NanoScience and Technology

NanoScience and Technology

Series Editors:
P. Avouris B. Bhushan D. Bimberg K. von Klitzing H. Sakaki R. Wiesendanger

The series NanoScience and Technology is focused on the fascinating nano-world, mesoscopic physics, analysis with atomic resolution, nano and quantum-effect devices, nanomechanics and atomic-scale processes. All the basic aspects and technology-oriented developments in this emerging discipline are covered by comprehensive and timely books. The series constitutes a survey of the relevant special topics, which are presented by leading experts in the field. These books will appeal to researchers, engineers, and advanced students.

Applied Scanning Probe Methods I
Editors: B. Bhushan, H. Fuchs, and S. Hosaka

Nanostructures
Theory and Modeling
By C. Delerue and M. Lannoo

Nanoscale Characterisation of Ferroelectric Materials
Scanning Probe Microscopy Approach
Editors: M. Alexe and A. Gruverman

Magnetic Microscopy of Nanostructures
Editors: H. Hopster and H.P. Oepen

Silicon Quantum Integrated Circuits
Silicon-Germanium Heterostructure Devices: Basics and Realisations
By E. Kasper, D.J. Paul

The Physics of Nanotubes
Fundamentals of Theory, Optics and Transport Devices
Editors: S.V. Rotkin and S. Subramoney

Single Molecule Chemistry and Physics
An Introduction
By C. Wang, C. Bai

Atomic Force Microscopy, Scanning Nearfield Optical Microscopy and Nanoscratching
Application to Rough and Natural Surfaces
By G. Kaupp

Applied Scanning Probe Methods II
Scanning Probe Microscopy Techniques
Editors: B. Bhushan, H. Fuchs

Applied Scanning Probe Methods III
Characterization
Editors: B. Bhushan, H. Fuchs

Applied Scanning Probe Methods IV
Industrial Application
Editors: B. Bhushan, H. Fuchs

Nanocatalysis
Editors: U. Heiz, U. Landman

Roadmap of Scanning Probe Microscopy
Editors: S. Morita

Nanostructures – Fabrication and Analysis
Editor: H. Nejo

Applied Scanning Probe Methods V
Scanning Probe Microscopy Techniques
Editors: B. Bhushan, H. Fuchs, S. Kawata

Applied Scanning Probe Methods VI
Characterization
Editors: B. Bhushan, S. Kawata

Applied Scanning Probe Methods VII
Biomimetics and Industrial Applications
Editors: B. Bhushan, H. Fuchs

Bharat Bhushan
Satoshi Kawata (Eds.)

Applied Scanning Probe Methods VI

Characterization

With 195 Figures and 7 Tables

Springer

Editors:

Professor Bharat Bhushan
Nanotribology Laboratory for Information
Storage and MEMS/NEMS (NLIM)
W 390 Scott Laboratory, 201 W. 19th Avenue
The Ohio State University, Columbus
Ohio 43210-1142, USA
e-mail: Bhushan.2@osu.edu

Satoshi Kawata
Osaka City University, Graduate School
of Science, Department of Mathematics
Sugimoto 3-3-138, 558-8585 Osaka, Japan
e-mail: skawata@skawata.com

Series Editors:

DOI 10.1007/11776314
ISSN 1434-4904

ISBN 978-3-642-07212-3 e-ISBN 978-3-540-37319-3

Springer is a part of Springer Science+Business Media

springer.com

Softcover reprint of the hardcover 1st edition 2007

Cover: WMX Design, Heidelberg
Printed on acid-free paper 2/3100/YL - 5 4 3 2 1 0

Preface

The scanning probe microscopy field has been rapidly expanding. It is a demanding task to collect a timely overview of this field with an emphasis on technical developments and industrial applications. It became evident while editing Vols. I–IV that a large number of technical and applicational aspects are present and rapidly developing worldwide. Considering the success of Vols. I–IV and the fact that further colleagues from leading laboratories were ready to contribute their latest achievements, we decided to expand the series with articles touching fields not covered in the previous volumes. The response and support of our colleagues were excellent, making it possible to edit another three volumes of the series. In contrast to topical conference proceedings, the applied scanning probe methods intend to give an overview of recent developments as a compendium for both practical applications and recent basic research results, and novel technical developments with respect to instrumentation and probes.

The present volumes cover three main areas: novel probes and techniques (Vol. V), charactarization (Vol. VI), and biomimetics and industrial applications (Vol. VII).

Volume V includes an overview of probe and sensor technologies including integrated cantilever concepts, electrostatic microscanners, low-noise methods and improved dynamic force microscopy techniques, high-resonance dynamic force microscopy and the torsional resonance method, modelling of tip cantilever systems, scanning probe methods, approaches for elasticity and adhesion measurements on the nanometer scale as well as optical applications of scanning probe techniques based on nearfield Raman spectroscopy and imaging.

Volume VI is dedicated to the application and characterization of surfaces including STM on monolayers, chemical analysis of single molecules, STM studies on molecular systems at the solid–liquid interface, single-molecule studies on cells and membranes with AFM, investigation of DNA structure and interactions, direct detection of ligand protein interaction by AFM, dynamic force microscopy as applied to organic/biological materials in various environments with high resolution, noncontact force microscopy, tip-enhanced spectroscopy for investigation of molecular vibrational excitations, and investigation of individual carbon nanotube polymer interfaces.

Volume VII is dedicated to the area of biomimetics and industrial applications. It includes studies on the lotus effect, the adhesion phenomena as occurs in gecko feet, nanoelectromechanical systems (NEMS) in experiment and modelling, application of STM in catalysis, nanostructuring and nanoimaging of biomolecules for biosensors, application of scanning electrochemical microscopy, nanomechanical investigation of pressure sensitive adhesives, and development of MOEMS devices.

As in the previous volumes a distinction between basic research fields and industrial scanning probe techniques cannot be made, which is in fact a unique factor in nanotechnology in general. It also shows that these fields are extremely active and that the novel methods and techniques developed in nanoprobe basic research are rapidly being transferred to applications and industrial development.

We are very grateful to our colleagues who provided in a timely manner their manuscripts presenting state-of-the-art research and technology in their respective fields. This will help keep research and development scientists both in academia and industry well informed about the latest achievements in scanning probe methods. Finally, we would like to cordially thank Dr. Marion Hertel, senior editor chemistry, and Mrs. Beate Siek of Springer for their continuous support and advice without which these volumes could have never made it to market on time.

July, 2006

Prof. Bharat Bhushan, USA
Prof. Harald Fuchs, Germany
Prof. Satoshi Kawata, Japan

Contents – Volume VI

Contents – Volume V

Contents – Volume VII

Contents – Volume I

Contents – Volume II

Contents – Volume III

Contents – Volume IV

List of Contributors – Volume VI

Asa H. Barber
Queen Mary, University of London, Department of Materials
Mile End Road, London E1 4NS, UK
e-mail: a.h.barber@qmul.ac.uk

Lilia A. Chtcheglova
Institute for Biophysics, Johannes Kepler University of Linz
Altenbergerstr. 69, A-4040 Linz, Austria
e-mail: lilia.chtcheglova@jku.at

Sidney R. Cohen
Chemical Research Support, Weizmann Institute of Science
Rehovot 76100, Israel
e-mail: Sidney.cohen@weizmann.ac.il

Andreas Ebner
Institute for Biophysics, Johannes Kepler University of Linz
Altenbergerstr. 69, A-4040 Linz, Austria
e-mail: andreas.ebner@jku.at

Hermann J. Gruber
Institute for Biophysics, Johannes Kepler University of Linz
Altenbergerstr. 69, A-4040 Linz, Austria
e-mail: hermann.gruber@jku.at

Norihiko Hayazawa
Nanophotonics Laboratory
RIKEN (The Institute of Physical and Chemical Research)
2-1 Hirosawa, Wako, Saitama, 351-0198, Japan
e-mail: hayazawa@riken.jp

Peter Hinterdorfer
Institute for Biophysics, Johannes Kepler University of Linz
Altenbergerstr. 69, A-4040 Linz, Austria
e-mail: peter.hinterdorfer@jku.at

Ferry Kienberger
Institute for Biophysics, Johannes Kepler University of Linz
Altenbergerstr. 69, A-4040 Linz, Austria
e-mail: ferry.kienberger@jku.at

Kei Kobayashi
International Innovation Center, Kyoto University, Katsura, Nishikyo
Kyoto 615-8520, Japan
e-mail: keicoba@iic.kyoto-u.ac.jp

Tadahiro Komeda
Institute of Multidisciplinary Research for Advanced Materials (IMRAM)
Tohoku University
2-1-1, Katahira, Aoba, Sendai, 980-0877 Japan
e-mail: komeda@tagen.tohoku.ac.jp

Andrzej J. Kulik
Ecole Polytechnique Fédérale de Lausanne, EPFL – IPMC – NN
1015 Lausanne, Switzerland
e-mail: Andrzej.Kulik@epfl.ch

Piotr Laidler
Institute of Medical Biochemistry, Collegium Medicum Jagiellonian University
Kopernika 7, 31-034 Kraków, Poland
e-mail: mblaidle@cyf-kr.edu.pl

Małgorzata Lekka
The Henryk Niewodniczański Institute of Nuclear Physics
Polish Academy of Sciences, Radzikowskiego 152, 31–342 Kraków, Poland
e-mail: Malgorzata.Lekka@ifj.edu.pl

Thomas Mueller
Veeco Instruments, 112 Robin Hill Road, Santa Barbara, CA 93117, USA
e-mail: tmueller@veeco.com

Theeraporn Puntheeranurak
Institute for Biophysics, Johannes Kepler University of Linz
Altenbergerstr. 69, A-4040 Linz, Austria
e-mail: theeraporn.puntheeranurak@jku.at

Yuika Saito
Nanophotonics Laboratory
RIKEN (The Institute of Physical and Chemical Research)
2-1 Hirosawa, Wako, Saitama, 351-0198, Japan

Yasuhiro Sugawara
Department of Applied Physics, Graduate School of Engineering,
Osaka University, Yamada-oka 2-1, Suita, Osaka 565-0871, Japan
e-mail: sugawara@ap.eng.osaka-u.ac.jp

Neil H. Thomson
Molecular and Nanoscale Physics Group, University of Leeds
EC Stoner Building, Woodhouse Lane, Leeds, LS2 9JT, UK
e-mail: n.h.thomson@leeds.ac.uk

Kohei Uosaki
Division of Chemistry, Graduate School of Science, Hokkaido University
N10 W8, Sapporo, Hokkaido, 060-0810, Japan
e-mail: uosaki@pcl.sci.hokudai.ac.jp

H. Daniel Wagner
Dept. Materials and Interfaces, Weizmann Institute of Science
Rehovot 76100, Israel
e-mail: daniel.wagner@weizmann.ac.il

Hirofumi Yamada
Department of Electronic Science & Engineering, Kyoto University
Katsura, Nishikyo, Kyoto 615-8510, Japan
e-mail: h-yamada@kuee.kyoto-u.ac.jp

Ryo Yamada
Division of Material Physics, Graduate School of Engineering Science
Osaka University, Machikaneyama-1-3, Toyonaka, Osaka, 060-0810, Japan
e-mail:yamada@molectronics.jp

List of Contributors – Volume V

Maria Allegrini
Dipartimento di Fisica "Enrico Fermi", Università di Pisa
Largo Bruno Pontecorvo, 3, 56127 Pisa, Italy
e-mail: maria.allegrini@df.unipi.it

Yasuhisa Ando
Tribology Group, National Institute of Advanced Industrial Science and Technology
1-2 Namiki, Tsukuba, Ibaraki, 305-8564, Japan
e-mail: yas.ando@aist.go.jp

Bharat Bhushan
Nanotribology Laboratory for Information Storage and MEMS/NEMS (NLIM)
W 390 Scott Laboratory, 201 W. 19th Avenue, Ohio State University
Columbus, Ohio 43210-1142, USA
e-mail: bhushan.2@osu.edu

Daniel Ebeling
Center for NanoTechnology (CeNTech), Heisenbergstr. 11, 48149 Münster
e-mail: Daniel.Ebeling@uni-muenster.de

Pietro Guiseppe Gucciardi
CNR-Istituto per i Processi Chimico-Fisici, Sezione di Messina
Via La Farina 237, I-98123 Messina, Italy
e-mail: gucciardi@its.me.cnr.it

Sadik Hafizovic
ETH-Zurich, PEL, Hoenggerberg HPT H6, 8093 Zurich, Switzerland
e-mail: hafizovi@phys.ethz.ch

Andreas Hierlemann
ETH-Zurich, PEL, Hoenggerberg HPT H6, 8093 Zurich, Switzerland
e-mail: hierlema@phys.ethz.ch

Hendrik Hölscher
Center for NanoTechnology (CeNTech), Heisenbergstr. 11, 48149 Münster
e-mail: Hendrik.Hoelscher@uni-muenster.de

Lin Huang
Veeco Instruments, 112 Robin Hill Road, Santa Barbara, CA 93117, USA
e-mail: lhuang@veeco.com

Hideki Kawakatsu
Institute of Industrial Science, University of Tokyo
Komaba 4-6-1, Meguro-Ku, Tokyo 153-8505, Japan
e-mail: kawakatu@iis.u-tokyo.ac.jp

Kay-Uwe Kirstein
ETH-Zurich, PEL, Hoenggerberg HPT H6, 8093 Zurich, Switzerland
e-mail: kirstein@phys.ethz.ch

Christine Kranz
School of Chemistry and Biochemistry, Georgia Institute of Technology
311 Ferst Dr., Atlanta GA 30332-0400, USA
e-mail: Christine.Kranz@chemistry.gatech.edu

Boris Mizaikoff
School of Chemistry and Biochemistry, Georgia Institute of Technology
311 Ferst Dr., Atlanta GA 30332-0400, USA
e-mail: Boris.Mizaikoff@chemistry.gatech.edu

Salvatore Patanè
Dipartimento di Fisica della Materia e Tecnologie Fisiche Avanzate
Università di Messina, Salita Sperone 31, I-98166 Messina, Italy
e-mail: patanes@unime.it

Craig Prater
Veeco Instruments, 112 Robin Hill Road, Santa Barbara, CA 93117, USA
e-mail: cprater@veeco.com

Elisa Riedo
Georgia Institute of Technology, School of Physics
837 State Street, Atlanta, GA 30332-0430, USA
e-mail: elisa.riedo@physics.gatech.edu

Tilman E. Schäffer
Institute of Physics and Center for Nanotechnology, University of Münster
Heisenbergstr. 11, 48149 Münster, Germany
e-mail: tilman.schaeffer@uni-muenster.de

Udo D. Schwarz
Department of Mechanical Engineering, Yale University
P.O. Box 208284, New Haven, CT 06520-8284, USA
e-mail: Udo.Schwarz@yale.edu

Yaxin Song
Nanotribology Lab for Information Storage and MEMS/NEMS (NLIM)
The Ohio State University
650 Ackerman Road, Suite 255, Columbus, Ohio 43202, USA
e-mail: Song.220@osu.edu

Chanmin Su
Veeco Instruments, 112 Robin Hill Road, Santa Barbara, CA 93117, USA
e-mail: csu@veeco.com

Robert Szoszkiewicz
Georgia Institute of Technology, School of Physics
837 State Street, Atlanta, GA 30332-0430, USA
e-mail: robert.szoszkiewicz@physics.gatech.edu

Sebastiano Trusso
CNR-Istituto per i Processi Chimico-Fisici, Sezione di Messina
Via La Farina 237, I-98123 Messina, Italy
e-mail: trusso@its.me.cnr.it

Cirino Vasi
CNR-Istituto per i Processi Chimico-Fisici, Sezione di Messina
Via La Farina 237, I-98123 Messina, Italy
e-mail: vasi@its.me.cnr.it

Justyna Wiedemair
School of Chemistry and Biochemistry, Georgia Institute of Technology
311 Ferst Dr., Atlanta GA 30332-0400, USA
e-mail: Justyna.Wiedemair@chemistry.gatech.edu

List of Contributors – Volume VII

Flemming Besenbacher
Interdisciplinary Nanoscience Center (iNANO), University of Aarhus
DK-8000 Aarhus C, Denmark
e-mail: fbe@inano.dk

François Bessueille
LSA, Université Lyon I, 43 Boulevard du 11 Novembre 1918
69622 Villeurbanne Cedex, France
e-mail: francois.bessueille@univ-lyon1.fr

Bharat Bhushan
Nanotribology Laboratory for Information Storage and MEMS/NEMS (NLIM)
W 390 Scott Laboratory, 201 W. 19th Avenue, Ohio State University
Columbus, Ohio 43210-1142, USA
E-mail: bhushan.2@osu.edu

Malte Burchardt
Faculty of Mathematics and Sciences
Department for Pure and Applied Chemistry
and Institute of Chemistry and Biology of the Marine Environment (ICBM)
Carl von Ossietzky University of Oldenburg
D-26111 Oldenburg, Germany
e-mail: malteburchardt@gmx.de

Abdelhamid Errachid]
Laboratory of NanoBioEngineering, Barcelona Science Park
Edifici Modular, C/Josep Samitier 1–5, 08028-Barcelona, Spain
e-mail: aerrachid@pcb.ub.es

Horacio D. Espinosa
Department of Mechanical Engineering, Northwestern University
2145 Sheridan Rd., Evanston, IL 60208-3111, USA
e-mail: espinosa@northwestern.edu

Yanxia Hou
Ecole Centrale de Lyon, STMS/CEGELY
36 Avenue Guy de Collongue, F-69131 Ecully Cedex, France
e-mail: yanxiahou24@yahoo.com

Nicole Jaffrezic-Renault
Ecole Centrale de Lyon, STMS/CEGELY
36 Avenue Guy de Collongue, F-69131 Ecully Cedex, France
e-mail: Nicole.Jaffrezic@ec-lyon.fr

Changhong Ke
Department of Mechanical Engineering, Northwestern University
2145 Sheridan Rd., Evanston, IL 60208-3111, USA
e-mail: c-ke@northwestern.edu

Keun-Ho Kim
Department of Mechanical Engineering, Northwestern University
2145 Sheridan Rd., Evanston, IL 60208-3111, USA
e-mail: kkim@nualumni.edu

Jeppe Vang Lauritsen
Interdisciplinary Nanoscience Center (iNANO)
Department of Physics and Astronomy
University of Aarhus, DK-8000 Aarhus C, Denmark
e-mail: jvang@phys.au.dk

Huiwen Liu
Nanotribology Laboratory for Information Storage and MEMS/NEMS (NLIM)
The Ohio State University
650 Ackerman Road, Suite 255, Columbus, Ohio 43202, USA
e-mail: Huiwen.Liu@seagate.com

Claude Martelet
Ecole Centrale de Lyon, STMS/CEGELY
36 Avenue Guy de Collongue, F-69131 Ecully Cedex, France
France
e-mail: Claude.Martelet@ec-lyon.fr

Nicolaie Moldovan
Department of Mechanical Engineering, Northwestern University
2145 Sheridan Rd., Evanston, IL 60208-3111, USA
e-mail: n-moldovan@northwestern.edu

Martin Munz
National Physcial Laboratory (NPL), Quality of Life Division
Hampton road, Teddington, Middlesex TW11 0LW, UK
e-mail: martin.munz@npl.co.uk

Michael Nosonovsky
Nanomechanical Properties Group
Materials Science and Engineering Laboratory
National Institute of Standards and Technology
100 Bureau Dr., Mail Stop 8520, Gaithersburg, MD 20899-8520, USA
e-mail: michael.nosonovsky@nist.gov

Sascha E. Pust
Faculty of Mathematics and Sciences
Department for Pure and Applied Chemistry
and Institute of Chemistry and Biology of the Marine Environment (ICBM)
Carl von Ossietzky University of Oldenburg
D-26111 Oldenburg, Germany
E-mail: sascha.pust@uni-oldenburg.de

Robert A. Sayer
Nanotribology Lab for Information Storage and MEMS/NEMS (NLIM)
The Ohio State University
650 Ackerman Road, Suite 255, Columbus, Ohio 43202, USA
e-mail: Sayer.11@osu.edu

Heinz Sturm
Federal Institute for Materials Research (BAM), VI.25
Unter den Eichen 87, D-12205 Berlin, Germany
e-mail: heinz.sturm@bam.de

Ronnie T. Vang
Interdisciplinary Nanoscience Center (iNANO)
Department of Physics and Astronomy
University of Aarhus, DK-8000 Aarhus C, Denmark
e-mail: rtv@inano.dk

Gunther Wittstock
Faculty of Mathematics and Sciences
Department for Pure and Applied Chemistry
and Institute of Chemistry and Biology of the Marine Environment (ICBM)
Carl von Ossietzky University of Oldenburg, D-26111 Oldenburg, Germany
e-mail: gunther.wittstock@uni-oldenburg.de

11 Scanning Tunneling Microscopy of Physisorbed Monolayers: From Self-Assembly to Molecular Devices

Thomas Müller

11.1 Introduction

Driven by visions of future nanoscale engineering, recent years have witnessed a surge in research activity concerned with the organization and properties of molecules and nanoscale objects adsorbed on surfaces and at interfaces. Fundamental questions are being addressed in two broad areas. First, as bottom-up assembly of future nanoscale devices is expected to complement existing (lithographic) top-down techniques, the driving forces for two-dimensional self-organization of molecular films are being examined. Second, device properties, commonly associated with macroscale objects, are being transposed into the quantum-confined world of individual molecules and nanoparticles.

Monolayers physisorbed on the basal plane of graphite can serve as ideal model systems for studies of self-assembly and molecular device properties. Highly ordered adsorbate monolayers on graphite substrates have been generated for a wide range of experimental conditions and molecular species. The dominant dispersion and electrostatic interactions on this inert substrate often afford sufficient adsorbate mobility to aid in the self-assembly of thermodynamically favored monolayer structures. A combination of theoretical and experimental approaches, including proximal probes, diffraction-based techniques, and thermal desorption, has been employed to interrogate adsorbate structures and dynamics, and ultimately unravel the delicate balance of forces driving the self-assembly process.

Scanning tunneling microscopy (STM) [1–3] has played a prominent role in elucidating the spatial orientation and conformation of individual adsorbate molecules. The high resolution achieved with relative ease in STM images sets this technique apart from other SPMs and has allowed numerous STM studies to address the structure and dynamics of self-assembled monolayers in exquisite detail [4–6]. The elucidation of driving forces for the self-assembly process has been helped further by the ability to perform STM studies of self-assembly under experimental conditions ranging from idealized ultrahigh vacuum (UHV) environments to more "practical" air/solid and liquid/solid interfaces [7].

The understanding and control of charge transport across individual molecules constitutes an important focal point of current research aimed at single-molecule device properties. Due to its ability to interrogate electron transport with submolecular resolution, and with theoretical models relating STM image contrast to electronic structure [8–15], STM is uniquely positioned as an important tool in this area. Recent years have witnessed a synthesis where ambient STM studies have been designed to interrogate charge transport in molecular devices while building on previous work to self-assemble the species of interest into ordered arrays [16–20].

This chapter is organized as follows. First, a discussion of the STM tunneling junction sets the stage and provides the basis for interpreting images presented in later sections. The image contrast observed in the presence of a weakly adsorbed molecule is seen to arise from geometric and electronic contributions. Sections 11.3 and 11.4 focus largely on the geometric factor by reviewing STM studies elucidating the driving forces for the self-assembly of functionalized hydrocarbons on the basal plane of graphite. The structure and dynamics of the observed ordered, lamellar monolayers, in some cases with unexpected symmetry properties, are seen to result from a set of competing geometric requirements. The examples of postassembly modifications and template generation discussed in the subsequent section may suggest avenues for expanding the set of tools available for the control of nanoscale structure. The final section of this chapter is concerned with recent studies exploiting both geometric and electronic contributions to STM image contrast to address molecular device properties in carefully tailored model systems self-assembled on graphite.

11.2 Source of Image Contrast: Geometric and Electronic Factors

As shown by Tersoff and Hamann, the tunneling current between a metallic tip and a metallic surface is proportional to the surface local density of states (LDOS) at the Fermi level, E_F, evaluated at the location of the tip [8, 9]. While this simple result is subject to certain approximations, such as small bias, low temperature, and a spherical tip represented solely by an s-wave function, the interpretation of STM images in terms of the surface LDOS at (or near) E_F has met with overwhelming success.

The present chapter is mainly concerned with surfaces decorated with relatively weakly interacting molecules. In contrast to the assumptions underlying Tersoff and Hamann's theory, larger bias voltages (typically on the order of 1 V) are employed for imaging such systems, thus expanding somewhat the energy range of surface states involved in tunneling. In addition, when considering the density of states of the overall system, the molecular orbitals associated with the adsorbate must be taken into account [4, 10, 13] as illustrated in Fig. 11.1.

In constructing a diagram as shown schematically in Fig. 11.1, the energies of the adsorbate states need to be placed with respect to the energies of the substrate surface states. Shown in Fig. 11.1 are the frontier orbitals of the molecular adsorbate, i.e. the highest occupied molecular orbital, HOMO, and the lowest unoccupied molecular orbital, LUMO. As a first approximation, the vacuum levels can be "pinned" or "lined up", as a free electron does not care where it came from. As indicated in Fig. 11.1, the Fermi level of the substrate and the HOMO of a free molecule are referred to the vacuum level by the workfunction, ϕ, and ionization potential, IP, respectively. However, a quantitative treatment will have to take into account shifts (and broadening) of energy levels due to effects such as the electronic coupling, Γ,

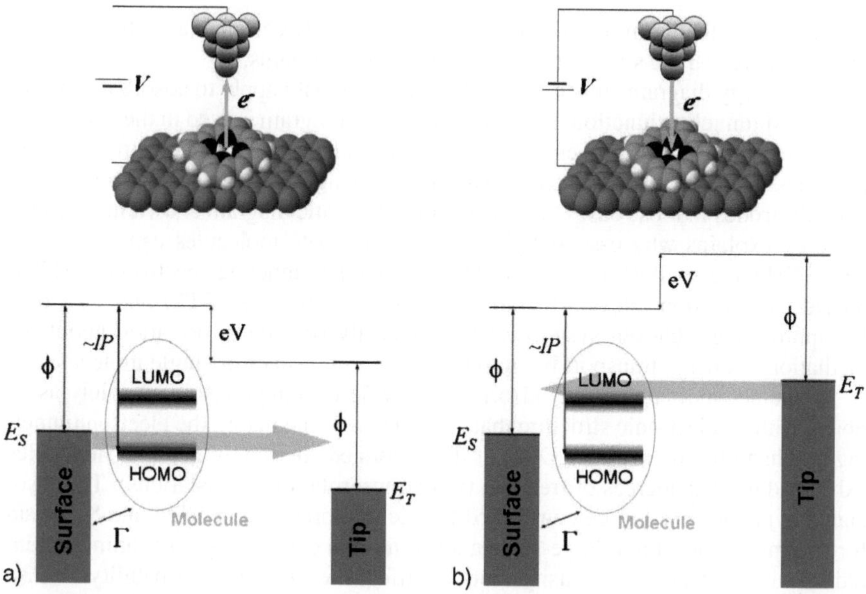

Fig. 11.1. Energy diagram for an STM junction when applying a negative (**a**) or positive (**b**) sample bias. The position of molecular frontier orbitals (HOMO and LUMO) and Fermi levels of substrate (E_S) and tip (E_T) are indicated. While molecular energy levels are broadened and shifted due to electronic coupling (Γ) to substrate states (and other phenomena), their approximate position can be estimated by pinning the vacuum level of the isolated molecule to that of the substrate

between adsorbate and substrate states, the presence of any solvent (typically lowering the energy of ions), and the electric field in the vicinity of the STM tip (typically $\sim 10^7$ V/cm).

At first it may seem that the adsorbate does not contribute a significant number of states to the overall system, especially not near the Fermi level. In the case of Xe on Ni, not a single adsorbate state lies within several eV of the Fermi level. Nevertheless, Xe atoms on Ni have been imaged successfully [21]. The presence of Xe has a significant influence on the LDOS near the Fermi level as evaluated at the position of the STM tip. Simply due to its spatial location even this most insulating adsorbate can serve as an "antenna" (i.e. at least a better antenna than vacuum) for the transport of electrons across the tunneling junction. Thus, STM images of insulating adsorbates can provide a topographic map of the topmost adsorbate atoms [22].

When the surface adsorbate exhibits electronic states near the Fermi level (i.e. for a moderate HOMO–LUMO gap), then the electron transmission may be dominated by individual molecular orbitals. For negative sample bias, electrons from filled surface states tunnel to the tip. As illustrated in Fig. 11.1a, mediation of this process is dominated by the HOMO of the adsorbate, if the HOMO is situated near the Fermi level. Conversely, electrons tunnel into empty surface states for positive sample bias. As illustrated in Fig. 11.1b, the LUMO may play a dominant role in this case if it is located near the Fermi level. Thus, with appropriately chosen bias voltages, STM images can reveal the shape of individual molecular orbitals.

The energy diagram shown in Fig. 11.1 does not only apply to adsorbate species in an STM tunneling junction. Essentially the same diagram is used in the analysis of single-molecule device studies performed with break junctions and electromigration junctions. The only difference is the assumption of significant electronic coupling to both electrodes in those cases. Therefore, the schematic diagram shown in Fig. 11.1 not only explains why essentially insulating adsorbate molecules can be imaged with STM but also illustrates how STM and scanning tunneling spectroscopy (STS) provide information relevant to molecular device properties. STM and STS data for appropriately chosen systems can intrinsically provide information about the mediation of charge transport by individual molecular states of single molecules.

In the above treatment the adsorbed molecule is being considered solely as an object with an electronic structure that plays a role in mediating the electron tunneling by changing the overall LDOS of the decorated surface. In general, molecules exhibit additional degrees of freedom (vibrations, rotations, translations). The associated dynamics need to be considered for the interpretation of STM and STS data for several reasons. First, large-scale nuclear motion entails a corresponding spatial redistribution of electron density that can affect the tunneling probability [4, 23]. Second, vibrational and rotational degrees of freedom are associated with spectral fine structure. In experiments performed at or near room temperature, thermal broadening reduces the energy resolution of tunneling spectra sufficiently to prevent detection of such fine structure. However, this is not the case in UHV STM experiments performed a few degrees above absolute zero where vibrational structure has been observed due to its influence on both, the elastic and inelastic tunneling probability [24].

11.3 Two-Dimensional Self-Assembly: Chemisorbed and Physisorbed Systems

In recent STM studies of two-dimensional self-assembly, several types of systems have received considerable interest, including thiols chemisorbed on Au(111) [6] and alkane derivatives physisorbed on the basal plane of graphite [4]. The chemisorption of thiols on gold is an irreversible process driven by the formation of the strong Au–S bond. Intermolecular interactions and the structure of the Au(111) surface also play a role during the self-assembly. Dense monolayers are formed, where only the thiol headgroup is in contact with the gold surface. In the case of alkane thiols, a commensurate $(\sqrt{3} \times \sqrt{3})$R30° overlayer structure is formed with additional superstructures depending on chain length and functionalization.

Figure 11.2 shows an example of an ambient high-resolution STM image where the ordered structure of a thiol layer can be seen clearly [25]. The monolayer was prepared by exposing a gold substrate to an ethanol solution containing a simple alkane thiol (decanethiol) and a bifunctionalized alkane derivative (11-mercaptoundecanoic acid). Both molecular species of interest were incorporated in the monolayer and can be distinguished in the STM image. Thus, model systems for nanoscience can be embedded in a sea of simple alkane thiols and studied by STM.

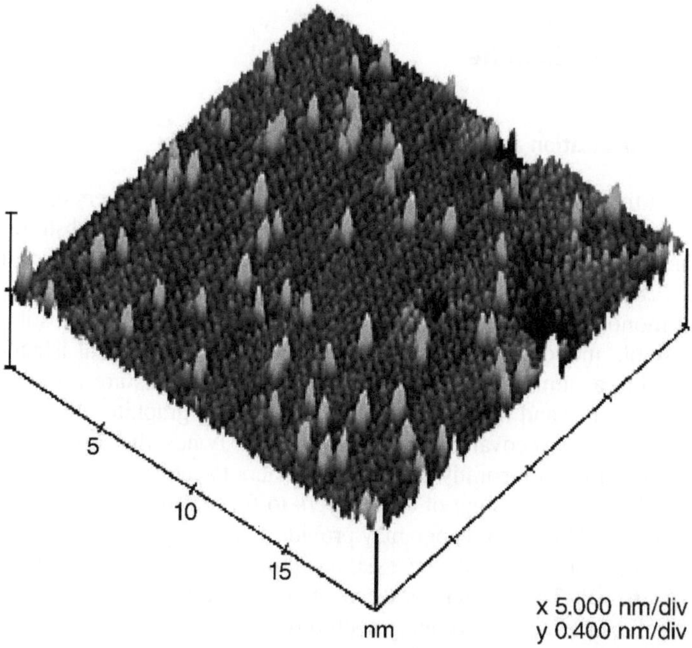

Fig. 11.2. Low-current STM topographic images of mixed self-assembled monolayers on Au(111) formed from a solution of 11-mercaptoundecanoic acid and decanethiol (1:9) at 65 °C with a bias voltage of 1.02 V and a set point current of 1.10 pA. Z bar is 0.4 nm. Used with permission from Lingyan Li, Shengfu Chen, and Shaoyi Jiang, Langmuir 2003, 19 (8), 3266–3271. Copyright (2003) American Chemical Society

Alkane thiols are subject to oxidation in air and high-resolution STM studies are often performed in UHV, even when the preparation takes place by immersing the substrate in a solution. The present chapter is mainly concerned with physisorbed monolayers self-assembled on the basal plane of graphite, which are frequently generated and probed under ambient conditions at the liquid/solid interface. In contrast to thiols on gold, physisorbed monolayers on graphite are not based on the irreversible formation of a chemical bond between a surface atom and a specific adsorbate functional group. Numerous types of organic molecules permit the formation of weakly adsorbed monolayers on graphite that exist in equilibrium with a supernatant solution. Often, either long alkyl chains or functional groups (or both) are present with both the associated intermolecular and molecule–substrate interactions providing key driving forces for the self-assembly process. The following sections of this chapter will provide several examples of STM work on such self-assembled systems, including molecules that were custom-tailored to exhibit both the desired self-assembly and the molecular device properties of interest.

11.4 Self-Assembly on Graphite

11.4.1 Alkane Functionalization and Driving Forces for Self-Assembly

Alkanes and their derivatives form weakly physisorbed monolayers on graphite [4, 26]. Dispersion interactions between alkyl chains and the graphite substrates provide adsorption energies of $\sim 1-3$ kcal/mol per methylene group, depending on chain length [27, 28] and functionalization, [29] making desorption barriers substantially higher for monolayer than for multilayer adsorption sites [27, 29]. Substituted alkanes form stable monolayers (i.e. as opposed to three-dimensional island growth) on graphite in vacuum although no directed chemical bonds are formed between adsorbate molecules and individual carbon atoms of the graphite substrate. Consistent with the absence of covalent adsorbate–substrate bonds, the adsorption energies are accompanied by comparatively smaller barriers to lateral motion. High lateral mobility enables the annealing of monolayers to form minimum energy structures in vacuum while solution exchange may provide an alternate path at the liquid/solid interface where ordered monolayers exist in equilibrium with a supernatant solution or melt [4, 26, 30]. Thus, conditions can often be found, both in vacuum and at the liquid/solid interface, for probing such ordered monolayers unimpeded by any formation of additional, ordered adsorbate layers.

Monolayers formed by substituted alkanes on graphite were among the first systems interrogated in STM studies of self-assembly from solution [4, 22, 31]. As can be seen in Figs. 11.3–11.10, close-packed lamellar structures are formed where the lamellae are composed of parallel alkyl chains each assuming an extended, all-

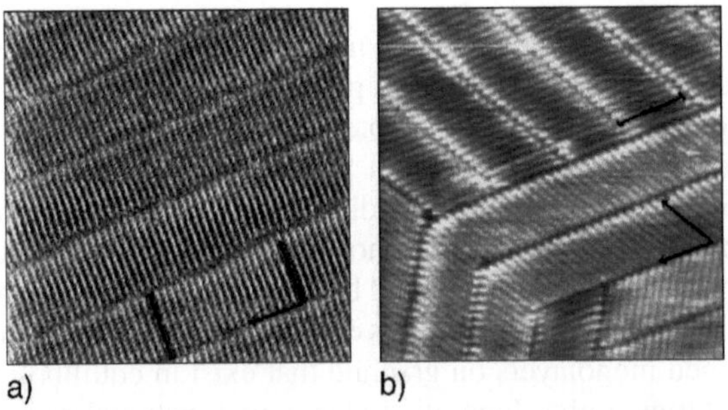

Fig. 11.3. (**a**) Constant height STM image (20 nm × 20 nm, sample bias −1.235 V, 166 pA tunneling current) of triacontane ($C_{30}H_{62}$) in phenyloctane physisorbed on graphite. One molecular length is represented by a *black bar*. The molecules are oriented with a 90° angle between the molecular axis and the direction of the lamellae. (**b**) Constant-current STM image (20 nm × 20 nm, sample bias −1.4 V, 80 pA tunneling current) of triacontanol ($CH_3(CH_2)_{29}OH$) in phenyloctane physisorbed on graphite. One molecular length is indicated by a *black bar*. The molecules are oriented with a 60° angle between the molecular axis and the direction of the lamellae as indicated on the image. Used with permission from Donna M. Cyr, Bhawani Venkataraman, and George W. Flynn, J. Phys. Chem. 1996, 100(32), 13747–59. Copyright (1996) American Chemical Society

trans conformation. In contrast to the dense self-assembled monolayers formed by alkane thiols on gold substrates, [6] the backbone axis of each molecule is oriented parallel to the plane defined by the substrate surface, thus bringing the entire alkyl chain into contact with the graphite substrate. Dispersion interactions with both the substrate and neighboring molecules play an important role in the formation of this structure.

In the case of alkanes a right angle is formed by the backbone and lamella axes (see Fig. 11.3a), while the introduction of terminal functional groups can lead to different lamella–backbone angles (see Fig. 11.3b) [4]. Alkane functionalization affects such structural parameters as the lamella–backbone angle since functional group interactions can exhibit a pronounced angular dependence (in addition to steric requirements) compared to the more isotropic alkane–alkane and alkane–graphite dispersion forces.

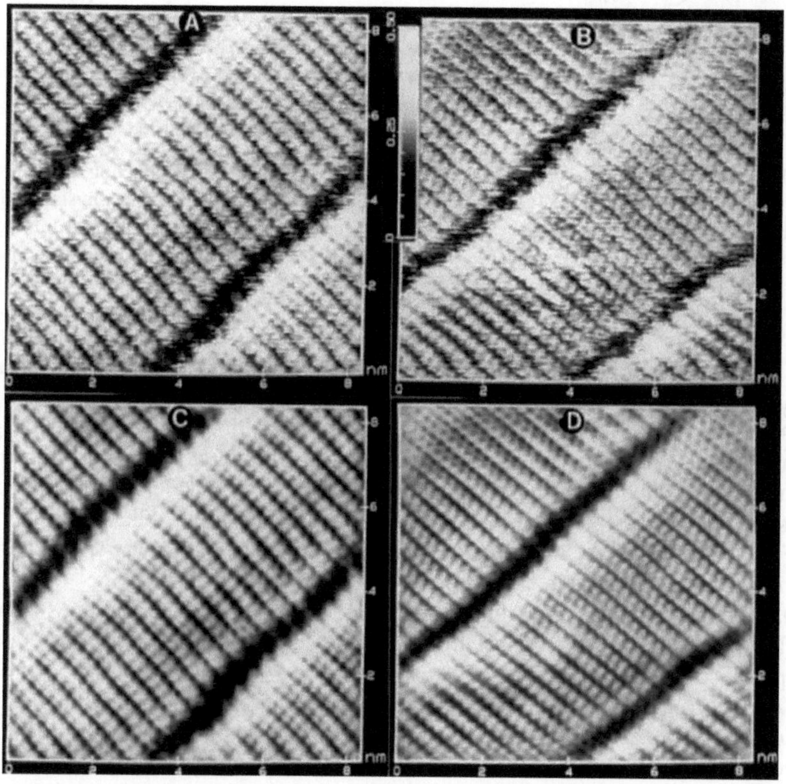

Fig. 11.4. Constant current STM image (sample bias −168 mV, 300 pA tunneling current) of hexatriacontane at the interface between a dodecane solution and the basal plane of graphite. Panels (**a**) and (**b**) show raw data acquired in two successive scans. Panels (**c**) and (**d**) show filtered versions of the STM images in (**a**) and (**b**), respectively. Used with permission from Weigen Liang, Myung-Hwan Whangbo, Aleksander Wawkuschewski, Hans-Joachim Cantow, and Sergei N. Magonov, Adv. Mater. 1993, 5, 817. Copyright 1993 VCH Verlagsgesellschaft mbH, D-69469 Weinheim

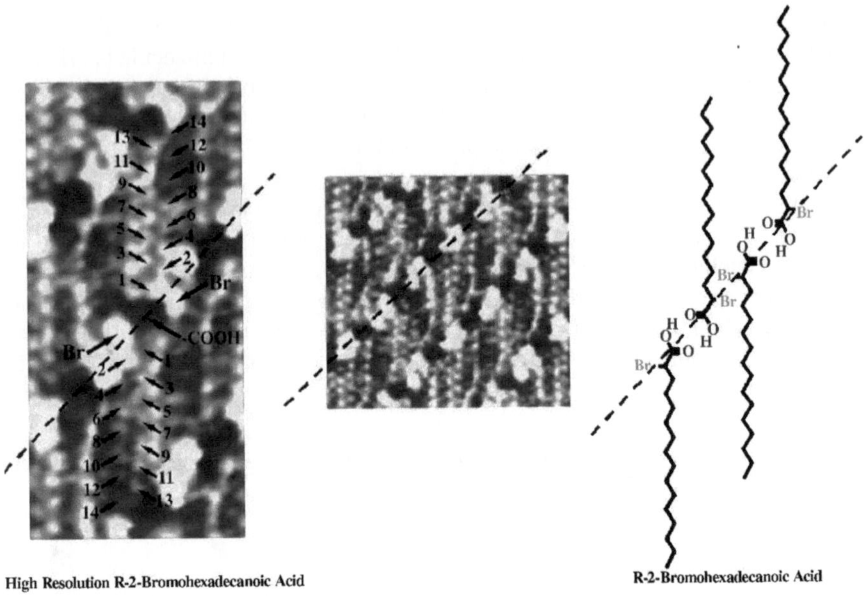

Fig. 11.5. Constant-current STM images (−1.4 V sample bias, 300 pA tunneling current) and a corresponding model of (*R*)-2-bromohexadecanoic acid in phenyloctane physisorbed on graphite. *Numbers* and *letters* in the image refer to individual hydrogen atoms and functional groups, while the *dashed line* emphasizes the 45° angle between lamella and molecular backbone. Reproduced with permission from www.columbia.edu/cu/chemistry/groups/flynn/r2bromo.html

The presence of functional groups has effects on the self-assembly that are more subtle than changing the lamella–backbone angle. Note that some of these STM images (e.g., see Figs. 11.4 and 11.5) clearly reveal the position of individual hydrogen atoms associated with the methylene groups of the carbon backbone. The orientation of individual methylene groups and thus of the whole carbon skeleton with respect to the surface plane can be determined. Consider first unsubstituted alkanes. As shown in Fig. 11.4, alkanes lacking functional groups can exhibit different spot patterns, even in successively acquired STM images. Apparently, the plane formed by the all-*trans* carbon backbone can assume a variety of angles with respect to the substrate surface plane [22, 31]. The distribution of angles is influenced by temperature and the presence of solvent [32, 33].

Adding to the more subtle effects, the spatial requirements of individual alkane molecules may depend on the angle assumed by the carbon backbone plane. In three-dimensional alkane crystals, lateral nearest-neighbor spacings of ~4.8 Å and ~4.2 Å are found parallel and perpendicular to the molecular carbon planes, respectively [34]. In STM studies of long-chain alkane monolayers at the liquid/solid interface, lateral spacings have been found to be closer to 4.2 Å [31]. If the molecular backbone axis is aligned with one of the high-symmetry directions of the underlying graphite substrate, a spacing of 4.26 Å is required for each alkane molecule to occupy an equivalent surface site. Such a commensurate monolayer structure is consistent with the absence of Moiré patterns in STM images of alkanes at the interface between an organic solution and the basal plane of graphite [31]. At first it may appear to be inconsistent with the substantially larger spacings reported by Herwig et al. [32] based on diffraction studies of vapor-deposited monolayers composed of similarly long alkanes (i.e. $C_{32}H_{66}$). However, Herwig et al. found that the presence of "solvent" (i.e. heptane adsorbed atop a monolayer formed by dotriacontane) leads to a contraction of the monolayer structure accompanied by a change in the distribution of angles assumed by the molecular backbone plane [32]. The observed shifts in diffraction maxima are consistent with a commensurate monolayer in the limiting case of full solvent coverage [32]. The sensitivity of alkane monolayer structures to the presence of additional alkane layers may also explain the existence of ordered alkane monolayers substantially above the bulk melting temperature in the presence of bulk alkane melt [30], while essentially no increased melting temperature (i.e. when compared to the bulk melting point) is found for analogous vapor-deposited alkane monolayers in vacuum [35]. Further variations in alkane monolayer structures exist as a function of alkane chain length [36–38]. Diffraction studies of short alkanes on graphite have revealed the existence of a herringbone structure for monolayers composed of some molecules with an even number of carbon atoms, with the transition to a rectangular structure depending on coverage and temperature [37].

In contrast, no comparable structural variation has been found for alkanoic acids. Close inspection of Fig. 11.5 clearly reveals a zig-zag spot pattern along the backbone axis, indicating that the molecular backbone plane is consistently oriented parallel to the surface plane. The high resolution achieved in such images also permits the identification of hydrogen-bonded acid dimers. Introduction of the hydrogen-bonding carboxylic headgroup has forced the entire molecular backbone plane into a "flat" orientation (parallel to the substrate surface). Lateral intermolecular spacings are expanded to $\sim$ 4.8 Å, and STM images of alkanoic acid monolayers often show Moiré effects due to the lack of commensurability [31]. Being the result of a geometric constraint imposed by headgroup interactions and headgroup steric requirements, the expanded lateral spacing might be expected to be accompanied by an energetic tradeoff due to the reduced alkyl–alkyl dispersion interactions. Indeed, thermal desorption data indicates greatly reduced incremental adsorption energies per methylene unit upon introduction of the alkanoic acid headgroup [29].

Unsubstituted alkanes and alkanoic acids embody limiting cases regarding (hindered) molecular rotation about the backbone axis within self-assembled monolayers on graphite. The structural variability observed for unsubstituted alkanes points to a shallow potential along several degrees of freedom (as long as a close-packed structure is maintained). In contrast, the carboxylic acid headgroup apparently forces the entire molecular backbone plane of alkanoic acids into a "flat" configuration. Terminal halogen substituents represent an intermediate case. Figure 11.6 shows UHV STM images of 1-halohexane monolayers acquired at $T_{substrate} = 80$ K. As indicated by the overlaid molecular models, an essentially identical, angled head-to-head configuration is found for all 1-halohexanes, with differences in image contrast ascribed to electronic factors. The molecular backbone plane is oriented parallel to the surface plane. The lack of commensurability following from the monolayer structural parameters is qualitatively consistent with the observed Moiré patterns in the STM images.

For 1-bromohexane, the results from energy minimizations and molecular dynamics simulations are in quantitative agreement with the experimentally determined monolayer structures. The theoretical studies indicate that electrostatic headgroup interactions are small in magnitude yet play an important role in the self-assembly due to their anisotropy. While alkyl-chain dispersion interactions drive the formation of a close-packed monolayer composed of all-*trans* extended molecules, headgroup

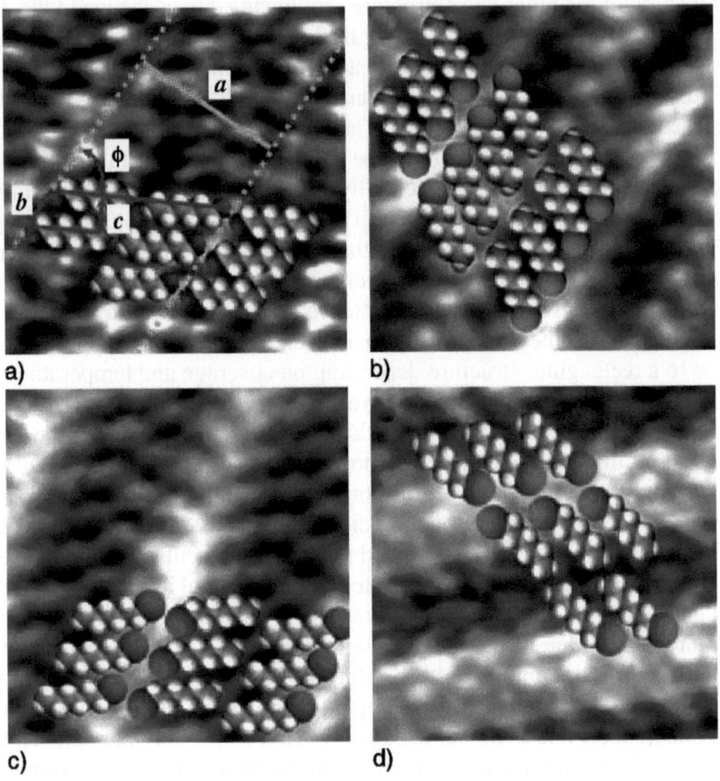

Fig. 11.6. High-resolution STM topographs of 1-halohexanes on HOPG. High-resolution (4 nm × 4 nm) constant-current STM topographs are shown for (**a**) 1-fluorohexane (2.0 V sample bias, 100 pA tunneling current), (**b**) 1-chlorohexane (1.9 V sample bias, 90 pA tunneling current), (**c**) 1-bromohexane (2.0 V sample bias, 100 pA tunneling current), and (**d**) 1-iodohexane (1.9 V sample bias, 85 pA tunneling current). To minimize distortion in these small-scale images acquired at relatively slow absolute scan speeds ($\sim$ 40 nm/s), a correction was applied for constant drift in the image plane, based on a careful analysis of measured adsorbate structural parameters as a function of absolute scan speed. Arrays of overlaid molecular models illustrate the proposed monolayer packing structure. The adsorbate molecules are seen to be in a fully extended all-*trans* configuration with their molecular backbone plane (i.e. including the carbon–halogen bond) parallel to the surface. The monolayer structural parameters (i.e. lamella spacing, a, neighbor spacing measured along the lamella direction, b, neighbor spacing measured perpendicular to the molecular backbone, c, and lamella–backbone angle, ϕ) highlighted in panel (**a**) are found to be similar for all 1-halohexanes, suggesting that the differences in image contrast are primarily electronic in nature. From Thomas Müller, Tova L. Werblowsky, Gina M. Florio, Bruce J. Berne, and George W. Flynn, Proc. Natl. Acad. Sci. USA 2005, 102(15), 5315–22. Copyright 1993–2005 by The National Academy of Sciences of the United States of America, all rights reserved

forces compete successfully to control both the angle between the lamella and backbone axes and the angle between the surface and backbone planes. Consistent with this concept of competing interactions, adsorption energies measured as a function of alkyl chain length and functionalization reveal the presence of energetic tradeoffs albeit of smaller magnitude than in the case of the (more strongly interacting) carboxylic acid substituent [29].

The STM images in Fig. 11.6 show what appears to be a robust self-assembly pattern for 1-haloalkanes. The same structure has been observed not only for different halogen substituents but also for different chain lengths [7]. In contrast, nearly rectangular structures with halogen substituents in a head-to-head configuration have been observed in ambient STM images of 1-bromoalkane monolayers at the liquid/solid interface (i.e. in the presence of solvent) [4, 5, 39]. Ongoing studies at Columbia University [7] are designed to elucidate the ability of solvent to shift the monolayer structure by tilting the balance among these interactions.

11.4.2 Expression of Chirality

The geometric requirement of close-packed monolayer structures allows such seemingly negligible changes as the addition of a single methylene group to a long alkyl chain to alter the self-assembly significantly. Panels a and b of Fig. 11.7 display STM images of 12-bromododecanoic acid and 11-bromoundecanoic acid, respectively [40]. Although only differing by one methylene group, the monolayer structures resulting for these two species are obviously very different. An additional geometric subtlety is associated with the transition from three to two dimensions. The bromoalkanoic acids under consideration are not chiral, i.e. they can be superimposed on their mirror images. However, in their all-*trans* extended conformation, bromoalkanoic acids contain two enantiotopic faces, i.e. faces that are interchanged by a symmetry plane or by an alternating axis of symmetry but not by a simple symmetry axis [41]. As a consequence, bromoalkanoic acids are prochiral. If adsorption to an achiral substrate prevents rotation about the backbone axis, as is the case on the basal plane of graphite, then two enantiomorphous adsorbed states can be formed [42]. Close inspection of Fig. 11.7 reveals that in the case of 11-bromoundecanoic acid, immobilization at the interface leads to the formation of chiral domains, which are the 2D analog of the enantiomerically pure 3D crystals (or conglomerates) formed when racemic mixtures spontaneously resolve during crystallization. Each chiral domain is populated by a single enantiomorph. Apparently, the nearest-neighbor interactions driving the self-assembly are most favorable in the case of identical enantiomorphs. The simple addition of a single methylene group changes geometric packing requirements, and therefore nearest-neighbor interactions, such that an alternating arrangement of opposite enantiomorphs is preferred. Thus, achiral domains (or 2D racemates) are found in the case of 12-bromododecanoic acid [40]. A third case, not found here, is that of nearest-neighbor interactions comparable in magnitude between identical vs. opposing enantiomorphs, leading to a 2D racemate composed of a random mixture of enantiomorphs.

The same effect is known for alkanoic acids without the terminal bromine substituents. In that case, chiral domains (or 2D conglomerates) are observed for an even

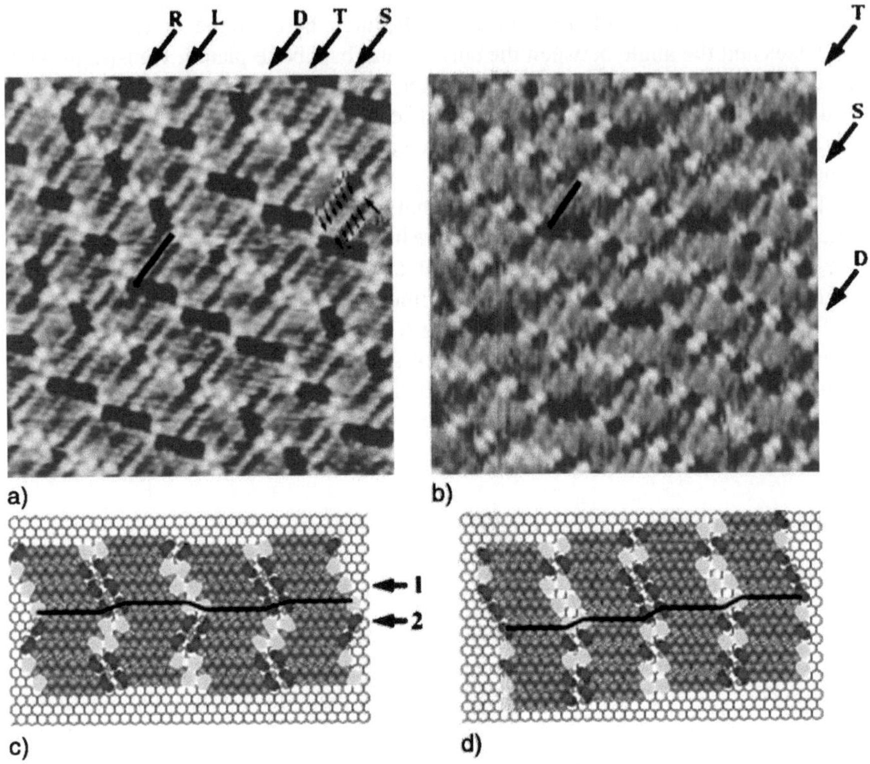

Fig. 11.7. Constant current STM images (12 nm × 12 nm, sample bias −1.4 V, 300 pA tunneling current) are shown of (**a**) 12-bromododecanoic acid and (**b**) 11-bromoundecanoic acid on graphite under phenyloctane solution. Black bars indicate a molecular length. *Capital letters S, D,* and *T* point to strips of single, double, and triple-twin structures, respectively. *Capital letters R* and *L* point respectively to the positions where the lower twin in a double or triple twin shifts right or left relative to the twin lying above it. The *small numbers 1–11* in (**a**) point to the positions of individual hydrogen and bromine atoms and Br refers to the bromine substituent. (**c**) and (**d**) show molecular models for the images displayed in (**a**) and (**b**), respectively. *Numbers 1* and *2* in (**c**) refer to rows, where the molecules shift right or left relative to the row above. Used with permission from Hongbin Fang, Leanna C. Giancarlo, and George W. Flynn, J. Phys. Chem. B 1998, 102(38), 7421–4 (1998). Copyright (1998) American Chemical Society

number of carbon atoms (including the acid group) [40]. An example of this is shown in Fig. 11.8. Panels (a) and (b) show the two opposite enantiomorphous domains formed by behenic acid ($C_{21}H_{43}COOH$, i.e. with an even number of carbon atoms). The mirror image relationship between the two types of domains is emphasized by the unit cells shown as (nonrectangular) parallelograms. As can be seen in panel (c), nonadecanoic acid (with an odd number of carbon atoms) forms a packing structure with a rectangular unit cell that is identical to its mirror image and represents a 2D racemate with an alternating arrangement of opposite enantiomorphs.

The expression of chirality has also been examined for self-assembled monolayers composed of several distinct molecular building blocks. When mixing chiral with

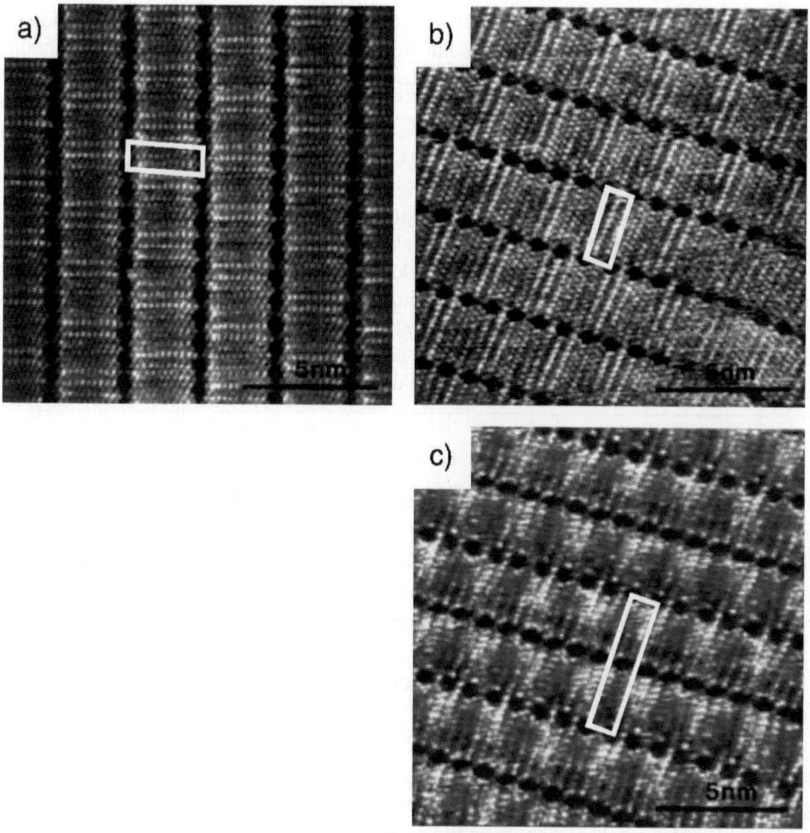

Fig. 11.8. Constant current STM images (~12 nm × 12 nm, sample bias −1.3 V, 800 pA tunneling current) are shown of (**a**,**b**) behenic acid and (**c**) nonadecanoic acid on graphite under phenyloctane solution. Unit cells are indicated by parallelograms (**a**,**b**) and a rectangle (**c**). Used with permission from Masahiro Hibino, Akinori Sumi, Hiroshi Tsuchiya, and Ichiro Hatta, J. Phys. Chem. B 1998, 102(23), 4544–7. Copyright (1998) American Chemical Society

achiral species, the chirality can be not only transferred to the two-dimensional array but also from the chiral to the achiral component. An example of this is shown in Fig. 11.9, where the chiral 5-[10-(2-methylbutoxy)-decyloxy]isophthalic acid (ISA) is self-assembled at the heptanol/graphite interface [44]. Chiral domains are formed composed of alternating ISA-heptanol lamellae stabilized by hydrogen bonding. The domains are chirally pure not only with respect to ISA but also with respect to the sense of rotation of the HOC bond angle of the coadsorbed heptanol molecules and the angle between the ISA and heptanol molecular backbone axes. The multicomponent self-assembly has extended the enantiomorphous character to heptanol [44]. Heptanol has become prochiral.

The opposite effect has also been observed. An achiral solute can form enantiomorphous domains that direct the self-assembly of a chiral cosolute. In that case the template created by the achiral species can afford the resolution of racemic mix-

Fig. 11.9. (**a**) Chemical structure of 5-[10-(2-methylbutoxy)-decyloxy]isophthalic acid (ISA). (**b**) Constant current STM image (11.7 nm × 11.7 nm) of an (*S*)-ISA/heptanol monolayer on graphite under heptanol solution. Note the angle between the heptanol and ISA backbones. This domain will be referred to as "positive". (**c**) Constant-current STM image (11.5 nm × 11.5 nm) of an (*R*)-ISA/heptanol monolayer on graphite under heptanol solution. Note the angle between the heptanol and ISA backbones. This domain will be referred to as "negative". (**d**) Proposed molecular model for the self-assembled monolayer giving rise to the STM image displayed in (**b**). (**e**) Proposed molecular model for the self-assembled monolayer giving rise to the STM image displayed in (**c**). (**f**) Proposed model emphasizing the molecular arrangement in a "positive" domain. (**g**) Proposed model emphasizing the molecular arrangement in a "negative" domain. Reproduced with permission from Steven de Feyter, Petrus C. M. Grim, Markus Rücker, Peter Vanoppen, Christian Meiners, Michel Sieffert, Suresh Valiyaveettil, Klaus Müllen, and Frans C. De Schryver, Angew Chem Int Ed 1998, 37(9), 1223–6. Copyright 1998 Wiley-VCH Verlag GmbH, D-69469 Weinheim

tures by separately depositing each enantiomer into chirally pure domains. Exactly this outcome can be seen in Fig. 11.10, where a self-assembled domain of hexadecanoic acid at the phenyloctane/graphite interface is shown, decorated with the chiral (*S*)-2-bromohexadecanoic acid also present in the solution. Domains of opposite chirality were found not to contain any coadsorbed (*S*)-2-bromohexadecanoic acid [45].

Note that in the absence of the achiral cosolute, 2-bromohexadecanoic acid forms a monolayer structure characterized by a different lamella–backbone angle (see Fig. 11.11) [5]. The formation of bromine dimers suggests that intermolecular interactions between bromine substituents play a role in driving this self-assembly [5,46]. Again, the 2D self-assembly leads to a spontaneous resolution of the racemic mixture, as intermolecular interactions drive the formation of chirally pure, enantiomorphous domains, composed of a single type of chirally pure, hydrogen-bonded acid dimers.

11.5 Beyond Self-Assembly

11.5.1 Postassembly Modification

Self-assembly on graphite has been probed for a wide variety of molecular species, some of which include functional groups that permit further, postassembly manipulation. As an example, an STM image and corresponding model are shown in Fig. 11.12a,b for an isophthalic acid derivative (ISA) self-assembled at the undecanol/graphite interface [26, 47]. An ordered, lamellar structure is observed, with alternating rows of close-packed ISA and solvent (undecanol) molecules, stabilized by dispersion interactions (involving alkyl chains and the graphite substrate), electrostatic forces (involving the polar carboxylic acid groups and the semimetallic graphite substrate), and hydrogen bonds (between the carboxylic acid and alcohol groups).

The isophthalic acid derivative under consideration belongs to the class of diacetylene compounds (R–C≡C–C≡C–R′) where polymerization is known to occur

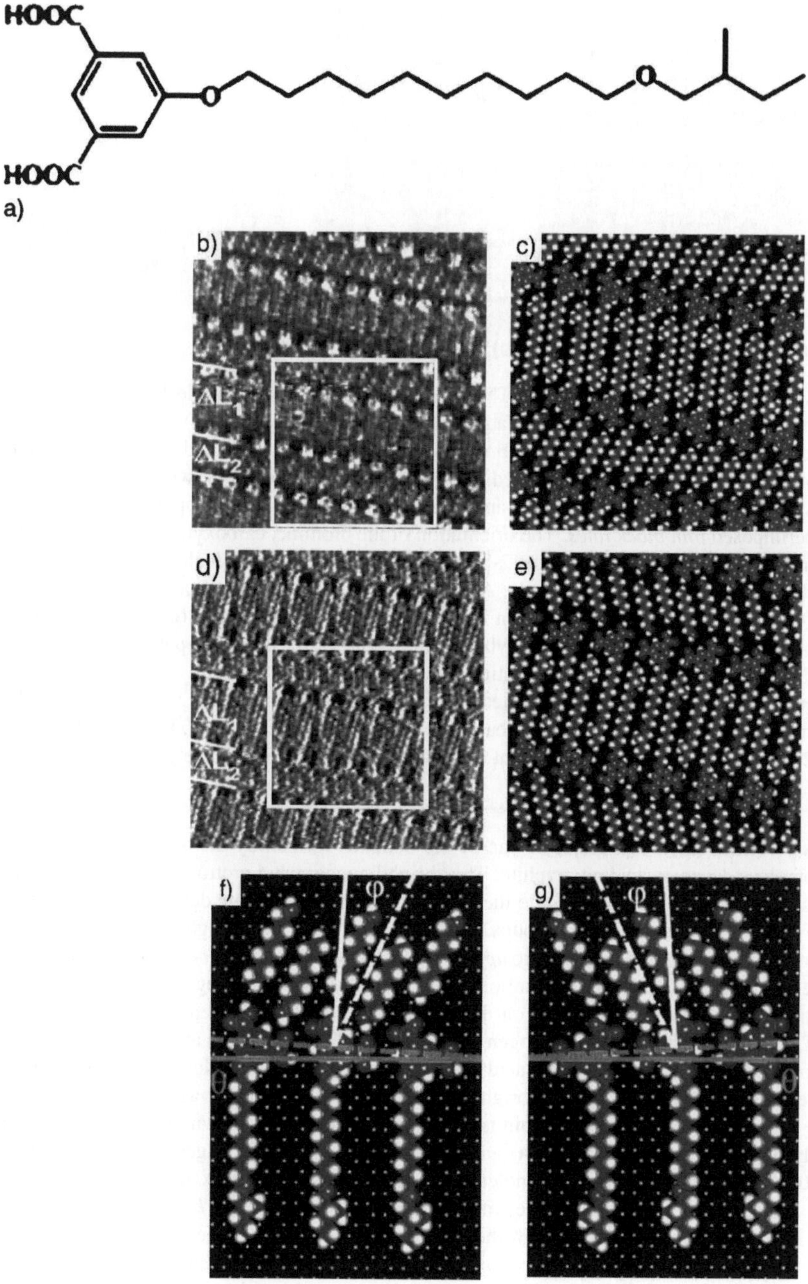
HOOC
HOOC
a)
b)
c)
d)
e)
f)
g)
ΔL1
ΔL2
φ
θ

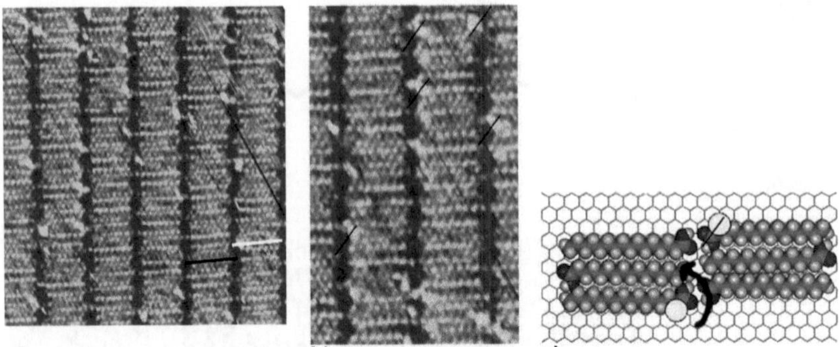

Fig. 11.10. (**a**) A constant current STM image (13 nm × 13 nm, sample bias −1.5 V, 300 pA tunneling current) of a domain of hexadecanoic acid interspersed with (*S*)-2-bromohexadecanoic acid is shown where a *solid black bar* depicts hexadecanoic acid and a *blue bar* depicts 2-bromohexadecanoic acid. An enlarged portion of (**a**) is presented in (**b**) where the bromine atom lies above and to the right or below and to the left of the carboxyl group as highlighted by the superimposed *thin black lines*. The orientation of the bromine, carboxyl, and alkyl chain on the chiral carbon is used to identify the 2-bromohexadecanoic acid molecules in this domain as the *S* chiral conformers of the molecule. A top view of a model of hexadecanoic acid interspersed by (*S*)-2-bromohexadecanoic acid on the graphite surface is shown in (**c**). The *thin black line* superimposed on a bromine/carboxylic combination shows the same pattern depicted in the STM image (**b**). An *arrow* demonstrating the counterclockwise direction of the brominecarboxyl group-alkyl group orientation identifies the brominated molecules as the *S* conformer. Used with permission from Dalia G. Yablon, Leanna C. Giancarlo, and George W. Flynn, J. Phys. Chem. B 2000, 104(32), 7627–35. Copyright (2000) American Chemical Society

Fig. 11.11. (**a**) STM topograph of one of the domains formed by the assembly of (*R*)/(*S*)-2-bromohexadecanoic acid on graphite. The "bright" topographic protrusions are assigned as the positions of the α-Br atoms, while the adjacent "dark" topographic depressions correspond to the hydrogen-bonding COOH groups. An alternating pattern of "bright"-"dark"-"bright" runs through the image from *lower left* to *upper right*, as denoted by the *black arrow*. This domain has been assigned as containing *R*-enantiomers exclusively. Two parallelograms denote a hydrogen-bonded molecular pair lying flat on the surface. The angle β formed by the molecular axis, OC, and the direction of the hydrogen bond, OD, has been measured as $53 \pm 5°$. (**b**) A second domain of 2-bromohexadecanoic acid molecules found at the phenyloctane/graphite interface. The alternating "bright"-"dark"-"bright" pattern now extends from *lower right* to *upper left* in the topographic image. This domain is the mirror image of that shown in (**a**) and is comprised only of the *S*-enantiomer. Both (**a**) and (**b**) are 12 nm × 12 nm images (−1.4 V sample bias, 300 pA tunneling current). (**c**) A molecular model of (*R*)-2-bromohexadecanoic acid organized on a graphite lattice based on the image shown in (**a**). The *black bars* denote a chiral pair of *R*–*R* molecules bonded through their carboxylic acid groups. The *yellow circles* represent Br, the *red balls* oxygen, the *green areas* carbon, and the *white spheres* hydrogen atoms. The alternating pattern seen in the STM image is also reproduced here. Further, as indicated by the parallelograms, the dimerized acid molecules occupy four rows of the graphite surface. (**d**) Molecular model representing the STM image of (**b**). Here, only *S*-enantiomers are depicted, and the *black bars* denote an *S*–*S* chiral pair. Again, the alternating Br–COOH–Br pattern is shown. Used with permission from Leanna C. Giancarlo and George W. Flynn, Acc. Chem. Res. 2000, 33(7), 491–501. Copyright (2000) American Chemical Society in (three-dimensional) solid crystals upon exposure to ultraviolet radiation. High-resolution STM studies of ISA monolayers with submolecular resolution can address the question whether the UV-induced polymerization reaction also occurs in two dimensions. The STM images and proposed model shown in Fig. 11.12a,b suggest that this might indeed be the case. As in three-dimensional crystals, the diacetylene (–C≡C–C≡C–) moieties of neighboring molecules are seen to occupy adjacent positions, presumably mimicking the transition state of the topochemical polymerization reaction. This is not the case for any additional ISA molecules present in the supernatant solution. Thus, it can be expected that UV irradiation will lead to selective polymerization of the interfacial layer only.

The results of UV illumination are shown in Fig. 11.12c,d. The clear changes in STM image contrast suggest that the intermolecular polymerization reaction has taken place as anticipated [26, 47]. Each individual lamella that used to consist of separate molecular entities has been transformed into a single conjugated polydiacetylene (=RC–C≡C–CR′=)$_n$ moiety with new mechanical and electronic

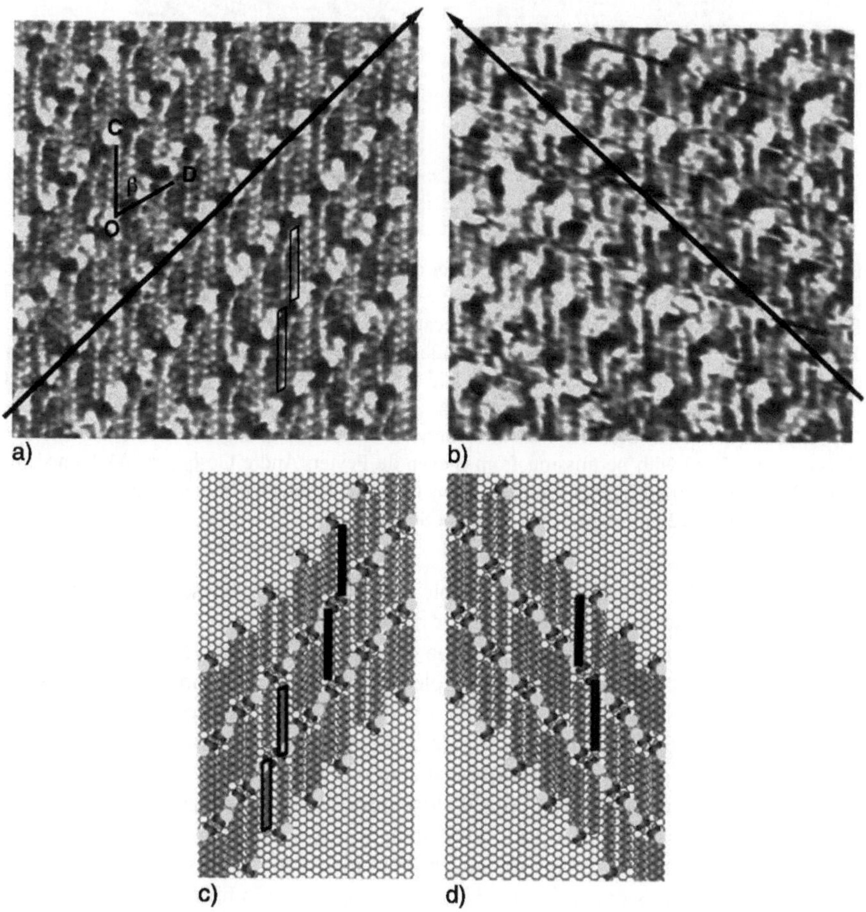

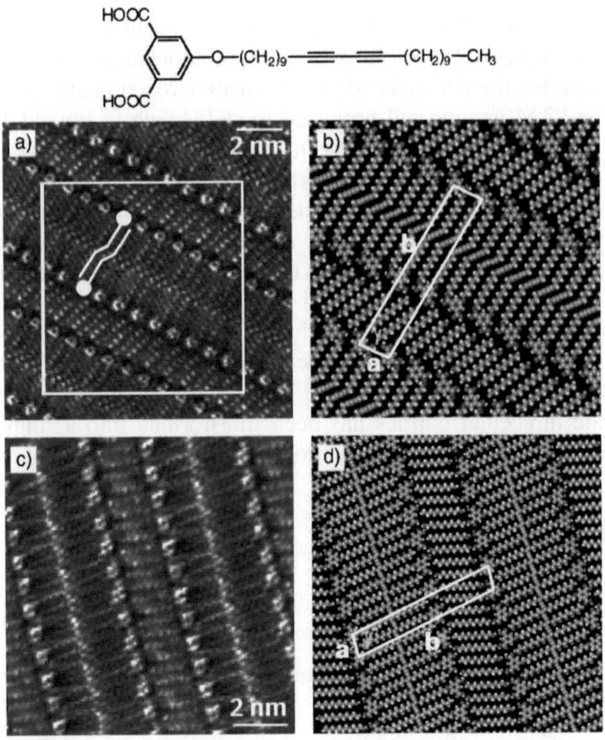

Fig. 11.12. The isophthalic acid derivative under consideration (ISA) is shown in the *top panel*. (**a**) Constant height STM image (sample bias −1.2 V, 1.0 nA tunneling current) of a mixed ISA/undecanol monolayer formed at the undecanol/graphite interface before UV illumination. *Yellow symbols* highlight a pair of ISA molecules. (**b**) Proposed molecular model for the array of molecules included in the *white box* overlaid on the STM image shown in (**a**). (**c**) Constant height STM image (sample bias −0.5 V, 1.0 nA tunneling current) of ISA after UV illumination. (**d**) Proposed molecular model for the polymerized monolayer giving rise to the STM image shown in (**c**). Used with permission from Steven de Feyter, Andre Gesquiere, Mohamed M. Abdel-Mottaleb, Petrus C., M. Grim, and Frans C. de Schryver, Acc. Chem. Res. 2000, 33(8), 520–31. Copyright (2000) American Chemical Society

properties. Due to their conjugated double and triple bonds, these molecularly thin "tapes" might serve as model systems for molecular wires.

In general, postassembly manipulation of interfacial monolayers can be induced through a variety of means. Aside from electromagnetic radiation, electrochemical control can be used to initiate chemical reactions. The application of brief voltage pulses through the STM tip can provide a means for highly localized nanoscale manipulation. As shown by Okawa et al. [48, 49] this strategy can be used successfully with dry monolayers of diacetylene derivatives self-assembled on graphite substrates. Figure 11.13 shows an STM image of a monolayer formed by 10,12-pentacosadiynoic acid (PCA) at the air/graphite interface [48, 49].

As in the previous example, the diacetylene derivative is found to form a monolayer structure that may be conducive to the topochemical polymerization reaction.

Again, the diacetylene (–C≡C–C≡C–) moieties of neighboring molecules occupy adjacent positions with the required mutual orientation. The result of applying a brief voltage pulse through the STM tip is illustrated in Fig. 11.14. A single row ($\sim$ 3 nm wide) of individual PCA molecules has been transformed into a conjugated polydiacetylene, thus demonstrating the targeted creation of a molecular wire from self-assembled precursors. One could envision the utilization of such a process for the targeted interconnection of coassembled molecular devices, model systems for which will be discussed in the next section.

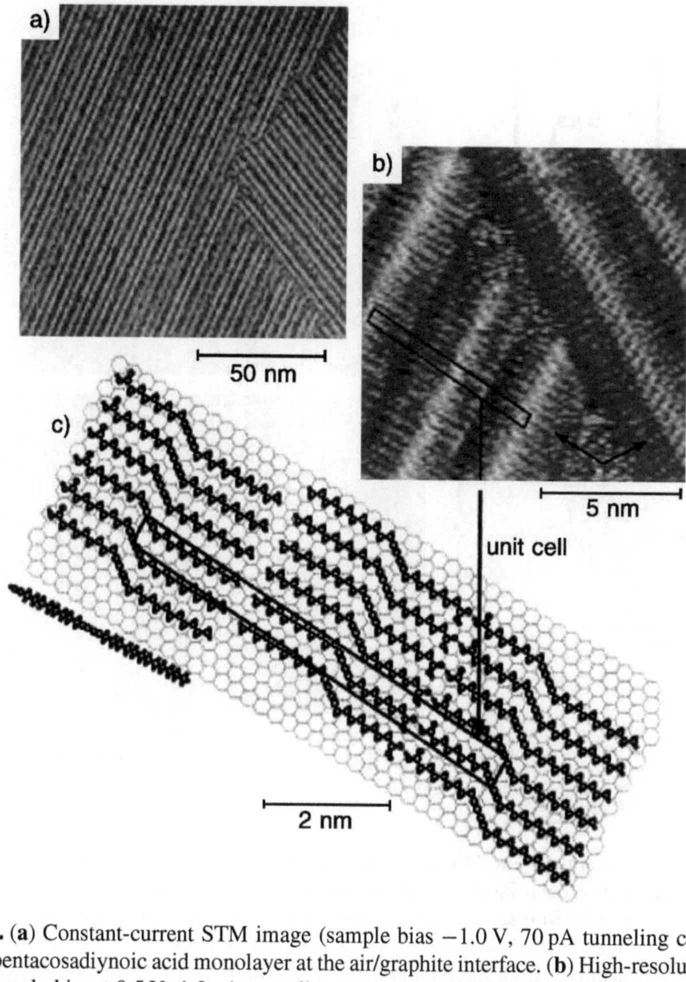

Fig. 11.13. (**a**) Constant-current STM image (sample bias -1.0 V, 70 pA tunneling current) of a 10, 12-pentacosadiynoic acid monolayer at the air/graphite interface. (**b**) High-resolution STM image (sample bias $+0.5$ V, 1.0 nA tunneling current) emphasizing the packing structure. As indicated by the *black arrows*, the molecular axis is aligned with high symmetry axes of the underlying graphite substrate. (**c**) Proposed molecular model for the monolayer giving rise to the STM images displayed in (**a**) and (**b**). *Black parallelograms* shown in (**b**) and (**c**) depict a unit cell of the monolayer structure. Reprinted Fig. 1 with permission from Y. Okawa and M. Aono, J. Chem. Phys. (2001), 115(5), 2317–22 (2001). Copyright (2001) by the American Physical Society

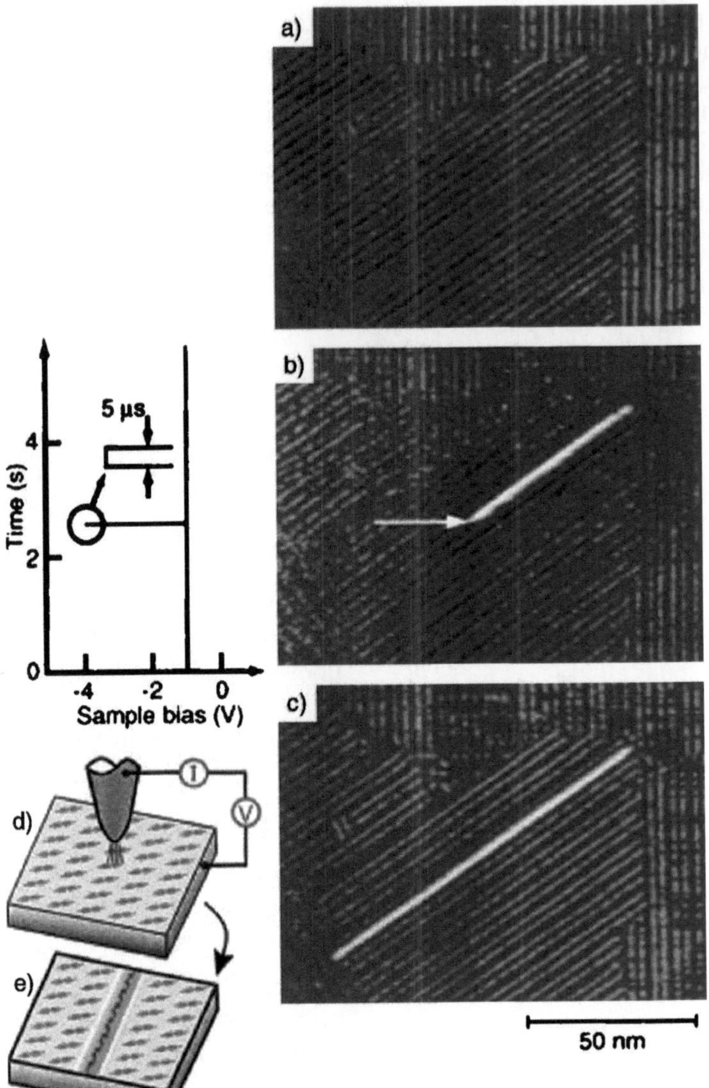

Fig. 11.14. (**a**) Constant-current STM image (sample bias −1.0 V, 70 pA tunneling current) of the original monomolecular layer of 10,12-pentacosadiynoic acid at the air/graphite interface. (**b**) Constant-current STM image of the same area as in (**a**) but with application of a bias voltage pulse during imaging while the STM tip passed the location indicated by the *white arrow*. The image was acquired from the *bottom to the top*. The contrast of a single-molecular row has changed drastically. (**c**) Constant-current STM image acquired immediately after the image shown in (**b**). The bright feature is seen to extend in both directions from the location where the voltage pulse was applied. (**d,e**) Diagrams illustrating the initiation of chain polymerization with the STM tip. Reprinted Fig. 4 with permission from Y. Okawa and M. Aono, J. Chem. Phys. (2001), 115(5), 2317–22 (2001). Copyright (2001) by the American Physical Society

STM is seen to be a powerful tool not only for local probing but also for local nanoscale manipulation. Molecular self-assembly is often cited as an essential ingredient for future nanoscale engineering. These studies demonstrate a path for augmenting the power of self-assembly through targeted postassembly manipulation.

11.5.2 Templates for Bottom-Up Assembly

In the quest for novel device architectures, controlling structure at the nanoscale is a key part of a broader challenge: Closing the gap between current bottom-up and top-down capabilities in order to achieve complete hierarchy of organization at all length scales. Meaningful control at multiple length scales further implies the ability to produce structures with a high degree of complexity. While self-assembly can lead to simple periodic structures, it does not necessarily preclude the generation of complexity, even before being augmented by postassembly modifications as discussed in the previous section. Some of the examples discussed above (see, e.g., Sect. 11.4.2) have employed monolayers formed by more than one component, in some cases leading to well-defined supramolecular structures. As shown by de Feyter et al. [50] self-assembly from multiple components at the solution/graphite interface can lead to monolayers composed of large heterocomplexes with unit cell parameters approaching $\sim$ 5 nm, which can serve to immobilize species that would otherwise be quite mobile.

In a host-guest, or template approach to multicomponent self-assembly, the hierarchy of (self-) organization is especially obvious. In this case one ingredient serves as a building block of a host (or template) structure that can exist by itself (i.e. conceptually before adding any guest), thus offering predictable results upon adding a variety of guest species. Several recent STM studies of self-assembly on graphite have implemented this strategy. Figure 11.15 shows STM images of self-assembled domains formed by 1,3,5-tris(10-carboxydecyloxy)benzene (TCDB) [51]. The self-assembly was formed in the presence of Cu phthalocyanine molecules. In some domains, the large tetragonal cavities in the TCDB network are filled with Cu phthalocyanine guest molecules. Apparently, the self-assembly of TCDB produces a template for organizing Cu phthalocyanine. The presence of the guest slightly distorts the host unit cell, which might account for the fact that most domains show either nearly complete or nearly absent decoration with guest molecules. Similar results are seen with coronene as guest species where up to two units can be included per TCDB cavity.

As the list of nanoscale objects (e.g., particles, rods, tubes) has been growing steadily, the organization and interconnection of multiple types of species is of obvious interest. Hoeppener et al. [20] have tailored an STM study of self-assembly on graphite to examine such a coassembly of multiple species. In their study, the self-assembled monolayer formed by octadecanoic acid at the phenyloctane/graphite interface has served as a template for organizing Au_{55} clusters. As seen in Fig. 11.16, the resulting STM images exhibit rows of features ascribed to the Au_{55} clusters, with distances between rows given by twice the lamella width of the (presumably underlying) alkanoic acid monolayer.

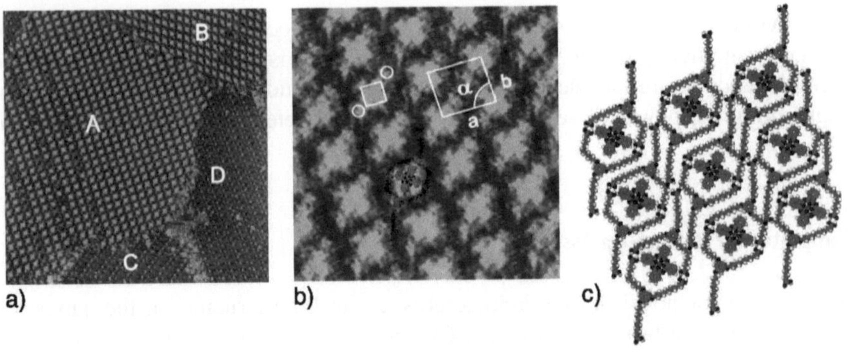

Fig. 11.15. (**a**) STM image (86.4 nm × 86.4 nm, sample bias 700 mV, tunneling current 853 pA) of TCDB networks and Cu phthalocyanine/TCDB host–guest architecture. The z-axis is 0.5 nm. (**b**) A higher-resolution STM image (13.2 nm × 13.2 nm) of the Cu phthalocyanine/TCDB host–guest architecture. (**c**) Suggested molecular model for the host–guest architecture. Used with permission from Jun Lu, Sheng-bin Lei, Qing-dao Zeng, Shi-Zhao Kang, Chen Wang, Li-jun Wan, and Chun-li Bai, J. Phys. Chem. B (2004), 108(17), 5161–5. Copyright (2004) American Chemical Society

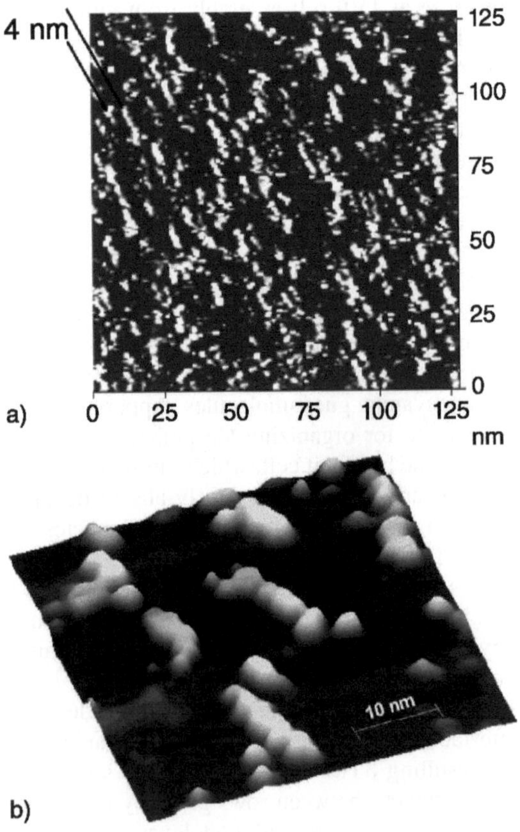

Fig. 11.16. (**a**) STM image of cluster rows formed by incorporating Au_{55} clusters into the molecular template measured at the solid/liquid interface of HOPG at a sample bias of ∼1 V and tunneling current of 80 pA. The distance of the cluster strands is ∼4 nm. This distance corresponds to the distance that is spanned by a bilayer consisting of two facing molecules. (**b**) The zoom shows some examples of highly resolved Au_{55} clusters. The bright strands consist of individual clusters that are arranged close to each other within the lamella that is provided by the fatty acid template and that mediates the clusters' positions. Used with permission from Stephanie Hoeppener, Lifeng Chi, and Harald Fuchs, Nano Letters (2002), 2(5), 459–63. Copyright (2002) American Chemical Society

11.6
Toward Molecular Devices

11.6.1
Ring Systems and Electronic Structure

STM studies of self-assembly on graphite have not been limited to simple substituted linear alkanes. Indeed, the success in the formation and interrogation of ordered monolayers for a wide range of substituted alkanes suggests strategies for the self-assembly of monolayers composed of ring systems. Large cycloalkanes have been found to self-assemble at the liquid/solid interface, forming structures analogous to those of unsubstituted n-alkanes [52]. In the case of more rigid, planar ring systems with extended π-electron systems, alkyl chains, functional groups, or both have been attached at their periphery to mediate self-assembly on the basal plane of graphite. Reports of self-assembled monolayers composed of such tailored ring systems include polyphenylenes, phthalocyanines, and porphyrins [53–55]. Figure 11.17 shows a high-resolution STM image of 21,23-Dihydro-5,10,15,20-tetrakis[4-(tetradecyloxy)phenyl] porphyrin (TTPP) at the air/graphite interface [55]. Clearly, an ordered monolayer structure is observed, presumably stabilized by the interdigitization of the 14-carbon long alkyl chains attached to each porphyrin. Consistent with this hypothesis, shorter-chain analogs were found to produce less ordered monolayer structures [55]. Ring systems such as trimesic acid and DNA bases lack alkyl chains at their periphery. Here the formation of intermolecular hydrogen bonds drives the formation of ordered monolayers at the basal plane of graphite [56–58]. The ability to form intermolecular hydrogen bonds is essential to the biological function of DNA bases. It has been suggested that the stabilization of self-assembled monolayers (composed of DNA bases) by intermolecular hydrogen bonding may have played an equally important role in the emergence of life [59].

The STM image shown in Fig. 11.17 is clearly dominated by bright features. Their size and four-fold symmetry suggest that they arise from porphyrin cores that lie flat on the graphite substrate. The porphyrin cores are seen to be consistently brighter than the alkyl chains at their periphery, as expected from the fact that the extended π-electron system of the porphyrin cores gives rise to occupied and unoccupied states that are much closer to the Fermi level than any electronic states with appreciable density on the alkyl chains [4, 13].

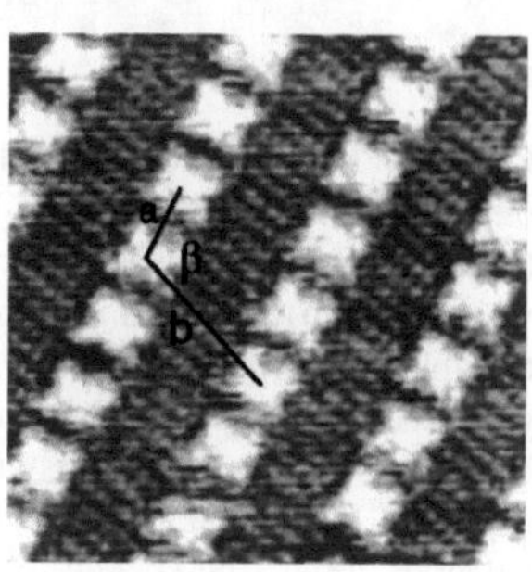

Fig. 11.17. Constant-current STM image (15 nm × 15 nm, sample bias 0.752 V, 1.032 nA tunneling current) of 21,23-Dihydro-5,10,15,20-tetrakis[4-(tetradecyloxy)phenyl] porphyrin on graphite. The porphyrin cores appear as bright features of four-fold symmetry. Hongna Wang, Chen Wang, Qingdao Zeng, Shandong Xu, Shuxia Yin, Bo Xu, and Chunli Bai, Surf. Interface Anal. 2001; 32:267. Copyright 2001 John Wiley and Sons Limited. Reproduced with permission

In a few cases the monolayers formed by ring systems without alkyl chains or functional groups have been interrogated under ambient conditions. Both very large polycyclic aromatic hydrocarbons [60] and porphyrins [61] form ordered monolayers at the liquid/solid interface. However, in the presence of a supernatant solution such monolayers often coexist with a significant number density of π-stacked aggregates and adsorbate multilayers [4, 60, 61], suggesting at most a weak selective stabilization of the adsorbate monolayer. The same species may be accessible to interrogation when vapor deposited under UHV conditions where the different adsorption procedure and absence of a supernatant solution changes both energies and kinetics. For ring systems that form stable monolayers both in UHV and at the liquid/solid interface, the presence of solvent can have subtle effects such as changing the phase relation with respect to the substrate [62].

Most studies of monolayers formed by ring systems lacking both, alkyl chains and functional groups at their periphery have been performed under UHV conditions. Figure 11.18 shows UHV STM images of an ordered monolayer formed by free base naphthalocyanine on the basal plane of graphite [14]. A single-point monolayer defect near the center of the image emphasizes the four-fold rotational symmetry of naphthalocyanine and aids in identifying the molecular orientation within the monolayer. Naphthalocyanine is seen to lie "flat" on the graphite substrate, i.e. with its π-electron system parallel to the surface plane. A seamless monolayer packing can be seen, apparently maximizing the number density of adsorbed molecules. However, substrate commensurability is maintained [14],

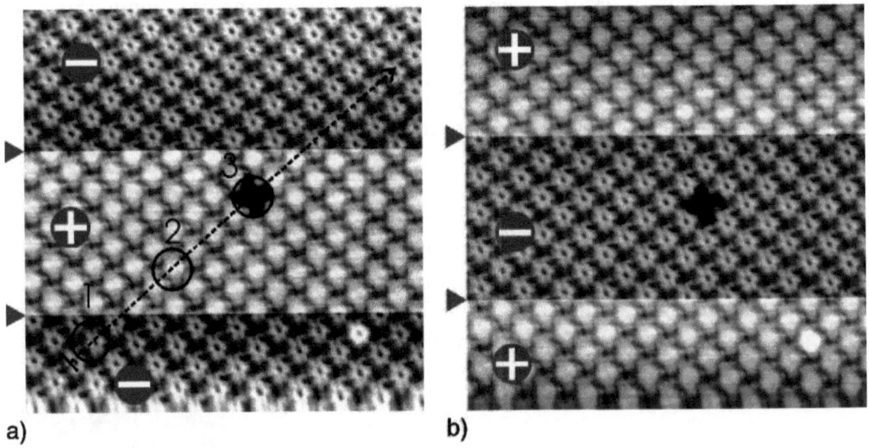

Fig. 11.18. High-resolution STM constant-current topographs of a naphthalocyanine monolayer on HOPG ($25 \times 55\,\text{nm}^2$, $|V| = 1.83$ V, 83 pA) obtained at 50 K; during the scans the polarity of the tunneling voltage was switched twice. In (**a**) scanning was started with negative sample bias, whereas in (**b**) it started with positive sample bias. The corresponding lines where switching took place are marked by *triangles* on the *left-hand side* and the actual sign of the sample bias is given within each part of the frame. Both images represent approximately the same sample area, but have a different order of switching. Used with permission from Markus Lackinger, Thomas Müller, T. G. Gopakumar, Falk Müller, Michael Hietschold, and George W. Flynn, J. Phys. Chem. B 2004, 108(7), 2279–84. Copyright (2003) American Chemical Society

indicating that the periodic corrugation of adsorbate–substrate interactions plays a significant role in the formation of this monolayer structure. Commensurable monolayer structures have been observed also for other planar ring systems with delocalized π-electron systems, even when intermolecular hydrogen bonds are present [58].

As indicated by "+" and "–" symbols overlaid on the images, the bias voltage polarity was switched during acquisition of the UHV STM images shown in Fig. 11.18. The images have sufficient resolution to exhibit details of the internal structure within individual molecules and reveal a distinct dependence of submolecular image contrast on bias voltage polarity. For negative sample bias, i.e. tunneling out of occupied sample states, the center of the naphthalocyanine appears dark. For positive sample bias, i.e. tunneling into unoccupied sample states, all molecules appear as bright humps, without any indication of a decrease in apparent height at the center. These differences are consistent with the frontier molecular orbitals calculated for gas-phase molecules [14].

High-resolution UHV STM images of porphyrin and phthalocyanine monolayers on noble metal surfaces have shown that characteristic changes in submolecular image contrast also occur when the central protons in the free base macrocycle are replaced by a transition-metal substituent [11, 12]. Here the contribution of metal d-electrons to the electronic structure near the Fermi level and thus the tunneling probability lends chemical sensitivity to STM images, which can be used to characterize supramolecular structures in multicomponent films [63].

11.6.2 Model Systems for Molecular Electronics

Recently, STM studies have included large molecules specifically tailored both to facilitate self-assembly and for probing electron transport at the nanoscale in model systems for molecular electronics. Following early suggestions by Aviram and Ratner [64] molecular analogs for current rectifiers (diodes) and transistors are being investigated where electron donor (D) and acceptor (A) units are linked covalently. STM constitutes an ideal tool for such studies, as the tunneling junction at the center of the STM technique intrinsically probes charge transport at the nanoscale. As illustrated in Fig. 11.1, STM studies of molecules adsorbed on surfaces intrinsically address the mediation of charge transport by molecular states and their coupling to one of the electrodes (i.e. the conducting substrate supporting the molecule). Essentially the same diagram as shown in Fig. 11.1 also applies for molecular conductance junctions (e.g., break or electromigration junctions). However, STM provides the added advantage of (lateral) scanning, thus combining the spectroscopic mode with an imaging mode. In molecular conductance junctions, the presence of a molecule has to be inferred from the spectroscopic data alone. The same is true for molecular orientation and bonding to the electrodes. In contrast, STM images can complement STS data acquired with a scanning tunneling microscope to ascertain the presence of the species of interest and determine its orientation and conformation. In addition, STS data can be acquired with the STM tip positioned over different parts of a given molecule, effectively mapping out the spatial variation of the conductance contributions associated with molecular orbitals.

Miura et al. [19] have investigated a large, symmetric D-A-D system where oligo p-phenylene vinylene (OPV) donor units are attached to a perylenediimide (PDI) acceptor moiety (see Fig. 11.19a). The STM images displayed in Fig. 11.19b,c,e reveal the formation of ordered monolayer domains at the liquid/solid interface. Dispersion interactions between the alkyl chains attached to the D-A-D triad may play a role in stabilizing the self-assembled structure as suggested by the interdigitated dodecyloxy chains in the molecular model displayed in Fig. 11.19d. The large molecular size and ordered monolayer structure observed in STM images permits the identification of donor and acceptor subunits (within individual molecules) so that their relative image contrast can be examined as a function of bias voltage. Bias-dependent imaging reveals brighter contrast (i.e. higher tunneling current) for the donor moieties at negative sample bias, while the acceptor moieties are brighter at positive sample bias (see Fig. 11.19e).

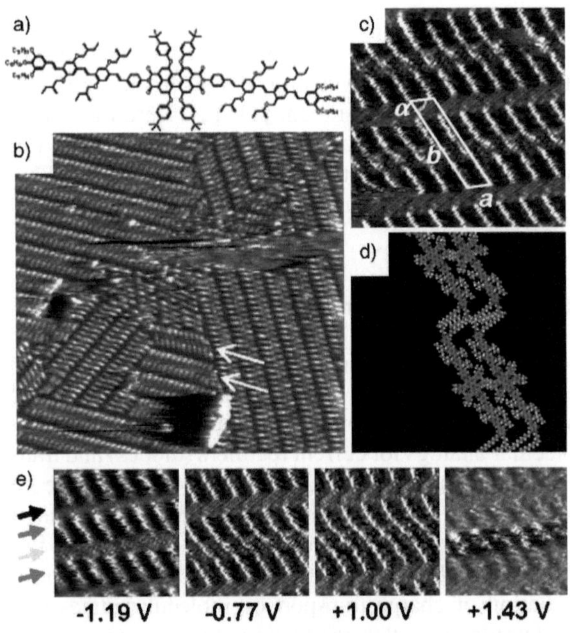

Fig. 11.19. (**a**) Chemical structure of the D-A-D triad. (**b**) Constant height STM image (70.2 nm × 70.2 nm, sample bias −0.96 V, 150 pA tunneling current) of a D-A-D monolayer at the interface between a 1-phenyloctane solution and the basal plane of graphite. *Arrows* indicate mirror-image-type packing. (**c**) High-resolution STM image (15.5 nm × 15.5 nm, sample bias −0.70 V, 400 pA tunneling current) with overlaid parallelogram indicating a monolayer unit cell. (**d**) Proposed molecular model reflecting the ordering in (**c**). (**e**) Bias-dependent imaging of the D-A-D triad (10.1 nm × 10.1 nm, 400 pA tunneling current, sample bias indicated below each image). In order *top*-to-*bottom*, the *arrows* on the *left* refer to alkyl chains, donor units, acceptor units, and donor units. Used with permission from Atsushi Miura, Zhijian Chen, Hiroshi Uji-i, Steven de Feyter, Magdalena Zdanowska, Pascal Jonkheijm, Albertus P. H. J. Schenning, E. W. Meijer, Frank Würthner, and Frans C. de Schryver, J. Am. Chem. Soc., 2003, 125(49), 14968–9. Copyright (2003) American Chemical Society

A simple physical picture has been proposed relating the observed bias-dependent charge transmission to the electronic structure of the D-A-D system [19]. Absorption spectra indicate that the donor and acceptor moieties are electronically decoupled or weakly coupled, allowing frontier orbitals of the D-A-D triad to be assigned as belonging to individual subunits. As the electron donor units give up electrons most easily (by definition), the donor HOMO constitutes the highest filled state of the overall molecule and has a dominant influence on the surface LDOS just below the Fermi level. Thus, the OPV donor moieties support a higher tunneling current at negative sample bias, where filled surface states are being probed. Conversely, the LUMO of the PDI acceptor moiety dominates the density of empty surface states just above the Fermi level, making the PDI unit more transmissive at positive sample bias. Thus, the bias-dependent charge transmission through this molecular device analog can be understood in terms of resonantly enhanced electron tunneling dominated by individual electronic states associated with molecular subunits [19].

Recent STM studies of model systems for molecular devices also include a prototypical single-molecule chemical-field-effect transistor where nanometer-sized charge-transfer complexes control current rectification through a covalently linked ring system [16]. At the center of these investigations is an electron-rich hexa-peri-hexabenzocoronene (HBC) core, surrounded by six electron-poor anthraquinone (AQ) subunits [16]. As in the previous example, individual molecular subunits can be identified in high-resolution STM images and the bias-polarity-dependent tunneling probability can be understood in terms of resonantly enhanced electron tunneling. Upon adding the electron-rich 9,10-dimethoxyanthracene (DMA) to the solution, DMA-AQ charge-transfer (CT) complexes are formed. New self-assembled domains that incorporate CT complexes are found to coexist with remaining domains free of DMA. As a local, high-resolution probe, STM allows the separate interrogation of individual molecules in these coexisting self-assembled structures. Current–voltage curves obtained atop the HBC cores are found to be asymmetric (rectifying) with the degree of asymmetry controlled by the presence (or absence) of the DMA-AQ-CT complex. The interfacial dipole associated with formation of the CT complex causes a relative shift between the adsorbate's electronic states and the Fermi level of the substrate [16, 17]. Thus, STM has been used to probe the charge transmission through a single molecule and its control by a chemically induced electric field effect.

The STM studies discussed in this section have combined self-assembly on graphite with detailed single-molecule conductance measurements by using specifically tailored molecules synthesized prior to deposition on the graphite substrate. Previous sections have covered examples for postassembly modification of self-assembled monolayers (e.g., topochemical polymerization of diacetylene derivatives) [47–49] on graphite and for self-assembled monolayers serving as templates for other species [20, 51]. Based on combinations of these ingredients one could envision powerful recipes for the targeted creation and interrogation of complex model systems for molecular devices with STM and self-assembly on graphite playing a central role.

11.7
Summary and Conclusions

Self-assembly on the basal plane of graphite can provide ideal model systems for the study of two-dimensional self-organization. A wide range of molecular species have been shown to form highly ordered monolayers on this inert substrate in vacuum and under ambient conditions. Due to its ability to provide high (submolecular) resolution locally in a variety of environments, STM has played a prominent role among experimental methods employed for studying the structure of self-assembled monolayers and dissecting the driving forces for their formation. For many functionalized alkanes, lamellar monolayer structures have been observed, with structural parameters depending on the balance of interactions associated with the alkyl chains and the functional group. In ambient studies of planar ring systems, monolayer formation is often driven by alkyl chains or functional group interactions. In contrast, UHV studies have been less dependent on substituents at the periphery of ring systems facilitating monolayer stabilization through specific intermolecular interactions.

The controlled bottom-up (self-) assembly at interfaces may become an important ingredient in future nanoscale engineering, particularly if the elements being assembled include single-molecule devices. The understanding and control of charge transport in model systems for such molecular devices constitutes an important focal point of current efforts in nanoscience. Due to its ability to interrogate the mediation of charge transport by individual molecular states, STM continues to be uniquely positioned as an important tool in this area. Research efforts reviewed in the present chapter include recent STM studies that have been designed to interrogate charge transport in molecular devices while building on previous work to self-assemble the species of interest into ordered arrays.

Simple models have been proposed relating the observed charge transport to the electronic structure of the adsorbed species. Future studies can be envisioned where the power of this approach is augmented further through combination with postassembly modification, molecular templates, and coassembly of multiple nanoscale components.

Acknowledgements. The author is grateful to Prof. George W. Flynn and Dr. Chunzeng Li for their support and insightful discussions.

References

1. Binning G, Rohrer H, Gerber C, Weibel E (1982) App Phys Lett 40:178
2. Binning G, Rohrer H, Gerber C, Weibel E (1982) Phys Rev Lett 49:57
3. Binning G, Rohrer H, Gerber C, Weibel E (1983) Phys Rev Lett 50:120
4. Giancarlo LC, Flynn GW (1998) Annu Rev Phys Chem 49:297
5. Giancarlo LC, Flynn GW (2000) Acc Chem Res 33:491
6. Poirier GE (1997) Chem Rev 97:1117
7. Florio GM, Werblowsky TL, Baker T, Müller T, Flynn GW (2005)
8. Tersoff J, Hamann DR (1983) Phys Rev Lett 50:1998
9. Tersoff J, Hamann DR (1985) Phys Rev B 31:805

10. Magonov SN, Whangbo MH (1996) Surface analysis with STM and AFM: experimental and theoretical aspects of image analysis. VCH, Weinheim, New York
11. Lu X, Hipps KW, Wang XD, Mazur U (1996) J Am Chem Soc 118:7197
12. Lu X, Hipps KW (1997) J Phys Chem B 101:5391
13. Claypool CL, Faglioni F, Goddard WA, III, Gray HB, Lewis NS, Marcus RA (1997) J Phys Chem B 101:5978
14. Lackinger M, Müller T, Gopakumar TG, Müller F, Hietschold M, Flynn GW (2004) J Phys Chem B 108:2279
15. Hipps KW, Scudiero L (2005) J Chem Educ 82:704
16. Jäckel F, Watson MD, Müllen K, Rabe JP (2004) Phys Rev Lett 92:188303
17. Jäckel F, Wang Z, Watson MD, Müllen K, Rabe JP (2004) Synth Met 136:269
18. Samori P, Yin X, Tchebotareva N, Wang Z, Pakula T, Jäckel F, Watson MD, Venturini A, Müllen K, Rabe JP (2004) J Am Chem Soc 126:3567
19. Miura A, Chen Z, Hiroshi U, De Feyter S, Zdanowska M, Jonkheijm P, Schenning APHJ, Meijer EW, Würthner F, De Schryver FC (2003) J Am Chem Soc 125:14968
20. Hoeppener S, Chi L, Fuchs H (2002) Nano Lett 2:459
21. Eigler DM, Weiss PS, Schweizer EK (1991) Phys Rev Lett 66:1189
22. Liang W, Whangbo MH, Wawkuschewski A, Cantow HJ, Magonov SN (1993) Adv Mater 5:817
23. Fang H, Giancarlo LC, Flynn GW (1999) J Phys Chem B 103:5712
24. Ho W (2002) J Chem Phys 117:11033
25. Li LY, Chen SF, Jiang SY (2003) Langmuir 19:3266
26. De Feyter S, Gesquiere A, Abdel-Mottaleb MMS, Grim PCM, De Schryver FC (2000) Acc Chem Res 33:520
27. Paserba KR, Gellman AJ (2001) J Chem Phys 115:6737
28. Paserba KR, Gellman AJ (2001) Phys Rev Lett 86:4338
29. Müller T, Flynn GW, Mathauser AT, Teplyakov AV (2003) Langmuir 19:2812
30. Magonov SN, Yerina NA (2003) Langmuir 19:500
31. Rabe JP, Buchholz S (1991) Science 253:424
32. Herwig KW, Matthies B, Taub H (1995) Phys Rev Lett 75:3154
33. Askadskaya L, Rabe JP (1992) Phys Rev Lett 69:1395
34. Small DM (1986) The physical chemistry of lipids: from alkanes to phospholipids. Plenum, New York
35. Hansen FY, Herwig KW, Matthies B, Taub H (1999) Phys Rev Lett 83:2362
36. Arnold T, Dong CC, Thomas RK, Castro MA, Perdigon A, Clarke SM, Inaba A (2002) Phys Chem Chem Phys 4:3430
37. Arnold T, Thomas RK, Castro MA, Clarke SM, Messe L, Inaba A (2002) Phys Chem Chem Phys 4:345
38. Krim J, Suzanne J, Shechter H, Wang R, Taub H (1985) Surf Sci 162:446
39. Giancarlo LC, Cyr DM, Muyskens K, Flynn GW (1998) Langmuir 14:1465
40. Fang H, Giancarlo LC, Flynn GW (1998) J Phys Chem B 102:7421
41. Eliel EL, Wilen SH (1994) Stereochemistry of organic compounds. Wiley-Interscience, New York
42. Wei Y, Kannappan K, Flynn GW, Zimmt MB (2004) J Am Chem Soc 126:5318
43. Hibino M, Sunni A, Tsuchiya H, Hatta I (1998) J Phys Chem B 102:4544
44. De Feyter S, Grim PCM, Rucker M, Vanoppen P, Meiners C, Sieffert M, Valiyaveettil S, Mullen K, De Schryver FC (1998) Angew Chem Intl Ed Engl 37:1223
45. Yablon DG, Giancarlo LC, Flynn GW (2000) J Phys Chem B 104:7627
46. Fang H, Giancarlo LC, Flynn GW (1998) J Phys Chem 102:7311
47. Grim PCM, De Feyter S, Gesquiere A, Vanoppen P, Rucker M, Valiyaveettil S, Moessner G, Mullen K, De Schryver FC (1997) Angew Chem Intl Ed Engl 36:2601

48. Okawa Y, Aono M (2001) J Chem Phys 115:2317
49. Okawa Y, Aono M (2001) Nature 409:683
50. De Feyter S, Miura A, Yao S, Chen Z, Würthner F, Jonkheijm P, Schenning APHJ, Meijer EW, De Schryver FC (2005) Nano Letters 5:7
51. Lu J, Lei SB, Zeng QD, Kang SZ, Wang C, Wan LJ, Bai CL (2004) J Phys Chem B 108:5161
52. Wawkuschewski A, Cantow HJ, Magonov SN, Möller M, Liang W, Whangbo MH (1993) Adv Mater 5:821
53. Wu P, Zeng Q, Xu S, Wang C, Yin S, Bai CL (2001) Chem Phys Chem 2:750
54. Qiu X, Wang C, Zeng Q, Xu B, Yin S, Wang H, Xu S, Bai CL (2000) J Am Chem Soc 122:5550
55. Wang H, Wang C, Zeng Q, Xu S, Yin S, Xu B, Bai CL (2001) Surf Interface Anal 32:266
56. Lackinger M, Griessl S, Heckl WM, Hietschold M, Flynn GW (2005) Langmuir 21:4984
57. Griessl S, Lackinger M, Edelwirth M, Hietschold M, Heckl WM (2002) Single Mol 3:25
58. Freund JE, Edelwirth M, Kröbel P, Heckl WM (1997) Phys Rev B 55:5394
59. Sowerby SJ, Heckl WM, Petersen GB (1996) J Mol Evol 43:419
60. Samori P, Severin N, Simpson CD, Müllen K, Rabe JP (2002) J Am Chem Soc 124:9454
61. Tao NJ, Cardenas G, Cunha F, Shi Z (1995) Langmuir 11:4445
62. Stecher R, Drewnick F, Gumpf B (1999)
63. Hipps KW, Scudiero L, Barlow DE, Cooke Jr. MP (2002) J Am Chem Soc 124:2126
64. Aviram A, Ratner MA (1974) Chem Phys Lett 29:277

12 Tunneling Electron Spectroscopy Towards Chemical Analysis of Single Molecules

Tadahiro Komeda

12.1 Introduction

The capability of scanning tunneling microscopy (STM) to provide real-space imaging of a surface with an atomic-scale resolution has revealed magnificent diversity of physical, chemical and electronic phenomena on the surface. With the use of the atomic-scale resolution, a chemical identification of a single molecule has been proposed since an early stage of the STM development.

For the realization of the spectroscopy of the tunneling current, scanning tunneling spectroscopy (STS) has shown that the tunneling current spectroscopy can reveal many important properties of the surface and the adsorbates. However for the reasons stated in the later part of this chapter, STS cannot be the ultimate solution of the chemical analysis of molecules.

Similar efforts towards chemical identification using a scanning probe have been made. Raman spectroscopy using a near-field scanning optical microscope (NSOM) is one of the successful results of chemical identification utilizing the detection of vibrational modes of adsorbates. Even with these development, the use of the tunneling current to unveil chemical properties has many advantages over other methods. The spatial resolution of the tunneling current enables the ultimate chemical analysis: single-molecule chemical analysis. The high current density can induce multiple excitation of electronic or vibrational modes, which has been enabled only with the use of short pulse lasers. It can, in principle, control chemical change of molecules in an atomic scale, and it may play a decisive role in the evolution of nanotechnology [1].

This chapter describes recent research which has utilized the scanning tunneling microscope as an electron source for the chemical analysis of molecules adsorbed on surfaces. Especially the detection of vibrational modes of the adsorbates for the determination of chemical species is focused upon. In Sect. 12.2, the characteristic features of tunneling electrons for the excitation of vibrations in adsorbates are explained in comparison with a conventional electron source. In Sect. 12.3, inelastic tunneling spectroscopy (IETS) with a STM setup is described. The technique can detect vibrational modes of molecules through inelastic tunneling (IET) processes. The excitation mechanism of STM-IETS is discussed and its application to molecules is examined. In Sect. 12.4, the manipulation of adsorbates by the injection of tunneling electrons is discussed; the phenomena include desorption, hopping, and chemical reactions which are caused by vibrational heating. In Sect. 12.5, the ability of action spectroscopy for the detection of vibrational modes is considered, in which the yield of the single-molecule manipulation is measured as a function of the energy of the tunneling electrons. Action spectroscopy can give supplemental information about the vibrational feature of adsorbates.

12.2 Vibrational Excitation Through Tunneling Electron Injection

12.2.1 Characteristic Features of the Scanning Tunneling Microscope as an Electron Source

The superior lateral resolution of STM has been confirmed by the vast number of STM images on various surfaces. The high resolution is a consequence of a high concentration of tunneling current. If 90% of the tunneling current of 1 nA is dosed in a single atom, the current density exceeds 10^6 A/cm^2. The convenient electron source can hardly give this much current density, and even it is realized, the heat generated by the current can easily melt the sample. The heating problem is minimum with the use of STM owing to the small injection area and rapid heat dissipation of electrons.

Thus, STM is not only a microscopy with an atomic resolution, it can also be regarded as an excellent source of electrons with which analysis of surface properties and modification of the surface characteristics might be made. At the same time it should be noted that the electrons obtained from the scanning tunneling microscope tip have different features from those obtained from conventional electron sources. An energy diagram for tunneling in a simplified 1D barrier is schematically shown in Fig. 12.1. Electrons in the energy range between the Fermi level (E_F) and $E_F - V_{bias}$ can tunnel from the tip to the unoccupied states of the substrate. If we assume a slow variation of the local density of state (LDOS) both for the substrate and for the tip, the tunneling current should be proportional to V_{bias} and the I–V curve shows ohmic behavior.

It should be pointed out that the energy of the electrons available from conventional electron sources is *above* the vacuum level of the sample; otherwise they cannot be emitted from the emitter. On the other hand, tunneling electrons can be injected to the states *below* the vacuum level; thus, the tunneling electron has a complementary nature in its energy to that obtained with the conventional electron sources.

At the same time there are several constricting conditions for the tunneling electrons. The first issue is the available energy range of the electrons. The tunneling barrier is close to the work function, which is approximately 4–5 eV for normal metals. If V_{bias} exceeds the work function, the triangle-shaped barrier becomes steep (Fig. 12.1b) [2], and the tunneling current shows a drastic increase. This is because the width of the tunneling barrier becomes thinner; and so-called Fowler–Nordheim (FN) tunneling occurs [3]. It had been generally accepted that the lateral spatial resolution is limited to rather large values of the order of 30 Å [4, 5]. Thus, the tunneling electrons with energy greater than 4–5 eV would not have an atomic-scale resolution and may not be classified as a STM electron source. It should

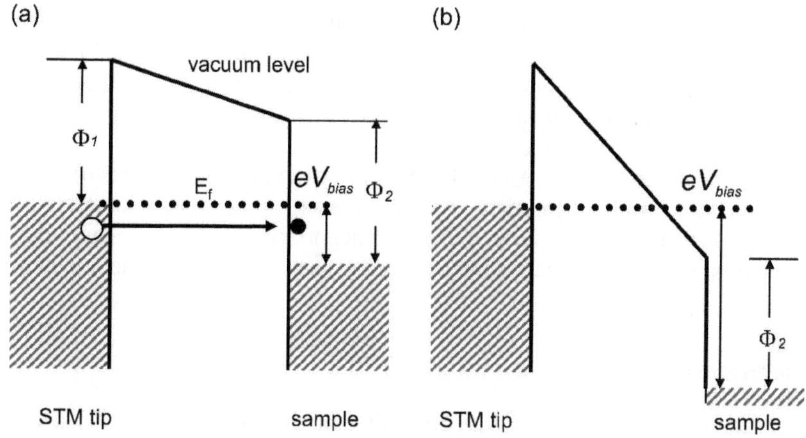

Fig. 12.1. Energy diagram for tunneling in a simplified 1D barrier: (**a**) normal tunneling and (**b**) Fowler–Nordheim tunneling. STM scanning tunneling microscope

be mentioned, in a special case, that FN tunneling electrons can give an atomic resolution. A recent observation of a diamond surface in the bias region of FN tunneling clearly revealed atomic-scale structures even though the bias was much larger than the electron affinity of the sample [6]. However, the high resolution was obtained by tuning the bias voltage to a resonant level which is not available for usual cases.

The next issue is the maximum current available for the tunneling current. Large current with the normal tunneling voltages means the reduction of the tip–substrate distance. It was argued that the tip–sample distance is approximately 1 Å for a tunneling gap of 1 MΩ. If the distance is further reduced, there appears a chemical bonding between the two and irreversible changes to the properties of the surface are made. The current expected for a 1-MΩ-gap distance and a maximum voltage of 5 V is apptoximately 5 μA, and the maximum available current should be close to this value [7–9].

Even with these restrictions, the tunneling current provided by a scanning tunneling microscope tip has a variety of unique characters, whose usage other than as the probe of microscopy has been envisioned for a long time, like the mode-selective vibrational excitations of a single molecule through inelastic electron tunneling process [10].

12.2.2 Electron-Induced Vibrational Excitation Mechanism

First let us examine how the tunneling electrons can excite vibrational modes of adsorbates and how this unique electron source can be utilized for chemical analysis. The high-resolution electron energy loss spectroscopy (HR-EELS) technique is a well-established surface analysis method which can detect vibrational modes of molecules. HR-EELS is an all-electron measurement technique and the energy range allowed for the normal tunneling is similar to the one used in HR-EELS measurements which can cover all the range of the vibrational energy of molecules. Thus, if vibrational spectroscopy with a similar resolution as HR-EELS is realized with the use of electrons from the scanning tunneling microscope tip, it enables the combination of real-space imaging and chemical analysis with an atomic resolution. I first give a brief review of the HR-EELS measurement.

The mechanism of the vibrational excitation with this technique has been investigated in detail both by theoretical and experimental studies. The interaction of the incoming electron with the surface can be categorized into the following classes: (1) dipole excitation, (2) impact scattering, and (3) negative ionic resonant scattering.

12.2.2.1
Dipole Excitation

In this excitation mechanism, it is assumed that the incoming electrons form a time-varying electric field which interacts with the dipole created by the vibration of the adsorbate molecules [11]. The interaction is thus through an electrostatic interaction which is of long-range nature. The excitation process occurs when the electron is far away from the molecule (typically 100 Å) for an electron with a few electronvolts of kinetic energy and the momentum transfer is in general small. The selection rule of dipole scattering is derived from the matrix element in the Golden rule $|\langle F|\gamma \cdot E|I\rangle|^2$, where I is the initial vibrational state and F is its final state. The coupling Hamiltonian is γE, where γ is the dipole moment operator due to the intra-adsorbate nuclear motion with the vibrational mode and E is the local electric field at the adsorbate site. The vibrational modes where the matrix is finite are called the dipole-active modes. Also an image potential plays an important role in an additional selection rule: the dipole created perpendicular to the surface is reinforced by the image dipole and stronger vibrational excitation may be expected. On the other hand, the dipole moment parallel to the surface induces an image dipole which cancels the local dipolar field, which ends up with an weak quadrupole field. As the momentum transfer is limited, the distribution of the inelastic signal owing to the dipole excitation is mainly distributed in the specular direction of the incident electrons [12].

12.2.2.2
Impact Scattering

The second class of the excitation channel is characteristic of EELS and cannot be seen in the similar vibration detection method of IR spectroscopy [13]; namely, the impact scattering channel in which the dipole selection rule is no longer valid and dipole-inactive hindered modes can be detected. Experimentally it is observed or becomes dominant when the scattering angle of inelastic electrons is large where the intensity of the dipole component is drastically decreased. The excitation of this mode is considered to occur in the proximity of the surface or adsorbate compared with the dipole scattering case. The incoming electron interacts with the ion core in the range of several angstroms from the center; the interaction is short range and shows a clear difference from the dipole scattering. The interaction is not so simple as the dipole scattering and multiple scattering should be considered.

Theoretical treatment of the mechanism of the excitation of this channel is done by the combination of the vibrational excitation followed by the formation of a deformed potential which scatters the incoming electron with different momentum and energy. Thus, if we assume a event in which a particular vibrational quantum (energy $\hbar\omega_s$; vibrational motion coordinate Q_s) is emitted and the initial electron is scattered from k^I into k^s in momentum, the following matrix element M should be considered [12]:

$$M(k^I, k^s; +s) = \langle n_s + 1 | f(k^s, k^I; R) | n_s \rangle = \sqrt{(n_s+1)} \sqrt{\frac{\hbar}{2M\omega_s}} \frac{\partial f}{\partial Q_s}, \quad (12.1)$$

where

$$\frac{\partial f}{\partial Q_s} = \sum_{i\alpha} \left(\frac{\partial f}{\partial R_{i\alpha}} \right)_0 \frac{\xi_{i\alpha}}{\sqrt{M_i}}, \quad (12.2)$$

n_s is the number of vibrational quanta, f is the scattering amplitude at the nucleus position R, M is a reduced mass, and $\xi_{i\alpha}$ denotes the amplitude of the displacement of nucleus i in Cartesian direction α where a quantum of vibrational motion is excited.

12.2.2.3 Negative Ion Resonance

This excitation channel is well known in the vibrational excitation of gas-phase molecules. The incident electron is temporally trapped in the unoccupied electronic state of a molecule. The existence of external charge in a molecule can cause a geometrical rearrangement; including the elongation/contraction of the distance between nuclei. Such kinetic movement of nuclei can induce a vibrational mode, which is called a shape resonance.

It is well established that the excitation of vibrational modes of molecules is greatly enhanced in this process for gas-phase molecules [14]. An example can be seen in vibrational excitation of gas-phase O_2 by electrons, in which the differential vibrational cross section shows its maximum near 9.5 eV. The observation is interpreted as the resonant contributions from the negative ion of O_2, and is evidence for the resonant enhancement of vibrational excitations [15].

However, there are a very limited number of reports for successful observation of vibrational excitation via negative ion resonance for molecules on surfaces by HR-EELS. One of the probable reasons is the short lifetime of the negative ionic state for the adsorbed molecules compared with that for gas-phase molecules, which is due to the increased decay path by the formation of chemical bonding. Experimentally the lifetime of the negative ionic state can be evaluated from the spectra of inverse photoemission spectroscopy. The intrinsic broadening of the corresponding feature reflects the lifetime of the ionic state, where the width of the peak (Γ) is connected with the lifetime (τ) as $\Gamma\tau \sim \hbar$. It has been reported that $\Gamma \sim 0.4$ eV for free CO molecules, and that $\Gamma \sim 1.5-2.5$ eV when chemisorbed on Ni [16] and Cu [17].

One of few reports of negative ion resonance is on physisorbed molecules (N_2) on evaporated silver film [18], in which the lifetime of the resonance is relatively long owing to the weak coupling between the molecule and the surface, which enhances the excitation of vibrational modes

Another mechanism which makes the observation of the negative ion resonance for adsorbates difficult is a shift of energy of the molecule's electronic state. The resonant state is formed by the hybridization between the gas molecule's discreet states (highest occupied molecule orbital,HOMO, and lowest unoccupied molecule orbital LUMO) and the valence/conduction band of the substrate. The $2\pi^*$ LUMO state is located approximately 2 eV above the vacuum level in the gas phase, but its resonant state is located approximately 2 eV above the Fermi level in the case CO adsorbed on Cu(111); the energy level is shifted approximately 4 eV downwards upon adsorption. The electrons from conventional sources are provided through a vacuum; it is not possible to inject the electrons directly to the states below the vacuum level. This is certainly part of the reason why the negative ion resonance cannot be seen in electron-induced vibrational spectroscopy studies.

12.3 IET Process of Vibrational Excitation

IETS is an all-electronic spectroscopy. Historically the successful observations of molecule vibrational modes were reported soon after the invention of tunneling devices such as the Esaki diode [19]. By measuring current and voltage across a metal–insulator–metal device whose interface contains a target molecule, it is possible to extract vibrational and electronic information about the metals (magnons and phonons), the insulator, and the molecules (Fig. 12.2a). Pioneering works by Jaklevic and Lambe [20] clearly showed its capability of detecting vibrational features of molecules buried in the interface. It has been successfully applied to problems in surface chemistry and catalysis. Combined data of conventional IETS results are summarized in detail in the reviews in [21,22].

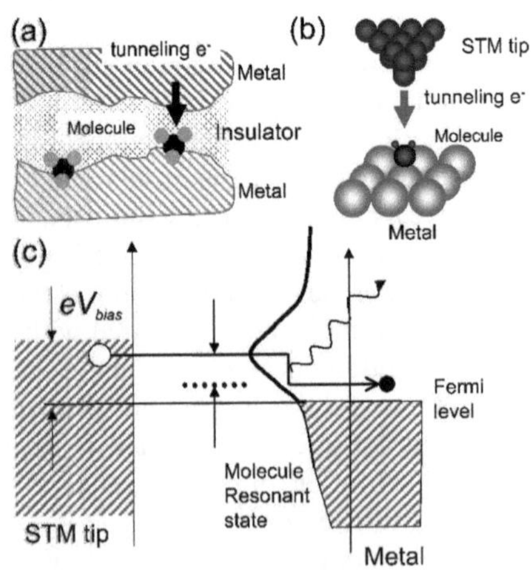

Fig. 12.2. Inelastic tunneling spectroscopy (IETS) measurement on molecules with (**a**) conventional IETS with a metal–insulator–metal tunneling junction (molecules buried in its interface) and with (**b**) a STM set-up. (**c**) A tunneling electron emitted from a STM tip is trapped in a molecule-induced resonant state. The electron is dissipated into the metal substrate after a short time, during which process the electron could excite a phonon or the molecule's vibration with a energy of $\hbar\omega$

Mazur and Hipps [23] have summarized the advantages of IETS as follows. First, IETS has ultrahigh sensitivity. Fewer than 10^{13} molecules are needed to provide a spectrum. Second, overtone and combination bands are exceptionally weak; thus, it is easier to identify fundamentals in IETS than in IR or Raman spectroscopy. Third, optically forbidden transitions may be observed as strong bands. It is possible to obtain adsorbate spectra in the "IR opaque" regions of the oxide spectrum.

In spite of such unique abilities to identify vibrational modes of molecules, it is difficult to say that IETS studies play a main role in the vibrational studies of surface science. The main reason is that, in spite of high sensitivity for the identification of molecules in the interface, there are few methods to characterize the properties of the molecules in the interface. Though vast information has been accumulated in the IETS studies, one can say surface science and IETS have developed in a separate manner. But the most critical problem of characterization of molecules in the tunneling gap might be solved with the use of a STM setup (Fig. 12.2b).

12.3.1 Basic Mechanism of Vibrational Excitation in the IET Process

The excitation of a vibrational mode of a molecule in the scanning tunneling microscope junction is schematically shown in Fig. 12.2c. The model is drawn based on the resonant tunneling model in which an electron first tunnels from the tip to a molecule's resonant state and is then dissipated into the substrate. The properties of the resonant state can be seen on the STM images of adsorbates. An example of a STM image of a CO molecule adsorbed on a Pd(110) surface is shown in Fig. 12.3. Though it has been investigated that a CO molecule shows an upright bonding configuration, isolated CO molecules are imaged with twofold symmetry (Fig. 12.3a) and fourfold symmetry (Fig. 12.3b) depending on their electronic state [24]. This is certainly due to the nature of the resonant state of CO molecules.

For the electrons tunneling through molecules, in addition to the elastic electron path, there appear processes in which a tunneling electron loses its energy by exciting a vibrational mode. Since the maximum energy which can be lost in the inelastic

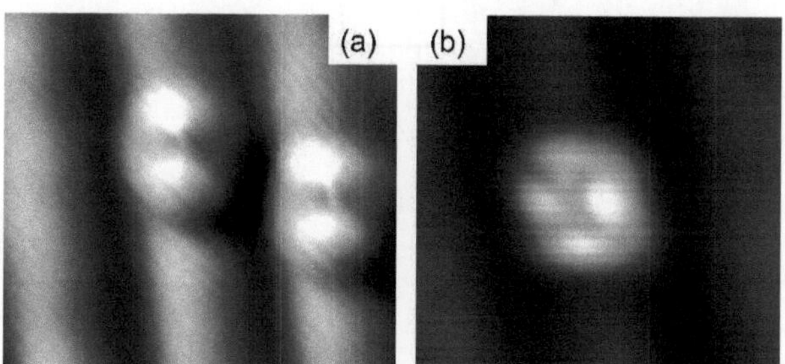

Fig. 12.3. STM images of CO molecules adsorbed on a Pd(110)-c(2 × 4):O surface: (**a**) the CO at the bridging site and a nodal plane between two bright spots is clearly observed, and (**b**) the CO at the terminal site in which the internal structures of the CO molecule appear as four bright spots. The tunneling conditions are $V_{\text{sample}} = -0.05$ V (**a**) and $V_{\text{sample}} = -0.2$ V (**b**) at a tunneling current of 1 nA

process is eV_{bias}, the process occurs only when the energy relation of $eV_{\text{bias}} > \hbar\omega$ is satisfied. The appearance of the new channel with the inelastic process contributes to the increase of the conductance ($\mathrm{d}I/\mathrm{d}V$) for $V > 0$. The relations of the I–V, $\mathrm{d}I/\mathrm{d}V$, and $\mathrm{d}^2I/\mathrm{d}V^2$ curves are schematically shown in Fig. 12.4. In the $\mathrm{d}^2I/\mathrm{d}V^2$ curve, two peaks with Gaussian shape are expected at the symmetric positions in energy with reverse polarities in the intensity [21,23].

The question is what is the mechanism for the excitation of a vibrational mode during an inelastic process. IETS is an all-electron spectroscopy as is HR-EELS; therefore the three processes discussed for the HR-EELS method should be considered in IETS process – namely, dipole scattering, impact scattering, and resonant negative ion formation. At the same time it should be noted that there exist distinct differences in the properties of incident electrons between the two methods. First, the interaction at a long distance of approximately 100 Å, which is a major part of dipole scattering of HR-EELS, cannot be realized in the IET process, where the electron source is located within 10 Å. Second, the electrons tunnel into and interact with the states located between E_f and the vacuum level (if electrons are injected into a sample). Owing to the lowering of the LUMO level from that in the gas phase when the molecules are adsorbed on surfaces, gas-induced resonance states are often observed in this energy range. It is highly possible for tunneling electrons to occupy the resonant state in the tunneling process.

In spite of the data accumulated for conventional IETS measurements, there does not seem to be a complete theoretical explanation for the mechanism of this measurement technique. Theoretical work on inelastic electron tunneling has been presented for the planar metal–oxide–metal junction [21], and it has been argued that IET is dominated by dipole coupling in most cases [25,26].

However, as discussed already, the resonant coupling via the formation of a temporary negative ion is favored in the IET process, and Persson and Baratoff have

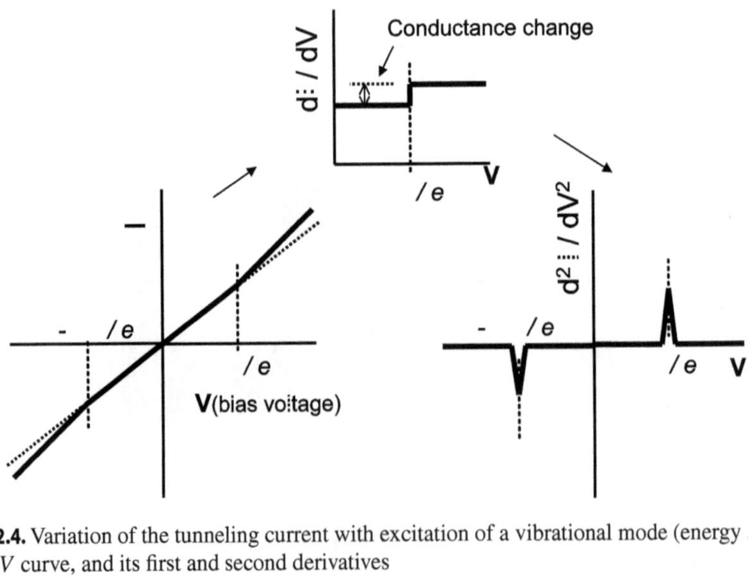

Fig. 12.4. Variation of the tunneling current with excitation of a vibrational mode (energy $\hbar\omega$) in an I–V curve, and its first and second derivatives

developed a theory which stresses the importance of resonant coupling. They showed that the contribution from the resonant coupling is equal to or larger than the long-range dipole coupling [27]. In addition it was claimed that the resonant coupling is important even when the resonance is centered well away from the Fermi energy, i.e., in a nonresonant condition (impact scattering).

Intriguingly it is claimed that the elastic channel suffers an effect of backscattering of elastic electrons by exciting and reabsorbing a vibrational mode at the threshold energy, which might reduce the intensity of an elastic component. Experimentally we can only measure the sum of the elastic and the inelastic component. Thus, the shape of the IETS spectra is not always protruded; it can be dipped or flat even if the inelastic component gives a positive contribution for the conductance change. Most conventional IETS experiments have been performed on molecules adsorbed on inert Al_2O_3 oxide and in these cases one would not expect any adsorbate-induced resonance states to occur in the vicinity of the Fermi level. This has been attributed as the main reason why most of the observations of the tunneling conductance show a positive feature [27]. One of a few examples in which such a dip is observedis methyl isocyanide molecules that are adsorbed on alumina-supported rhodium particles, and this can be evidence of the importance of the resonant coupling process [28]. Persson [27] claimed that such behavior is characteristic of negative ion resonance and that neither dipole nor impact scattering induces such behavior.

12.3.2
IETS with the Setup of STM

Spectroscopy of the tunneling current has been an important part of STM research. The investigation started just after the success of STM imaging [29, 30]. The STS technique utilizes the character of the tunneling current which depends on the magnitude of the LDOS of the tip and the substrate; dI/dV and more precisely $dI/dV/(I/V)$ can express the distribution of the LDOS well [31–33]. Since STS is a well-developed technique, it should be clarified why the STM-IETS technique is necessary in addition.

STS gives information on electronic states of the surface: LDOS as a function of energy, which has been investigated using photoemission and inverse photoemission techniques. It gives LDOS information with an atomic-scale resolution. However, there are several drawbacks in applying this technique to the study of a molecule adsorbed on a surface. One of the reasons is related to the range of the bias voltage allowed for the normal tunneling experiment. As mentioned already, this is in the range of 4–5 eV. The interesting electronic states included in this energy range are, for example, the band gap of semiconductors or the superconducting gap. As has been shown by photoemission spectroscopy and inverse photoemission spectroscopy experiments, interesting features expected for adsorbates often appear out of this range.

Second, it is quite difficult to make chemical identification only from the STS spectra, such as the normal photoemission in the valence band region. The appropriate interpretation of the spectra in the valence band region is possible only with a theoretical calculation.

Again it should be stressed that vibrational spectroscopy with the use of a tunneling current is a very adequate technique for chemical analysis for the following reasons: (1) the energy range allowed for the normal tunneling spectroscopy can cover all the range of the energy of the vibrational modes of molecules; (2) characteristic frequencies of internal vibrational modes of gas-phase molecules serve as excellent "fingerprints" that can be sought in the signal derived from a solid surface covered with unknown chemical species. Typically the sensitivity of IR absorption spectroscopy (IRAS) and EELS has been estimated to be approximately 0.005 monolayers; it should be magnificent to reduce the required number of molecules to one for chemical analysis.

12.3.3 Instrumentation of IETS with the Use of STM

An example of the experimental setup used in STM-IETS is shown in Fig. 12.5. Conventional STS measurements can be executed by taking the I–V curve followed by numerical differentiation of the curve. This procedure is possible because the natural width of the electronic states is of the order of 1 eV and there is no special technical difficulty in taking the numerical derivative of the I–V curve. The vibrational features, on the other hand, have a quite narrow peak width of the order of 1 mV, which makes it quite difficult to obtain $\mathrm{d}^2I/\mathrm{d}V^2$ by numerical differentiation. Instead, lock-in-amplifier detection using a voltage modulation method is required. The measurement of the lock-in-amplifier signal requires a relatively long acquisition time of the order of approximately 100 s to get an single spectrum. The change of relative location between the tip and the molecule is critical to obtain decent spectroscopy; stability is required for both lateral and horizontal directions. Much attention should be paid to both the mechanical stability of the scanning tunneling microscope head and the reproducibility of the control electronics [34,35].

An example of IETS measurement is shown in Fig. 12.6, panels a–d, where the variations of bias voltage, tunneling current, $\mathrm{d}I/\mathrm{d}V$, and $\mathrm{d}^2I/\mathrm{d}V^2$ are displayed as a function of time. The data were obtained for a *trans*-2-butene (C_4H_8) molecule

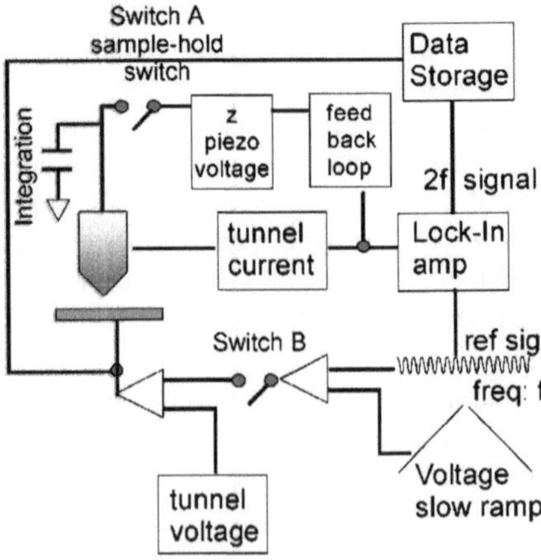

Fig. 12.5. Setup of STM-IETS. IETS spectra are measured by turning off the feedback circuit (*Switch A*) and applying a slow-ramping voltage in which a sine-shaped modulation voltage is superimposed (frequency approximately 797 Hz) to the sample bias by closing *Switch B*. The differentiation of the tunneling current is executed by the use of a lock-in amplifier

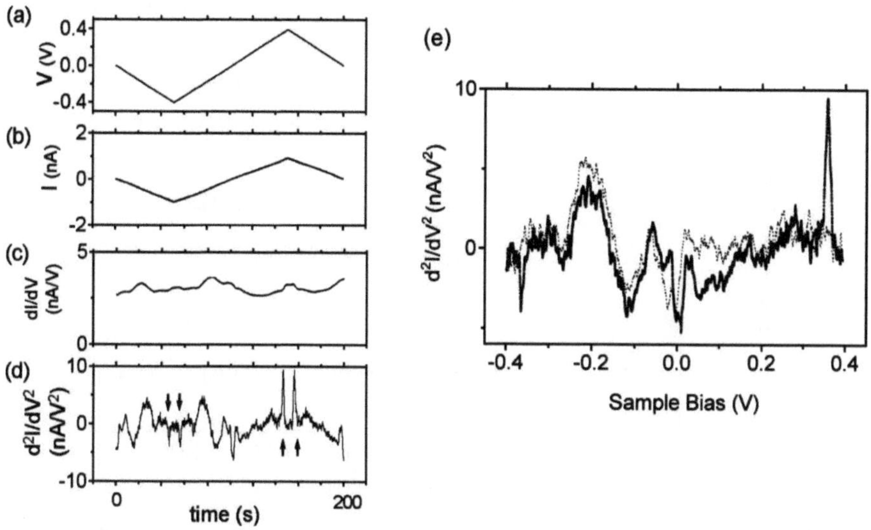

Fig. 12.6. Variation of sample bias (**a**), tunneling current (**b**), the output of the lock-in amplifier of the harmonic (**c**), and the second-harmonic component (**d**) as a function of the time in the STM-IETS measurement of a *trans*-2-butene molecule on Pd(110). All graphs are the average of 16 cycles of the measurements (see text). The *arrows* indicate sharp features corresponding to the ν(C–H) vibrational mode. (**e**) The d^2I/dV^2 vs V spectra obtained for a *trans*-2-butene molecule (*solid line*) and on the bare Pd(110) surface (*dotted line*)

adsorbed on Pd(110). The curves of d^2I/dV^2 vs V are shown in Fig. 12.6e by comparing the spectra obtained on a bare metal substrate (dotted line) and on an adsorbate (solid line). Sharp features can be identified at the sample bias of -363 mV and 360 mV only in the spectrum obtained for the molecule. The former appears as a dip and the latter is a peak which satisfies the conditions for the assignment to a vibration-induced feature. The intensity of the positive peak is approximately 9.5 nA/V^2 and the full width at half maximum is approximately 12 mV. Though there are several vibrational modes, they are not as obvious as the ν(C–H) component [36].

12.3.4 Examples of STM-IETS Measurements

The measurements of IETS spectra with the setup of STM (STM-IETS) started with the pioneering works by Ho's group, even though theoretical consideration was reported in advance [37]. One of the first successful observations of inelastic electron tunneling spectra can be seen for an isolated acetylene (C_2H_2) molecule adsorbed on a Cu(100) surface [38]. The results showed an increase in the tunneling conductance at 358 mV, resulting from the excitation of the C–H stretching mode, ν(C–H). An isotopic shift to 266 mV was observed for deuterated acetylene (C_2D_2) (Fig. 12.7). The conductance change of ν(C–H) which corresponds to the proportion of the inelastic channel is 4.2%. It is interesting to compare this with the conventional IETS data obtained on hydrocarbons. In the conventional IETS spectra ν(C–H) is

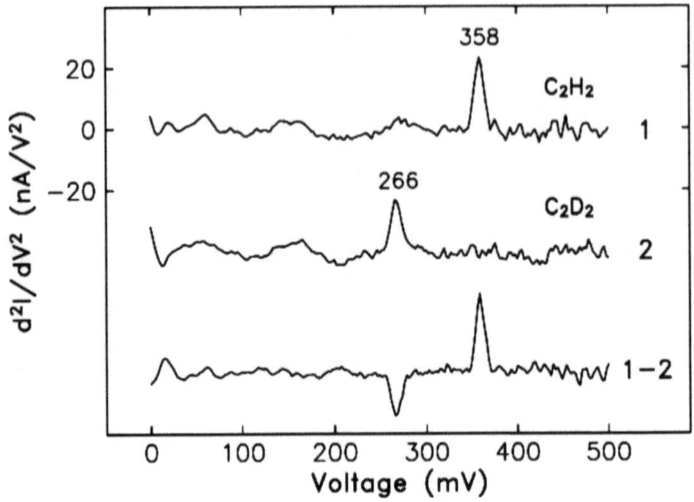

Fig. 12.7. Background difference d^2I/dV^2 spectra for C_2H_2 (*1*) and C_2D_2 (*2*), taken with the same STM tip, show peaks at 358 and 266 mV, respectively. Spectrum *1–2* corresponds to spectrum *2* subtracted from spectrum *1*. (From Stipe et al. [38])

measured also as the strongest feature; a conductance change of 1.8% was observed for a complex silane molecule containing aminopropyltrimethoxy [23,39].

The measurement of IETS spectra of CO molecules followed the success for a acetylene molecule; vibrations of four isotopes of CO on Cu(001) and Cu(110) at 8 K have been measured by STM-IETS [40]. The low-energy mode observed at 36 mV was assigned to the hindered rotation mode (R-mode), which exhibits strong intensity with a conductance change of 15%. The C–O stretch is detected at 256 mV, which is close to the energy obtained using the HR-EELS technique. However, the conductance change is small (approximately 1.5%) and it was claimed that the peak intensity is close to the detection limit of STM-IETS at this stage. It is intriguing to note the detected intensity of the ν(C–O) mode is weak, which is one of the standard vibration modes for the HR-EELS measurements, which make it clear that the detection mechanism of the vibrational modes with the use of IETS is different from that with HR-EELS. The hindered translational mode (T-mode) and the M–CO stretching mode are not observed [40]. It is interesting to see whether the appearance and disappearance of specific vibrational modes in STM-IETS can be related to other properties of these modes. Let us examine their relaxation time (τ), which indicates how long the vibrational mode lasts once it is excited [41]. Experimental results of τ are 2 ps for the CO stretch mode [42], and a similar value for the R-mode [43]. On the other hand, $\tau \sim 40$ ps for T-mode [44] and $\tau > 10$ ps for the CO–Pt(111) stretch have been reported [45]. It might be speculated that a vibrational mode with a short τ can give a strong intensity of IETS features. It might be considered that a short τ indicates a strong coupling between the vibrational mode and the electronic excitation whose reverse process corresponds to the IETS process. However, these arguments contains speculations, and it should be examined carefully with a theoretical calculation.

Other "normal" Gaussian-shaped IETS features (increase of conductance change) have been reported for pyridine [46], in which an interesting comparison between benzene and pyridine were made. The benzene ring lies down in the former case, but is in an upright position in the latter case. The IETS spectra show a ν(C–H) feature only for the pyridine molecule; no ν(C–H) feature is detected for the benzene molecule.

A similar change of IETS peak intensity can be seen for benzene adsorbed on a Cu(110) surface. It is known that a benzene molecule has a bonding configuration of ring-flat on the Cu surface, but injection of electrons apparently causes the dehydrogenation of the benzene and changes the bonding to the upright configuration. Figure 12.8a shows a comparison of IETS spectra obtained on the Cu(110) substrate (spectrum I), on the benzene molecule (spectrum II), and on the benzene molecule after the dosing of the tunneling electrons (spectrum III). Peaks are observable at the energies of 372 mV and −380 mV in spectrum III obtained for the reacted product of the benzene molecule. With the isotope shift for the deuterated benzene molecule as shown in Fig. 12.8b, which makes the peak shift to 275 mV, these observations indicate that the ν(C–H) vibrational mode appears only after the bonding configuration of the benzene molecule changes into the upright position.

As mentioned before, there are no established selection rules for IETS, but it has been learned that optically active modes have strong intensities in IETS spectra compared with the optically forbidden modes [21]. The C–H stretching mode should have a surface normal component of the dipole moment when the benzene ring is in an upright position both for pyridine and for dehydrogenated benzene. However, before concluding that these observations support the dipole coupling

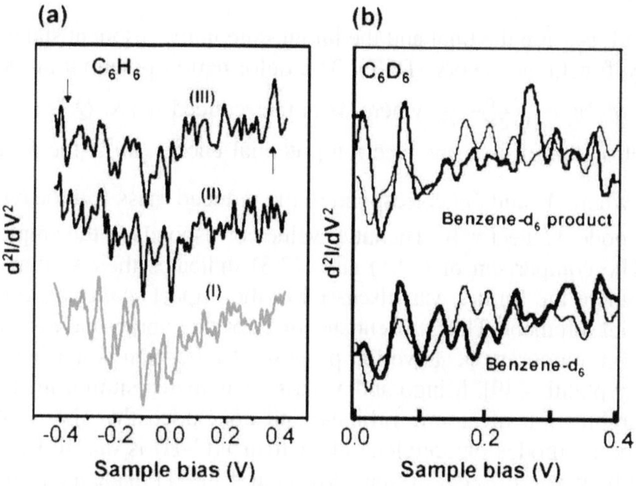

Fig. 12.8. (**a**) Vibrational spectra (dI^2/dV^2 vs V plot) obtained on a clean Cu(110) surface (*I*), a benzene molecule (*II*), and its reacted product (*III*) by STM-IETS taken at a gap resistance of 1 GΩ and V_{rms} = 5-mV modulation at 398 Hz. The spectra are the averages of 20 scans for the bias range between −0.4 and 0.4 V. (**b**) STM-IETS spectra of the deutrated benzene molecule (C_6D_6) and its reacted produce. In both spectra, the spectrum of the Cu substrate is superimposed by a *thin line*

process for the dominant vibrational excitation mechanism, the following issues should be examined. First, as the bonding configuration is changed to upright, some of the C–H groups are more isolated from the substrate than in the case when they lie flat. The expected longer relaxation time with a weak interaction with the metal might induce a higher efficiency of the excitation of the vibrational modes. Second, there should be a significant change in the electronic structure when the dehydrogenation with the change of bonding configuration takes place. The coupling of vibrational modes and the newly appeared electronic states should be considered in terms of the contribution to the vibrational excitation.

12.3.5 Theoretical Treatment of STM-IETS Results

The IETS feature of the ν(C–H) mode observed for the acetylene molecule, which has a much stronger intensity compared with other modes, has attracted attention and theoretical calculations have been performed. For example, the mechanism of the appearance of strong intensity was discussed by Lorente et al. [47] and Mingo and Makoshi [48]. Basically both groups calculated the matrix elements in which the initial electronic state i makes a transition to the final state f because of the presence of a deformation potential as illustrated in Fig. 12.2c [49]. The deformation potential is the change of the one-electron potential by the movement of nucleus accompanying the vibrational mode of the molecule.

More precisely Lorente et al. [47] obtained the following formula:

$$|\langle\varphi_\lambda|\delta v|\varphi_\mu\rangle|^2 \,. \tag{12.3}$$

Here $\langle\varphi_\lambda|$ and $|\varphi_\mu\rangle$ are the final and the initial state in the resonant states calculated with density functional theory (DFT). The deformation potential δv is expressed in the form of $\delta v = \sqrt{\frac{\hbar}{2M\omega}}\frac{\partial v}{\partial Q}$, where M is the reduced mass, Q is the coordinate of the vibration, and v is a one-electron potential energy. δv is the product of $\frac{\partial v}{\partial Q}$ and $\sqrt{\frac{\hbar}{2M\omega}}$, where M and $\hbar\omega$ correspond to the reduced mass and the energy of the vibrational mode, respectively. The latter value corresponds to the amplitude of the vibration. The comparison of (12.1) and (12.3) indicates their similar theoretical treatment, where the former was discussed in the impact scattering calculation of HR-EELS measurement. Thus, the enhancement of the vibrational excitation by the formation of a temporary negative ion proposed by Persson is not included in the calculation explicitly [49]. Mingo and Makoshi calculated similar matrix elements with the combination of atomic orbitals, and concluded that the weak intensity of all vibrational modes of acetylene other than ν(C–H) is due to the destructive interference between different orbitals. Mii et al. [50, 51] made a calculation using the Green functional method on the basis of the Anderson Hamiltonian.

Let us see how they are related to the actual observation of IETS signals. Kim et al. [52] have performed IETS on *trans*-2-butene and 1,3-butadiene (C_4H_6) molecules on the Pd(110) surface at 4.7 K. The STM images, bonding configuration, and structure of the molecules are shown in Fig. 12.9a,c for *trans*-2-butene and in Fig. 12.9b,d for 1,3-butadiene. The former molecule has a C=C at the center

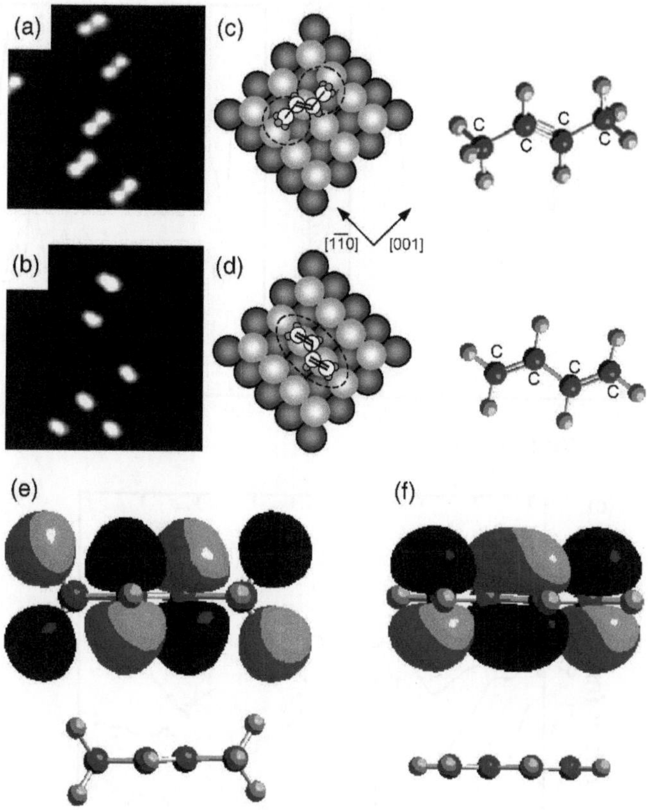

Fig. 12.9. The STM images, bonding configuration, and structure of the molecules (**a**),(**c**) for *trans*-2-butene and (**b**),(**d**) for 1,3-butadiene. The contour of the lowest unoccupied molecular orbital (LUMO) is shown (**e**) for *trans*-2-butene and (**f**) for 1,3-butadiene. Note some portion of the LUMO is located in the CH_3 group of *trans*-2-butene in (**e**)

of the molecule and two CH_3 groups attached at both carbon atoms. On the other hand, the butadiene molecule has two C=C. A clear positive conductance change was observed at 365 mV for *trans*-2-butene, corresponding to ν(C–H) (Fig. 12.10), whose peak strength is almost compatible with the one observed for acetylene on Cu(100) [38, 53]; however the feature is not observed for 1,3-butadiene in spite of the existence of C–H.

The origins of the difference in intensity of ν(C–H) for the two molecules can be considered in the following way. First is the difference in the electronic state of the molecules, especially the distribution of the molecular orbital at an energy close to the Fermi level. The molecular orbital at the LUMO level in the gas phase is shown for *trans*-2-butene and 1,3-butadiene in Fig. 12.9e,f, respectively. The LUMO level is expectedly shifted downwards upon adsorption and is responsible for the resonant state near the Fermi level. It can be seen that a portion of the LUMO exists at the C–H bond in the CH_3 group in *trans*-2-butene but no portion of the LUMO is located in the C–H bond in butadiene. If we follow the model suggested by Lorente

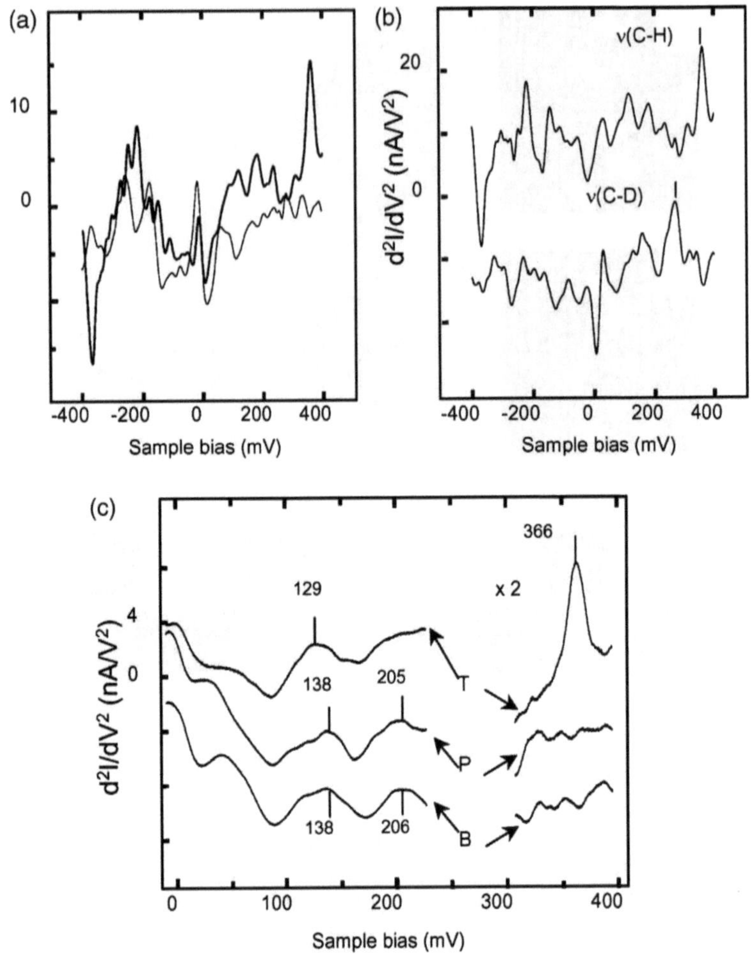

Fig. 12.10. (**a**) Vibrational spectra obtained for a clean Pd(110) surface (*thin line*) and *trans*-2-butene (*solid line*) by STM-IETS taken at a gap resistance of 1 GΩ with a 14.1-mV_{rms} alternating current modulation at 397 Hz. The spectra are averages of four scans from −400 to +400 mV. (**b**) Vibrational spectra for *trans*-2-butene (*upper line*) and trans-2-butene-d_8 (*lower line*) after background subtraction. The change in conductance for the peaks at ±366 mV (268 mV) is 6% (5%) at positive sample bias, and 7% (3%) at negative sample bias, for the C–H (C–D) stretching mode of *trans*-2-butene (*trans*-2-butene-d_8). (**c**) Vibrational spectra for *trans*-2-butene (*curve T*), reaction product (*curve P*), and 1,3-butadiene (*curve B*). The spectra were taken at a gap resistance of 200 MΩ within the range from 0 to 230 mV, in the region where the dehydrogenation does not occur, and at a gap resistance of 1.5 GΩ within the range from 300 and 400 mV

et al., the key factor which determines the IETS intensity should be the coupling strength shown in (12.3). It is reasonable to assume the space distribution of the deformation potential δv shows its maxima at around the bond directly related to the vibrational mode; C–H bond for C–H stretching mode for example. Then the matrix element of (12.3) should have a large component when the density of state of the electronic states φ_λ and φ_μ is spatially distributed at the bond connected to the vibration mode.

In addition it might be necessary to consider the strength of the bonding to the substrate as a factor which determines the IETS intensity. If the negative ion resonance is considered for the vibrational excitation mechanism, the probability of the excitation of vibrational modes should increase because the negative ion state stays longer. The weak interaction between the CH_3 and the substrate in *trans*-2-butene compared with the interaction between the CH_2 and the substrate for butadiene may indicate a longer resonance period, and enhances the excitation probability of ν(C–H).

Next we see examples where unusual behavior of the conductance change was observed at the energies of vibrational modes in which the shape of d^2I/dV^2 is not a Gaussian-type protrusion for $V_{bias} > 0$.

Hahn et al. [54] revealed two vibrational modes showing a *decrease* in conductance at ±82.0- and ±38.3-mV sample bias for a single oxygen molecule chemisorbed on the fourfold hollow sites of a Ag(110) surface at 13 K. Pascual et al. [55] have detected low-energy adsorbate–substrate (external or hindered) vibrational modes of benzene molecules adsorbed on a Ag(110) surface. These IETS features of vibrational modes are found to be strongly sensitive to the adsorption properties, and more interestingly the vibrational spectra of molecules close to kink sites exhibit a characteristic Fano-like line shape.

The dip and Fano-shaped features in IETS spectra should be attributed to the coupling of the inelastic component and the elastic component, where the latter occurs through second-order perturbation where the direct tunneling is coupled with the excitation and readsorption of a local vibrational mode. The appearance of these peaks was calculated theoretically by Lorente and Persson [56] and Persson [27]. Even though different mechanisms of impact scattering and the negative ion resonance formalism were used, respectively, they apparently reached a similar conclusion.

So far there are only a limited number of reports on the measurement of STM-IETS spectra compared with those for conventional IETS; however in many of them the STM-IETS features show unusual behavior which has been observed in few cases for conventional IETS. The following issues should be clarified prior to the complete understanding of STM-IETS.

First is the coupling between the individual resonant electronic states and the vibrational mode. Compared with the aluminum oxide substrate, the substrate of metals which were used for STM-IETS measurements can form an adsorbate-induced resonant state near the Fermi level. Thus, case-sensitive coupling between the resonant state and the vibrational mode should be considered. Especially the spatial distribution of the molecular orbital of the resonant state should have a large effect on the coupling with a vibrational mode; the coupling should be minimum when a smaller portion of the molecular orbital is distributed near the nuclei involved in the vibrational mode even if a deformation potential is created with the excitation of a vibrational mode.

Second, elastic component behavior at the energy of a vibrational mode should be more prominent than for the case of conventional IETS, which is assumingly related to the presence of a resonant state near the Fermi level [27].

12.3.6 IETS Mapping

The mapping of the intensity of the IETS feature can contribute to the understanding of the spatial distribution of functional groups and to the understanding of the basic mechanism of the IET process. The standard procedure for this measurement is to take the inelastic tunneling spectra at the grid points of a certain area. This method requires a long measurement time but gives an accurate result.

The results obtained for a C_2H_2 molecule on a Cu(100) surface with this method are shown in Fig. 12.11. The image corresponds to the mapping of the intensity of the ν(C–H) mode. The dashed line in Fig. 12.11a corresponds to the direction of the bond between two carbon atoms and a mirror plane can be defined at the point which bisects the two points. The mapping shows no node at the mirror plane and the ν(C–H) is excited most effectively at the middle of the two carbon atoms. This result can give important information on the selection rule of IET with a theoretical calculation [47].

Again (12.3) is used for the analysis where the initial state is directly related to the position of the electron injection. Two types of symmetry are considered: even (hereafter referred to as SS) or odd (AS) about the mirror plane. The DFT calculation indicates that a molecular orbital with higher LDOS has AS symmetry, which mainly derived from π^* states. There exist two vibrational modes of SS-ν(C–H) and AS-ν(C–H) at similar vibration energies, but the calculation indicates the AS mode gives a higher contribution to the IET process. Then the initial state should have SS symmetry in order to make the matrix element of (12.3) not diminish. This is consistent with the intensity mapping which shows no node at the mirror plane. A similar analysis was executed for an O_2 molecule on a Ag(110) surface where an O_2 molecule adsorbs with a flat-lying configuration [54]. The intensity mapping of the ν(O–O) mode shows a node at the mirror plane in the middle of two O atoms. The selection rule shown before explains the intensity distribution well in which the O–O stretching mode of SS symmetry and the electronic state of AS symmetry near the Fermi level were considered [47].

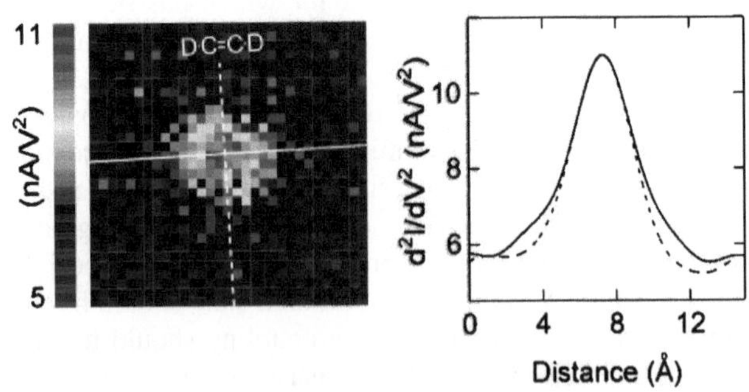

Fig. 12.11. Vibrational microscopy of DCCD. (**a**) Spatial map of d^2I/dV^2 at the energy of the C–D stretch mode (266 meV) showing localized excitation of the molecular vibration (14×14 Å^2). The *solid line* is the C–C bond axis along the [010] direction and the *dashed line* is along the [001] direction. (**b**) Smoothed cross sections of (**a**) taken along the [010] (*solid line*) and [001] (*dashed line*) directions. (From Lorente et al. [47])

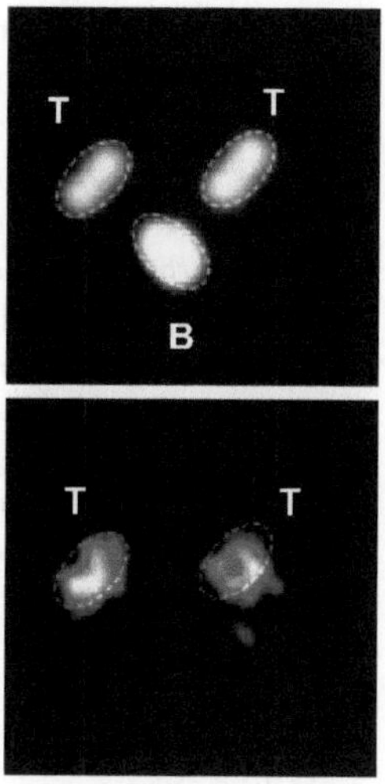

Fig. 12.12. Simultaneous observation of the topographic image (**a**) and the mapping of d^2I/dV^2 intensity (**b**) on the surface which contains both *trans*-2-butene (*T*) and butadiene (*B*) molecules. The tip scans the area with the feedback loop on (tunneling conditions of $V_{sample} = 360$ mV and $I_{tunnel} = 1$ nA, area 43×43 Å^2) The *color bar index* for the d^2I/dV^2 intensity corresponds to the range from 0 to 10 nA/V^2. No image processing including a Fourier transfer process was executed. (**c**) A reference image of d^2I/dV^2 mapping obtained at $V_{sample} =$ 300 mV which shows no obvious structures at the positions of the molecules

The mapping capability would be important in a future application to detect a functional group in a large organic molecule or in biomolecules. For that kind of purpose, a rapid mapping technique is required. Sainoo et al. [36] showed a rapid mapping example in which the surface morphology is observed with the feedback loop activated, while the IETS signal is obtained simultaneously with the modulation voltage superimposed on the sample voltage (Fig. 12.12). While the method is readily performable with conventional software, it clearly differentiated the molecules of *trans*-2-butene and butadiene through the mapping of the vibrational feature of ν(C–H) at 360 mV, demonstrating its capability of chemical identification in atomic scale.

12.4 Manipulation of Single Molecule Through Vibrational Excitation

The excitation of the vibrational modes of adsorbates by the tunneling electrons appears not only in tunneling spectroscopy (IETS) but also in the manipulation of adsorbates by the use of tunneling electrons. It has been shown that the injection of tunneling electrons into adsorbates can induce their motions and chemical reactions, many of which are related to the excitation of the vibrational modes in the mechanism. Thus, the analysis of the phenomena caused by tunneling electron injection can reveal the vibrational modes of adsorbates. In this section, examples of the manipulation of adsorbates through the excitation of vibrational modes are discussed.

12.4.1 Desorption via Vibrational Excitation

The desorption of adsorbates by the excitation of vibrational modes was discussed both from experimental and theoretical viewpoints in the early stage of research on atom manipulation, and is for the system of Xe atom transfer between a scanning electron microscope tip and a Ni(110) surface [57]. With applying a bias voltage of ±0.8 eV for 64 ms, Xe atoms were transferred in the same direction as the tunneling electrons flowed. The observed relation between the yield and the tunneling current is expressed as $I^{4.9}$. The vibrational excitation mechanism is attributed to a resonant tunneling process, which is via Xe 6s resonance. It is assumed that the tunneling electron can be regarded as temporarily hopping onto the Xe atom and forming a transient negative ion. The corresponding vibrational mode for the desorption should be metal–Xe z-motion which is estimated to have an energy of 2.5 meV. For the assumed activation barrier for the desorption in the range 10–20 meV, it reasonable to deduce the reaction order of approximately 4.9 corresponds to the number of ladders to overcome the barrier.

Further the probability of the desorption event was theoretically calculated [58]. The probability of the ladder climbing reaction decreases in accordance with the following equation:

$$\left(\frac{I}{e} f_{\text{in}} \tau\right)^n, \tag{12.4}$$

where n is the reaction order, I/e the number of dosed electron per second, f_{in} is the proportion of the inelastic component, and τ is the lifetime of the Xe–Ni vibration. For f_{in}, the vibrational excitation from $n = 0$ to $n = 1$ is considered as an inelastic channel. Numerical estimation was executed using the parameters $I = 200\,\text{nA}$, $f_{\text{in}} = 5 \times 10^{-4}$, $\tau = 25\,\text{ps}$, and the number of ladders $n \sim 4.9$, then $\left(\frac{I}{e} f_{\text{in}} \tau\right)^n \sim 10^{-9}$. This number indicates the degree of "vibrational heating". With the appropriate assumption for the pre-exponential factor (approximately 10^{13}), the observed frequency of the Xe switching can be reproduced.

One can notice that the probability of the event is sensitive both to the decay time τ and to the reaction order n. The relaxation time τ varies depending on the strength of the interaction between the adsorbates and the substrate [41]. The number of ladders is determined by the combination of the barrier height and the energy of the vibrational mode. For a high barrier and small vibrational energy, n becomes larger and the multiple excitation of the vibrational mode becomes very difficult.

The successful transfer for the Xe/Ni system is partly due to the long lifetime of the metal–Xe vibrational mode, which can be attributed to the physisorption-type bonding character. If stronger chemical bonding is there, the vibrational mode is quenched more quickly. Eigler et al. [57] claim that the transfer of Xe occurs in the direction of the electron flow; thus, if the mechanism of this phenomenon is exclusively through vibrational excitation, the system should have holeresonance than rather electron resonance as explained by an effective transient state [59].

In their calculation, the proportion of the inelastic component for the dipole moment was estimated to be 2×10^{-4} [59]. This value is almost compatible with the result obtained with the resonant coupling model [26], which stressed the importance of the resonant scattering process even in a physisorption system.

12.4.2 Vibration-Induced Hopping

The motion of adsorbates parallel to the surface, which is called hopping here, should be more commonly observed than desorption, since the activation barrier height for the former case is estimated to be one tenth of the latter. If one can excite sufficient quanta of the T-mode, which is the vibrational mode related to the lateral motion of adsorbates, the hopping of adsorbates can be induced.

Komeda et al. [60] studied the hopping of CO on a Pd(110) surface by the injection of tunneling electrons. The activation barrier and the energy of the vibrational modes are listed in Table 12.1. The T-mode can be excited if the tunneling electron energy exceeds 25 meV; however no hopping was observed even when the tunneling current was increased to 20 nA. The absence of the hopping might be due to small τ in the (12.4) which is expected from stronger bonding between CO and Pd(110) compared with Xe and Ni(110). Instead they observed that the hopping can be induced when the C–O stretching mode is excited.

An example of hopping of a CO molecule along the $[1\bar{1}0]$ direction is shown in Fig. 12.13. Four CO molecules are apparent in Fig. 12.13a as white protrusions; a CO molecule adsorbs at the bridge site on Pd(110) [61]. The molecule marked A in Fig. 12.13a was then dosed with tunneling electrons for 1 s at $V_s = 350\,\mathrm{mV}$ and $I_{\mathrm{tunnel}} = 7\,\mathrm{nA}$. Further imaging revealed that molecule A jumped three Pd lattice spacing to the left along the $[1\bar{1}0]$ direction (Fig. 12.13b), but no change was seen in the bonding position of other molecules.

The hopping probability as a function of sample bias is shown in Fig. 12.14a for voltages up to 550 mV. The probability shows a sharp increase beyond a threshold voltage which corresponds to the energy of the C–O stretching mode. The behavior near the threshold is shown in detail in Fig. 12.14b both for normal CO and for $C^{13}O^{16}$ molecules, in which the y-axis corresponds to the hopping probability per tunneling electron. In addition, the slope of the probability curve is shown in Fig. 12.14c. The probability shows a clear isotope shift, and the threshold energy can be assigned from Fig. 12.14c as 240 and 235 mV for the normal CO and $C^{13}O^{16}$, respectively. The excitation energy of the C–O stretching vibrational mode is very close to the threshold voltage of 240 mV [62], and the isotope shift of 5 mV to lower energy in the threshold energy is a reasonable value for $C^{13}O^{16}$ [40]. These findings suggest the involvement of C–O stretching excitation in the hopping of CO.

Table 12.1. Parameters related to the hopping of CO shown as a comparison between on Pd(110) surface and on Cu(110) surface. The original references are referred to in the text and also see [60, 67]

	CO on Pd(110)	CO on Cu(110)
Hopping barrier (meV)	180	97
Energy of C–O stretching mode (meV)	240	250
Energy of hindered translation mode (meV)	25	4
Anharmonic coupling (meV)	4.3	0.18
Number of ladders to overcome the barrier	7.2	24

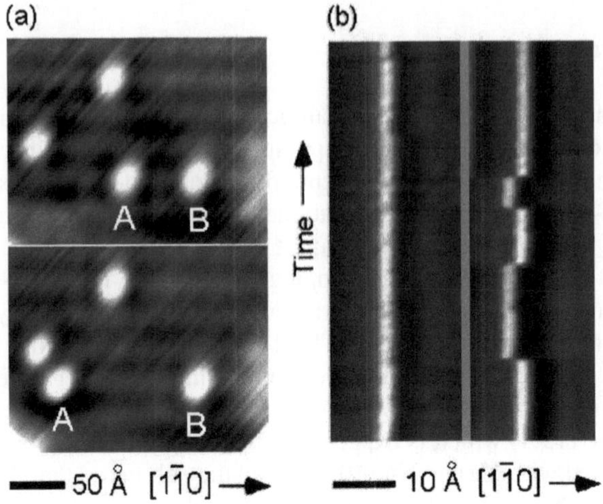

Fig. 12.13. (**a**) STM images (31×25 Å^2, $I_{\text{tunnel}} = 1$ nA, $V_s = 50$ mV) of four CO molecules on Pd(110). Before (*above*) and after (*below*) dosing tunneling electrons on the target molecule A. (**b**) One-dimensional dose-and-scan method images for an isolated CO on Pd(110). The sample bias voltage for electron dosing is (*left*) 200 and (*right*) 300 mV. In total 48 sequences were executed in both cases. Anisotropic hopping along the [1$\bar{1}$0] direction with a unit of a Pd lattice (2.75 Å) is clearly visible in the *right panel*

The observed phenomena are interpreted as (1) excitation of the ν(C–O) mode by tunneling electrons and (2) the decay of the ν(C–O) mode with an excitation of the T-mode to a higher quantum which can overcome the hopping barrier.

It has been argued that the C–O stretching mode decays through the creation of electron–hole pairs of the metal surface [63]; thus, the dumping of ν(C–O) to multiple excitation of T-modes is not the major process of vibrational relaxation, nonetheless it plays an important role in inducing surface phenomena. The mechanism is considered to be as follows.

First it should be noted that these two modes are "normal modes", and there is little intermixing between these two in the gas phase. However, with adsorption on a solid surface, there appears a coupling between normal modes which is often called anharmonic coupling. This has been studied intensively using IR and laser spectroscopy. One of the pioneering works was executed on CO molecules chemisorbed on a Pt(111) surface using an IR spectroscopy technique. The linewidth of the ν(C–O) mode for the on-top bonded molecules shows a strong temperature dependence, which is attributed to an anharmonic coupling to the T-mode [64].

Anharmonic coupling for stronger CO–metal bonding can be seen in the study of the lineshape of the ν(C–O) mode for CO chemisorbed at the twofold bridge site and the on-top site of Ni(111) over the temperature range from 80 to 300 K. The bridge-bonded CO undergoes pronounced broadening at higher temperatures, while the terminally bonded CO is only slightly broadened. The results are interpreted as the dephasing being brought about by rapid energy exchange between low-frequency modes of the substrate and low-frequency modes of the molecule which are anharmonically coupled to the high-frequency modes. The anharmonic coupling $\delta\omega$ is estimated 4.3 meV [65]

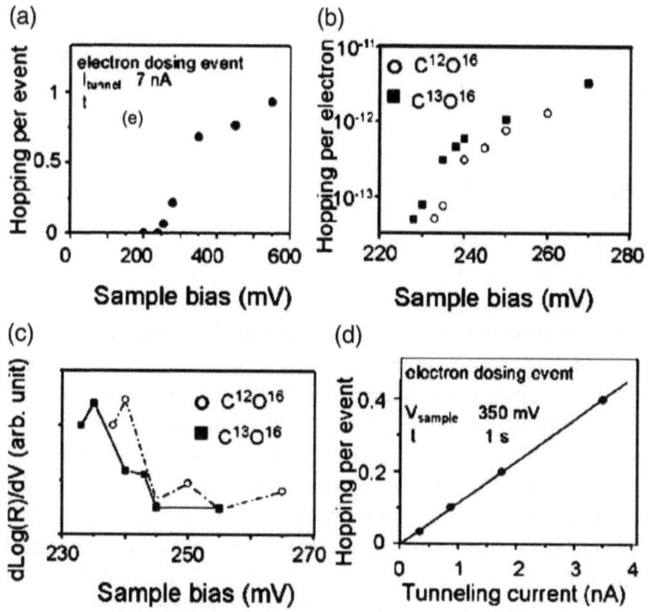

Fig. 12.14. (**a**) The hopping probability as a function of the bias voltage of dosed tunneling electrons ($I_{\mathrm{tunnel}} = 7\,\mathrm{nA}$, 1 s) examined for CO on a Pd(110) surface. (**b**) Expanded around $V_{\mathrm{s}} = 240\,\mathrm{mV}$; the y-axis is shown in a logarithmic scale and corresponds to the probability per electron. (**c**) The slope of the data in (**b**). (**d**) The hopping probability as a function of the tunneling current with the sample bias voltage of 350 mV. The *line* corresponds to the result of the least-squares fitting

On the other hand, the anharmonic coupling was found to be very small for CO on a Cu substrate, which was studied using fast laser spectroscopy [66]. The response of the stretch mode of CO on Cu(100) to resonant IR and nonresonant visible and UV pumping was measured on a picosecond timescale. The estimated anharmonic coupling $\delta\omega$ is 0.18 meV.

From the experimental results we can deduce that higher anharmonicity of ν(C–O) is observed as the bond strength of metal–CO gets stronger. Owing to the similarity of the electronic states, the anharmonicity on Pd is expected to be close to that on Ni, which is relatively strong. Thus, the comparison with the case on Cu(110) is an interesting issue. The results obtained for the Cu(110) surface showed no hopping for an electron energy up to 500 meV. Though the hopping barrier of CO is smaller on Cu(110), the anharmonic coupling is much weaker on the Cu surface, which can explain the lack of tunneling current induced hopping on the Cu surface.

Further theoretical consideration revealed that the difference of the anharmonic coupling constant is not the only reason for the difference observed on Cu and Pd surfaces [67]. It was found that the transition probability of the T-mode from 0 to n decays very rapidly as a function of $n^{1.5}\exp(-2n)$, where n is the number of the ladders required to overcome the hopping barrier. Though both the anharmonic coupling constant and the number of ladders contribute to the probability, the decrease of the yield with the increase of n is so drastic and the number of the required ladders is

24 for Cu case, which is much larger than 7.2 for the Pd(110) surface. The estimated probability is 10^{-9} for Pd and 10^{-26} for Cu; the latter excludes the observation on the timescale of the laboratory.

12.4.3 Vibration-Induced Chemical Reaction

There have been many efforts to control chemical reactions with nonthermal processes; special attention has been paid to mode-selective or site-selective chemistry, including an application to laser micromachining [68]. Most examples of such research have been performed by using lasers. It was described that controlling chemical reactions with light rests on the idea of exciting a vibration that becomes the reaction coordinate in subsequent chemistry [69,70]. Vibrational excitation techniques such as IR or stimulated Raman excitation of fundamental vibrations or vibrational overtone excitation of higher levels permit the preferential cleavage of a bond in a photodissociation. The key to bond-selected chemistry is the initial preparation of a suitable vibrational state followed by electronic excitation. Such experiments demonstrate bond-selected chemistry, permit detailed comparison with theory, and reveal general principles of vibrational state control of chemical reactions.

The use of STM for the investigation of mode-selective or site-selective chemical reactions has a variety of unique characteristics. First, the energy of the tunneling electrons can be tuned to excite a specified vibrational mode or an electronic state. In addition, the site of the electron dose can be selected with atomic resolution, which enables a site-selective excitation of a particular functional group of a molecule. Second, the reaction product can be characterized either with STM imaging or with spectroscopy, including IETS at a single-molecule level. Thus, the method can investigate a process whose reaction yield is so low that conventional methods cannot detect it.

A good example in which many of STM capabilities were used can be seen in an experiment in which two iodobenzene molecules are transformed into a biphenyl molecule. The reaction steps involve the separation of iodine from iodobenzene by using tunneling electrons, bringing together two resultant phenyls mechanically by lateral manipulation and, finally, their chemical association to form a biphenyl molecule mediated by excitation with tunneling electrons. The procedures in the report constitute an important step towards the assembly of individual molecules out of simple building blocks in situ on the atomic scale [71].

An example of chemical change of a molecule induced by vibrational excitation can be seen in the dissociation of an oxygen molecule on the Pt(111) surface. After dissociation, the two oxygen atoms are found one to three lattice constants apart. The dissociation rate as a function of current was found to vary as $I^{0.8\pm0.2}$, $I^{1.8\pm0.2}$, and $I^{2.8\pm0.3}$ for sample biases of 0.4, 0.3, and 0.2 V, respectively. These rates were explained using a general model for dissociation induced by intramolecular vibrational excitations via resonant inelastic electron tunneling [72]

An example of "molecule-to-molecule" conversion can be seen for the case of conversion of a *trans*-2-butene molecule into a 1,3-butadiene molecule adsorbed on the Pd(110) surface (Fig. 12.15). The reaction is induced by the multiple excitations of a vibrational mode in a ladder-climbing manner. The vibrational mode

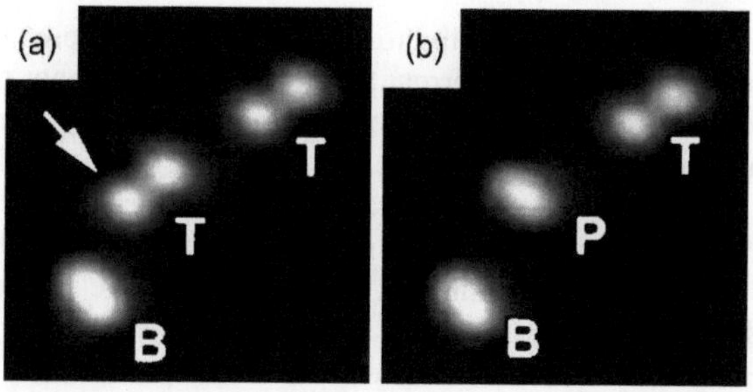

Fig. 12.15. STM images of the surface in which *trans*-2-butene and 1,3-butadiene molecule are coadsorbed (area 20 × 20 Å^2, $V_s = -200$ mV, $I_t = 0.86$ nA). The change of the image (**a**) before and (**b**) after dosing tunneling electrons on a target molecule of *trans*-2-butene (marked with an *arrow* in (**a**)). The *trans*-2-butene, reaction product, and 1,3-butadiene molecules are labeled T, P, and B, respectively

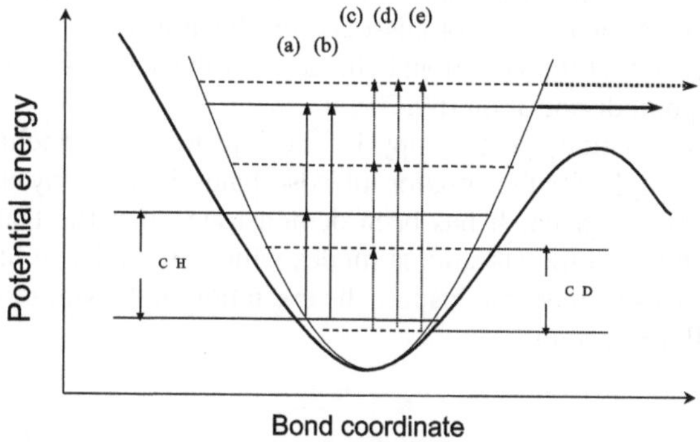

Fig. 12.16. Vibrational heating and dehydrogenation using a truncated one-dimensional harmonic oscillator. The potential well includes two C–H and three C–D vibrational states following the reaction orders determined from the experimental results. Representative types of vibrational transition for applied biases are considered as follows: (*a*) 370, 400, 600 mV and (*b*) 800 mV for C–H; (*c*) 400 mV, (*d*) 600, 800 mV, and (*e*) 1000 mV for C–D

was assigned to the C–H stretching mode which forms the ladder for the reaction. By exciting the ν(C–H) mode sequentially from $n = 0$ to $n = 1$ then from $n = 1$ to $n = 2$, the *trans*-2-butene molecule is converted to a butadiene molecule, which is a consequence of dehydrogenation from the methyl group of the former molecule (Fig. 12.16). For a deuterated molecule, the required number of ladders of the ν(C–D) mode becomes three. It was claimed that this is a first example where both the vibration-induced chemical reaction and the detection of the responsible vibrational mode were demonstrated simultaneously [52, 73].

12.5 Action Spectroscopy

As already discussed in Sect. 12.3, it has been demonstrated that IETS can detect vibrational modes of adsorbates with an atomic-scale spatial resolution; however, the observed vibrational modes are limited to some specific ones (e.g., C–H stretching mode). One of the reasons is that current STM instruments cannot produce a good enough signal-to-noise ratio to detect all the vibrational modes of adsorbates. Another reason is that the interference between the elastic and inelastic signals at the energy of a vibrational mode prohibits the observation of vibrational features.

In order to clarify these problems, it is important to combine the IETS methods with other measurement techniques which are sensitive to the excitation of vibrational modes.

As seen in Sect. 12.4, the motions and chemical reactions of adsorbates are induced by the injection of tunneling electrons, which are related to the excitation of vibrational modes. Action spectroscopy is a technique that is often used in the research areas of laser-driven photoionization and photochemical reaction of gas molecules; the yield of the reacted product is measured as a function of the energy of the incident light. A comparison of the absorption spectrum and the action spectrum yields both qualitative and quantitative information about the quantum state dependence of the reaction and the photodissociation dynamics, and the measurements reveal the influence that the vibrational character of the states has on the efficiency of reaction and photodissociation [69, 70].

In this section, action spectroscopy by the injection of tunneling electrons is discussed; an example for the rotation of adsorbates induced by the injection of tunneling electrons is given. It has been demonstrated that the features in action spectroscopy can reveal the vibrational modes which are invisible in IETS spectra. This technique gives further insights into the excitation of the vibrational modes and the features of IETS spectra.

12.5.1 Rotation of *cis*-2-Butene Molecules

For an example of action spectroscopy to reveal vibrational modes of adsorbates, the rotational motion of an adsorbate induced by tunneling electrons is examined here. If a molecule has multiple equivalent orientations at a single bonding site, the molecule shows rotation-like behavior. For example, an acetylene molecule has two diagonal sites on Cu(100) and the rotation of the molecule was observed when tunneling electrons were injected (Fig. 12.17). The activation barrier was determined by measuring the rotation rate as a function of temperature, which indicated an energy barrier of 169 ± 3 meV and a pre-exponential factor of $10^{11.8\pm0.2}$ [74].

The reversible rotation can be induced and monitored with tunneling electrons from a scanning tunneling microscope. Excitation of the C–H (C–D) stretch mode of C_2H_2 (C_2D_2) at 358 meV (266 meV) led to a 10-fold (60-fold) increase in the rotation rate. This increase is attributed to energy transfer from the C–H (C–D) stretch mode to the hindered rotational motion of the molecule [53].

More intriguing behavior can be observed for an isolated *cis*-2-butene molecule (CH_3–CH=CH–CH-$_3$) on Pd(110). A *cis*-2-butene molecule adsorbed on Pd(110) appears as an asymmetric oval in the STM image with a large bright region (head) and a small less bright region which can be seen in Fig. 12.18a. There are four equivalent orientations, labeled C_{UR}, C_{UL}, C_{DR}, and C_{DL}. An isolated *cis*-2-butene is π-bonded to an off-center position of the Pd atom, where the molecule is slightly

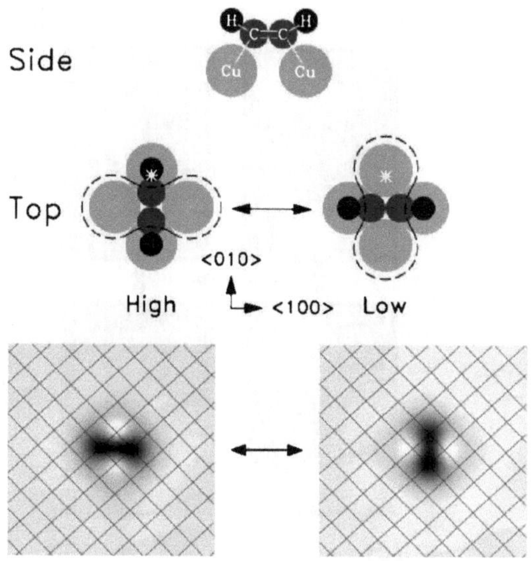

Fig. 12.17. Tunneling current induced rotation of an acetylene molecule on Cu(100). Two stable bonding sites in which the line connecting two carbon atoms are in the diagonal direction of the square lattice. (From Stipe et al. [53])

shifted towards the hollow site and the adsorption geometry is proposed as shown in Fig. 12.18b [75–77].

The motions of *cis*-2-butene with the injection of tunneling electrons can be categorized into two types [78]. They are extracted from the difference in the current changes and are simplified by defining the pairs of (C_{UR}, C_{UL}) and of (C_{DR}, C_{DL}); the pairs are indicated by arrows in Fig. 12.18a,b. C_{UR} and C_{UL} (C_{DR} and C_{DL}) are mirror images with respect to a plane parallel to [1–10], while C_{UR} and C_{DR} (C_{UL} and C_{DL}), are also mirror images with respect to a plane parallel to [001]. The motions with small changes in current, such as either between I_1 and I_2 or between I_3 and I_4 in Fig. 12.18c, correspond to the change between two orientations with the mirror plane parallel to [1–10]. This motion is named *low barrier* (LB) motion. The other motion is named *high barrier* (HB) motion with a larger current change between (I_1, I_2) and (I_3, I_4) in Fig. 12.18c, corresponding to the orientation change between two pairs for LB motion. The potential barrier for HB motion is higher than that of LB motion, which is clearly seen in action spectra.

12.5.2 Complimentary Information of Action Spectroscopy and IETS

The current variation shown in Fig. 12.18c indicates that it is possible to count the number of events per unit time. The result can be converted to the yield for a single electron injection. The yield of the HB and the LB motion as a function of the energy of the tunneling electrons can be plotted, and is considered as an action spectrum [79].

Figure 12.19 shows the action spectra of *cis*-2-butene, where the motion yield (number of molecular motions per injected electron) is indicated as a function of applied sample bias voltage at a chosen tunneling current. There are clear thresholds for

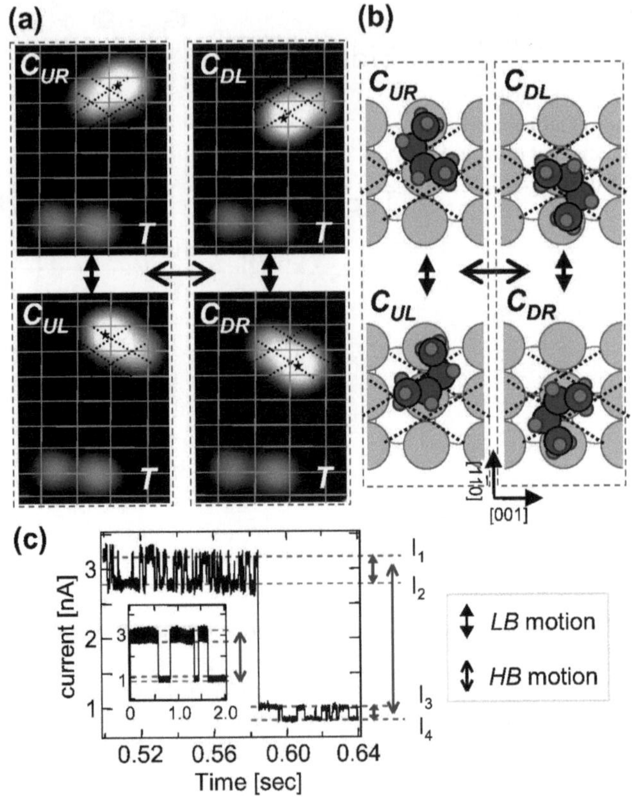

Fig. 12.18. (**a**) STM images of four equivalent orientations of a *cis*-2-butene molecule labeled C_{UR}, C_{DR}, C_{UL}, and C_{DL}, which are obtained by moving the molecule by injecting tunneling electrons. The suffix, UR, UL, DR, and DL, corresponds to the relative location of the head part, i.e., up-right, up-left down-right, and down-left, with respect to the center of the molecule in the STM images, respectively. A *trans*-2-butene molecule (T) was coadsorbed as a marker. The superimposed grid indicates the position of Pd atoms, which is determined from the STM image obtained with a molecular tip. The *four dotted lines* forming a rhombus correspond to the molecular long axes at each orientation. An *asterisk* represents the center of the large bright region of the molecule in each orientation, where the tunneling electrons were injected. (**b**) Proposed structure of *cis*-2-butene on Pd(110) at four equivalent adsorption orientations. (**c**) Current changes measured at a sample bias voltage of 170 mV with the tip fixed over the center of the head of a *cis*-2-butene molecule. The *inset* shows the wide time range of the same data. *HB* high barrier *LB* low barrier

both LB and HB motions of C_4H_8 in the upper spectra of Fig. 12.19a. For LB motion, the motion yield markedly increased at approximately 37 mV followed by a slight increase at approximately 115 mV. For HB motion, on the other hand, clear increases were observed at approximately 115 mV and approximately 360 mV. Comparison between C_4H_8 and fully deuterated *cis*-2-butene (C_4D_8) helps to assign the active vibrational modes to the aforementioned motions. In the action spectrum of C_4D_8, the increases of the motion yield for LB motion were observed at approximately

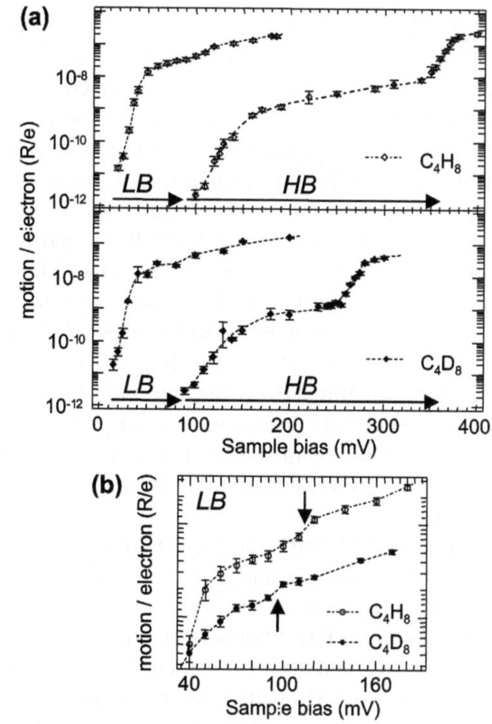

Fig. 12.19. (**a**) Action spectra for described motions of C_4H_8 (*upper*) and of C_4D_8 (*lower*). Data were taken under a fixed tunneling current of 3 nA for C_4H_8 and of 2 nA for C_4D_8. (**b**) Magnification of the action spectrum for LB motion at around the threshold energy. A slight increase in the yield was observed around 115 mV for C_4H_8 and 95 mV for C_4D_8, as indicated by *arrows*

31 mV and at approximately 95 mV. For HB motion, the increases of the motion yield appeared at approximately110 mV and at approximately 270 mV. The assignment of those modes is described next by making comparisons with STM-IETS, which gives us the vibrational signature of an individual molecule.

In order to compare them with the action spectra, IETS spectra were taken of C_4H_8 and C_4D_8 molecules. The significant features appear at ±358 mV for C_4H_8 and at ±268 mV for C_4D_8, which should be assigned to the vibrational peaks for $\upsilon(CH_3)$ and $\upsilon(CD_3)$, respectively. An additional feature with a very low intensity at about 36 (32) mV for $C_4H_8(C_4D_8)$ was present. No vibrational feature other than the two peaks was observable in the STM-IETS spectra.

Note that the vibrational energies for the *cis*-2-butene molecule are found in the following four regions: (1) metal carbon stretching mode at 20–40 meV, (2) CH bending and C–C stretching modes at approximately 100 meV, (3) C=C stretching mode, υ(C=C), at 160–180 meV, and (4) CH stretching mode at approximately 360 meV.

Compared with the HR-EELS studies, it was found that both the first threshold in the action spectrum for LB motion of C_4H_8 (C_4D_8) observed at 37 (31) mV and the small peak observed in the STM-IETS spectrum at 36 (32) mV correspond to the vibrational energy of υ(M–C). Thus, we conclude that the excitation of υ(M–C) directly couples with LB motion. For region 2, while the lower threshold for HB motion was found at approximately 110 mV both for C_4H_8 and for C_4D_8 (Fig. 12.19a), a clear isotopic shift is observed from 115 mV (C_4H_8) to 95 mV (C_4D_8) for LB motion in Fig. 12.19b. The latter case can be explained with the isotope shift of $\delta(CH_3)$ and $\delta(CD_3)$. On the other hand, υ(C–C) is expected to show a weak isotope shift. Thus, these two modes overlap in energy for C_4H_8 but are separated for C_4D_8, and both modes contribute to inducing molecular motion. For region 4, it is obvious that $\upsilon(CH_3)$ corresponds to both vibrational signals, observed at $\pm$358 (268) mV in the STM-IETS spectrum and at approximately 360 (approximately 270) mV in the action spectrum of HB motion of C_4H_8 (C_4D_8). However, υ(C=C) in region 3 was not visible in both the STM-IETS and the action spectrum. Similar behavior was observed in the hopping motion of C_2H_4 on Pd(110), where no response was observed in the C=C stretching mode region.

From the aforementioned analysis of the action spectra, it is clear that the vibrational modes υ(M–C), $\delta(CH_3)$, ν(C–C), and $\upsilon(CH_3)$ are excited via the inelastic electron tunneling. However, although the stretching mode and the bending mode of sp^3 CH in CH_3 groups were clearly observed in the action spectra, those for sp^2 CH at the C=C bond were not observed. Note that υ(=CH) and δ(=CH) for sp^2 CH appear in regions of different energies from those of sp^3 CH species, typically at 380 and at 80 meV.

Both the HOMO and the LUMO have significant parts at the π bonding and antibonding orbitals, and at the sp^3 hydrogen atoms, but not at sp^2 hydrogen atoms. Indeed, the action spectra revealed that the motion of *cis*-2-butene was actually enhanced for $\delta(CH_3)$ but was inactive for δ(CH), corresponding to the fact that the molecular orbitals of *cis*-2-butene have a distribution at sp^3 CH in methyl groups but not at sp^2 CH at the C=C bond. Because of the distribution of the electronic state, δ(CH) was not directly excited via the inelastic electron tunneling process even though the energy of the electron was sufficient for the excitation.

These findings indicate that the yield of the excitation of a vibrational mode strongly depends on the relation of the electronic state (distribution of LDOS) and the vibrational modes. The detection of a low-yield mode requires further development of the instruments for IETS, which realizes a better signal-to-noise ratio of the IET signals.

12.6 Conclusions

Detection of vibrational mode of adsorbates with tunneling electrons of a scanning electron microscope was discussed for the realization of chemical analysis with single-molecule resolution. First, recent reports on STM-IETS were introduced. IETS is an all-electron measurement technique which can detect vibrational modes of a molecule in the scanning electron microscope junction. Owing to the confinement of tunneling electrons, chemical analysis at a single-molecule level should be obtained with STM-IETS. Experimental techniques and examples of IETS spectra, including clear observation of the ν(C–H) mode in acetylene and *trans*-2-butene molecules, were introduced. At the same time, a comparison was made between STM-IETS and conventional IETS. The conventional IETS technique uses a metal–insulator–metal tunneling device and has been investigated for a long time. In the data accumulated byconventional IETS, almost all spectra show "normal" behavior of d^2I/dV^2 (increase of dI/dV for $V > 0$) for the peaks of the vibrational modes of molecules; thus, the excitation mechanism has been assigned to a dipole scattering. However, despite the limited number of spectra obtained with STM-IETS, many "unusual" spectra for vibrational modes in STM-IETS measurements can be seen. They include a dip in the spectra for $V < 0$ and the appearance of a Fano-shaped feature. The appearance of such features is certainly due to the increase of the importance of resonant states in STM-IETS measurements which are formed by molecules adsorbed on metal surfaces. On the other hand, they are hardly formed on oxide surfaces which are used for conventional IETS measurements. Thus, the resonant scattering model proposed by Persson and Baratoff [49], which assumes the tunneling to adsorbate-induced resonant states and the formation of a temporary negative ion, is seemingly appropriate to be the mechanism of vibrational excitation in STM-IETS. Nonetheless, the theoretical calculation by Lorente et al. using an impact scattering framework clearly simulated the strong intensity of the ν(C–H) mode of the acetylene molecule and the IETS mapping behavior. Systematic experimental and theoretical studies should be executed to fully understand the mechanism.

Since the basic mechanism of STM-IETS detection is not well established at this stage, a combination with other techniques that are sensitive to the excitation of vibrational modes can contribute to the development of STM-IETS. Manipulations of adsorbates with the excitation of vibrational modes can be seen for various surface phenomena, such as desorption, hopping, and chemical reactions. A case where a high-frequency vibrational mode is excited by tunneling electrons and decays by exciting multiple low-frequency vibrational modes to overcome an activation barrier was explained for hopping of CO on Pd(110). Also an example of chemical reaction from *trans*-2-butene to butadiene, which is caused by sequential excitation of the ν(C–H) mode ($n = 0$ to $n = 1$ followed by $n = 1$ to $n = 2$), was shown.

Action spectroscopy, in which the yield of the aforementioned single-molecule manipulation is measured as a function of the energy of the injected tunneling electrons, was discussed. The measurement of the action sepctrum for the case of a rotational motion of the *cis*-2-butene molecule revealed the vibrational modes which cannot be detected by STM-IETS measurements. The analysis of the action spectra has shown that the excitation probability of a vibrational mode strongly depends on the character of the LDOS with which the vibrational mode is involved. At the same time the features corresponding to these vibrational modes are missing in STM-IETS spectra even though it was proven that they are actually excited.

STM-IETS is a promising tool for the surface characterization with atomic-scale resolution. In order to make STM-IETS an established technique for chemical analysis of a single molecule, further development of instrumentation, including the improvement of the detection limit, as well as a theoretical study to determine the selection rules are necessary; further lowering of the measuring temperature should be one of the candidates.

References

1. Gimzewski JK, Joachim C (1999) Science 283:1683
2. Simmons JG (1963) J Appl Phys 24:1793
3. Fowler RH, Nordheim L (1928) Proc R Soc Lond Ser A 119:173

4. Becker RS, Golovchenko JA, Swartzentruber BS (1985) Phys Rev Lett 55:987
5. Caamano AJ, Pogorelov Y, Custance O, Mendez J, Baro AM, Veuillen JY, Gomez-Rodriguez JM, Saenz JJ (1999) Surf Sci 426:L420
6. Bobrov K, Mayne AJ, Dujardin G (2001) Nature 413:616
7. Durig U, Gimzewski JK, Pohl DW (1986) Phys Rev Lett 57:2403
8. Durig U, Zuger O, Pohl DW (1988) J Microsc 152:259
9. Durig U, Zuger O, Pohl DW (1990) Phys Rev Lett 65:349
10. Binnig G, Garcia N, Rohrer H (1985) Phys Rev B 32:1336
11. Sokcevic D, Lenac Z, Brako R, Sunjic M (1977) Z Phys B 28:273
12. Ibach H, Mills DL (1982) Electron energy loss spectroscopy and surface vibrations. Academic, New York
13. Li CH, Tong SY, Mills DL (1980) Phys Rev B 21:3057
14. Schulz GJ (1973) Rev Mod Phys 45:423
15. Wong SF, Boness MJW, Schulz GJ (1973) Phys Rev Lett 31:969
16. Himpsel FJ, Fauster T (1982) Phys Rev Lett 49:1583
17. Rogozik J, Scheidt H, Dose V, Prince KC, Bradshaw AM (1984) Surf Sci 145:L481
18. Demuth JE, Schmeisser D, Avouris P (1981) Phys Rev Lett 47:1166
19. Esaki L (1958) Phys Rev 109:603
20. Jaklevic RC, Lambe J (1966) Phys Rev Lett 17:1139
21. Hansma PK (1982) Tunneling spectroscopy. Plenum, New York
22. Walmsley DG, Tomlin JL (1985) Prog Surf Sci 18:247
23. Mazur U, Hipps KW (2002) In: Chalmers J, Griffiths P (eds) Handbook of vibrational spectroscopy, Vol. 1. Wiley, New York, p 812
24. Komeda T, Kim Y, Kawai M (2001) Riken Rev, p 26
25. Kirtley J, Soven P (1979) Phys Rev B 19:1812
26. Persson BNJ, Demuth JE (1986) Solid State Commun 57:769
27. Persson BNJ (1988) Phys Scr 38:282
28. Bayman A, Hansma PK, Kaska WC (1981) Phys Rev B 24:2449
29. Hamers RJ, Demuth JE (1988) Phys Rev Lett 60:2527
30. van Loenen EJ, Demuth JE, Tromp RM, Hamers RJ (1987) Phys Rev Lett 58:373
31. Feenstra RM (1994) Surf Science 300:965
32. Feenstra RM (1994) Phys Rev B 50:4561
33. Feenstra RM, Collins DA, Ting DZY, Wang MW, McGill TC (1994) Phys Rev Lett 72:2749
34. Lauhon LJ, Ho W (2001) Rev Sci Instrum 72:216
35. Stipe BC, Rezaei MA, Ho W (1999) Rev Sci Instrum 70:137
36. Sainoo Y, Kim Y, Komeda T, Kawai M (2004) J Chem Phys 120:7249
37. Baratoff A, Persson BNJ (1988) J Vac Sci Technol A 6:331
38. Stipe BC, Rezaei MA, Ho W (1998) Science 280:1732
39. Hipps KW, Mazur U (1989) Surf Sci 207:385
40. Lauhon LJ, Ho W (1999) Phys Rev B 60:R8525
41. Headgordon M, Tully JC (1992) Phys Rev B 46:1853
42. Morin M, Levinos NJ, Harris AL (1992) J Chem Phys 96:3950
43. Hirschmugl CJ, Williams GP, Hoffmann FM, Chabal YJ (1990) Phys Rev Lett 65:480
44. Persson BNJ (1991) Phys Rev B 44:3277
45. Ryberg R (1990) J Electron Spectrosc Relat Phenom 54–55:65
46. Lauhon LJ, Ho W (2000) J Phys Chem A 104:2463
47. Lorente N, Persson M, Lauhon LJ, Ho W (2001) Phys Rev Lett 86:2593
48. Mingo N, Makoshi K (2000) Phys Rev Lett 84:3694
49. Persson BNJ, Baratoff A (1987) Phys Rev Lett 59:339
50. Mii T, Tikhodeev SG, Ueba H (2002) Surf Sci 502:26
51. Mii T, Tikhodeev SG, Ueba H (2003) Phys Rev B 68:205406

52. Kim Y, Komeda T, Kawai M (2002) Phys Rev Lett 89:126104
53. Stipe BC, Rezaei MA, Ho W (1998) Phys Rev Lett 81:1263
54. Hahn JR, Lee HJ, Ho W (2000) Phys Rev Lett 85:1914
55. Pascual JI, Jackiw JJ, Song Z, Weiss PS, Conrad H, Rust HP (2001) Phys Rev Lett 86:1050
56. Lorente N, Persson M (2000) Phys Rev Lett 85:2997
57. Eigler DM, Lutz CP, Rudge WE (1991) Nature 352:600
58. Walkup RE, Newns DM, Avouris P (1993) Phys Rev B 48:1858
59. Persson BNJ, Baratoff A (1987) Phys Rev Lett 59:339
60. Komeda T, Kim Y, Kawai M, Persson BNJ, Ueba H (2002) Science 295:2055
61. Raval R, Haq S, Harrison MA, Blyholder G, King DA (1990) Chem Phys Lett 167:391
62. Kato H, Kawai M, Yoshinobu J (1999) Phys Rev Lett 82:1899
63. Persson BNJ, Persson M (1980) Solid State Commun 36:175
64. Ryberg R (1989) Phys Rev B 40:8567
65. Trenary M, Uram KJ, Bozso F, Yates JT Jr (1984) Surf Sci 146:269
66. Germer TA, Stephenson JC, Heilweil EJ, Cavanagh RR (1994) J Chem Phys 101:1704
67. Persson BNJ, Ueba H (2002) Surf Sci 502:18
68. Sarkar SK, Mittal JP (1999) Mater Sci Forum 301:1
69. Crim FF (1996) J Phys Chem 100:12725
70. Crim FF (1999) Acc Chem Res 32:877
71. Hla SW, Bartels L, Meyer G, Rieder KH (2000) Phys Rev Lett 85:2777
72. Stipe BC, Rezaei MA, Ho W, Gao S, Persson M, Lundqvist BI (1997) Phys Rev Lett 78:4410
73. Kim Y, Komeda T, Kawai M (2002) Surf Sci 502–503:7
74. Lauhon LJ, Ho W (1999) J Chem Phys 111:5633
75. Sainoo Y, Kim Y, Fukidome H, Komeda T, Kawai M, Shigekawa H (2002) Jpn J Appl Phys Part 1 41:4976
76. Sainoo Y, Kim Y, Komeda T, Kawai M (2004) J Chem Phys 120:7249
77. Kawai M, Komeda T, Kim Y, Sainoo Y, Katano S (2004) Philos Trans R Soc Lond Ser A 362:1163
78. Sainoo Y, Kim Y, Komeda T, Kawai M, Shigekawa H (2003) Surf Sci 536:L403
79. Sainoo Y, Kim Y, Okawa T, Komeda T, Shigekawa H, Kawai M (2005) Phys Rev Lett 95:246102

13 STM Studies on Molecular Assembly at Solid/Liquid Interfaces

Ryo Yamada · Kohei Uosaki

13.1 Introduction

Much interest has been shown in the structure of a solid/liquid interface because it plays important roles in many fields such as surface chemistry, electrochemistry and life science [1, 2]. Molecules are subjected to inhomogeneous force at the interface because of the unsymmetrical nature of the interface structure. The imbalance in forces results in the formation of unique structures. Much progress has been made in studies on structures of solid/liquid interfaces, especially in the field of electrochemistry [3–6].

Modern in situ surface analytical methods have revealed the existence of well-ordered monolayers at a solid/liquid interface spontaneously formed in a variety of systems [5–9]. This spontaneous ordering of molecules is called self-assembly (SA), and monolayers formed in this manner are called self-assembled monolayers (SAMs).

The construction of well-defined surfaces and interfaces is one of the key steps in material science because the electronic property and chemical reactivity of materials are influenced by characteristics of surfaces. A single monolayer is known to change surface characteristics and, therefore, surface structures should be controlled with a precision of a single molecular thickness. SA is a useful technique to achieve the design of surfaces with a desired level of precision.

In this chapter, structures formed by molecular assembly at metal/liquid interfaces are discussed. The scanning tunneling microscope (STM) is the most powerful tool for observing structures of monolayers at a solid/liquid interface because it provides direct information on the molecular order and defect structures in situ *when the imaging is successful*. We briefly describe how to perform STM measurement in liquids in the following section. Readers interested in the mechanism of STM should consult [10–13].

Many molecules have been reported to form ordered SAMs at a solid/liquid interface. We discuss several significant examples in this chapter. Review articles for other representative SAMs such as liquid crystals [14] and organic molecules in electrochemical environments are available [8, 9].

13.2 STM Operations in Liquids

13.2.1 Instruments

When a liquid is not conductive, such as liquid crystals and alkanes, STM measurements are very similar to those in an ultra-high-vacuum (UHV) environment and air. The tip of the STM is simply inserted into liquids [15–17]. Liquids of low vapor pressure are preferred to avoid drying up.

When a liquid is conductive, electrochemical control of the tip of the STM and a substrate is required [18–22]. A typical setup for the electrochemical STM (EC-STM) is shown in Fig. 13.1. The electrochemical potential of the tip and that of the substrate are independently controlled by using a bi-potentiostat. Typically, a small reversible hydrogen electrode and Ag/AgCl, fitted to a tiny STM cell, are used as reference electrodes. Cl^- should be carefully separated when Ag/AgCl is used, since even a very small residual amount of Cl^- can change the mobility of metal atoms on electrodes [23–28]. A gold oxide wire is useful when chemical reactions do not occur on it in the solution [29]. Pt and Ag wires are also used as a quasi-reference. The area of the counter electrode is not necessarily larger than that of a working electrode because EC-STM measurements are usually carried out with a small flow of electrochemical current.

Current due to an ionic conduction and electrochemical reactions should be smaller than tunneling current. Current other than from the tunneling process is proportional to the tip area exposed to the solution and, hence, can be decreased by the insulation of the side wall of the tip. Only the very end of the tip is required for STM measurement since several atoms can hold the whole tunneling current [10–13].

Insulation of the tip is carried out by dipping the whole tip into molten or softened polymers such as apiezon wax, nail polish and polypropyrene [30]. The very end of the tip spontaneously sticks out when the polymers are hardened. A leak current due to ionic and electrochemical current through the bare tip-end is easily reduced to less than 0.1 nA. Careful coating results in a leak current smaller than 10 pA, which is required to observe bulky molecules.

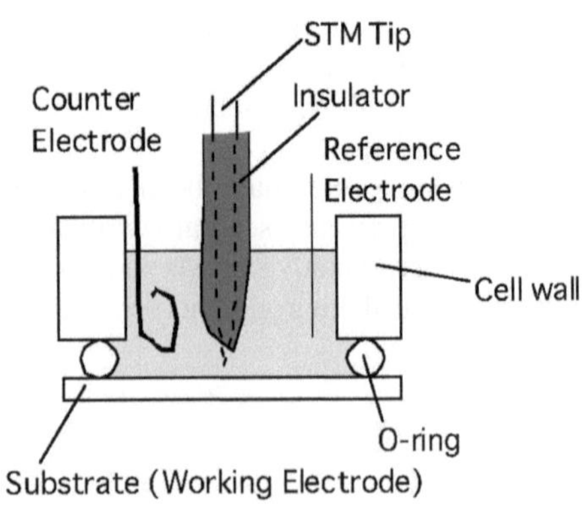

Fig. 13.1. Schematic drawing of the setup of an electrochemical scanning tunneling microscope (STM). The dimension of the cell is typically approximately 1–2 cm. The top of the cell is opened to introduce a STM tip. The cell wall is usually made from Teflon or other similar materials. A Teflon-coated O-ring is used to seal the solution

It should be noted that STM measurements sometimes alter electrochemical reactions of interest [31]. For example, the tip can disturb the diffusion of the reactant and change the local concentration. Overlapping of the electrical double layer between the tip and the substrate can change the local structure of the double layer [32]. The strong electric field between the tip and the substrate and/or the localized charge induced by the tip can change the local potential of the substrate [33, 34]. These effects are intentionally used to fabricate nanostructures via localized electrodepositions and dissolution [35].

Environmental control is required to remove an unfavorable gas such as oxygen from a solution. A liquid cell has to be opened to introduce an STM tip, and the whole microscope is sealed in a small chamber. The chamber is also useful to reduce evaporation of liquid.

13.2.2
Preparation of Substrates

An atomically flat and ordered surface is required. A (111)-oriented gold thin film is the most widely used substrate since it is stable in ambient conditions and can be easily formed on various substrates, such as mica [36–39], a glass slide and silicon [7, 40, 41], by thermal vacuum evaporation. A polished surface of a gold polycrystal is also used as a substrate [42]. Grains having atomically flat (111) terraces are grown when the substrate is heated at around 300 °C during the deposition. Flame annealing is also employed to obtain flat and wide (111) terraces [7].

An adhesive layer is usually required between a gold film and the substrate since adhesion of a gold film to glass and silicon is weak. Thin films of Cr and Ti are typically used as adhesive layers. It should be noted that Cr is known to migrate to the surface of the gold film [43–46]. An organic monolayer of mercaptotrimethyoxysilane is used as a molecular-adhesive layer to avoid metal contamination of gold films [47–50].

(111)-oriented films of other materials such as silver can also be grown by thermal evaporation or sputtering with heating of the substrate during the deposition process [40]. Oxidation of the surface is a serious problem when handling the surface in air. Flame annealing and chemical etching are employed to remove an oxide layer. The surface should be treated in an inert gas such as Ar and N_2.

Single crystals are required for studies on specific crystal faces other than (111). The so-called Cravilier- or flame-melting method is employed to grow small (a few millimeters in diameter) single crystals [51,52]. When a wire made of polycrystalline metal, such as gold, silver, platumum, palladium or rhodium, is melted by a flame, a small molten metal bead dangles on the edge of the wire. The molten metal bead is solidified to a single crystal by very gently cooling it. Small facets of (111) and (100) surfaces appear on the single-crystal bead. The orientation of the single crystal can be determined from the positions of the facets. A surface with a desired orientation is obtained by cutting the crystal. The facet itself is a very flat surface and is a good substrate for STM measurement. Electrochemical epitaxitial growth on gold surfaces is useful to prepare well-defined surfaces of various metals, such as platinum [53], palladium [54–56], cobalt, nickel and iron [57].

13.3 Surface Structures of Substrates

13.3.1 Introduction

A clean surface sometimes shows a unique structure that is different from what is expected from its bulk structure. Deformations of the surface originate from the imbalance of forces acting on the surface. These deformations are called reconstructions [58, 59]. Gold surfaces are known to show complex reconstructions. In this section, structures of clean Au(111) and Au(100) surfaces are explained because these surfaces are the most widely used substrates for studying molecular assemblies.

13.3.2 Structures of Au(111)

A clean Au(111) surface shows a reconstructed structure that is described as $22 \times \sqrt{3}$ [60–64]. It consists of a uniaxial compression of surface atoms by about 4% in the [110] direction. As a result, 23 atoms are located in the 22 atomic spaces of the (1×1) structure.

STM images of the reconstructed Au(111) surface are shown in Fig. 13.2 [61]. A paired stripe structure is shown in Fig. 13.2a. The brighter region of the stripe corresponds to the bridge site where the registry of the surface atoms is changing from the face centered cubic to hexagonal close packed sites as shown in Fig. 13.2b. An STM image of a large area (Fig. 13.2c) shows that the direction of the bridge site regularly changes and it forms a herringbone structure. The regular turning of the stripe is a result of the formation of stress domains [65–67]. The stress acting on the gold surface is determined by the balance between a repulsive interaction of the filled *d* shells and compressive stress of the *s* and *p* electrons [68].

13.3.3 Structures of Au(100)

The Au(100) surface shows a reconstructed structure that is described as (5×20) [62, 63, 69–71]. Since the surface atoms form a quasi-hexagonal arrangement, this structure is sometimes called hex-Au(100) phase. Figure 13.3a,b shows STM images of the hex-Au(100) phase obtained in UHV [71]. An image of a large area (Fig. 13.3a) shows a stripe structure running parallel to the step. A magnified image (Fig. 13.3b) shows that the stripes are attributed to the corrugation of the gold surface. Subtle height corrugation parallel to the stripe is also seen in Fig. 13.3b. As shown in the model in Fig. 13.3c, these features are due to the difference in the registry of the gold atoms of the top layer with respect to that of the second layer.

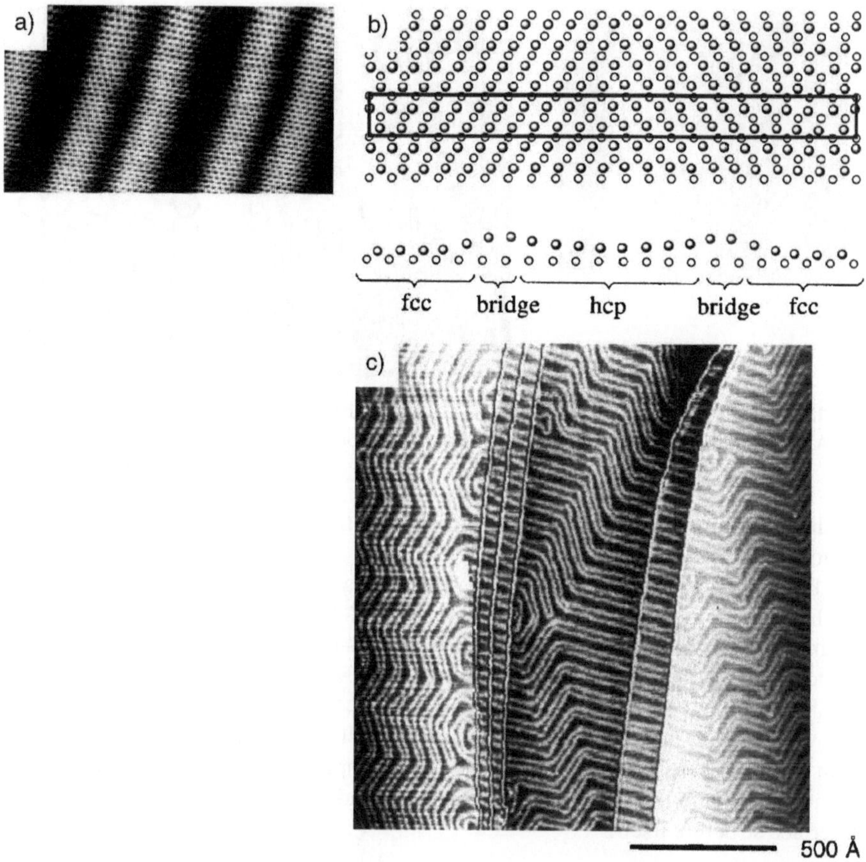

Fig. 13.2. (**a**) A STM image of a reconstructed Au(111) surface. (**b**) A model of a $22 \times \sqrt{3}$ structure; *fcc* face-centered cubic, *hcp* hexagonal close-packed. (**c**) A STM image of a reconstructed Au(111) surface in large scale. ((**a**),(**c**) Reprinted from [61]; (**b**) Reprinted from [64])

13.4 SA of Organic Molecules

13.4.1 Introduction

Formation of organic monolayers is an attractive way to modify or *functionalize* surfaces because various combinations of functional group can be attached and integrated in a molecule. SA of organic molecules at a solid/liquid interface is a useful and reliable way to construct a well-ordered monolayer.

SA accompanied by chemisorption provides robust SAMs, and this method is used to construct functional monolayers on solid surfaces [3, 72, 73]. We discuss the molecular assembly of alkanethiols in solutions on a Au(111) surface as an example of chemisorbed organic molecular assembly because this system has been widely

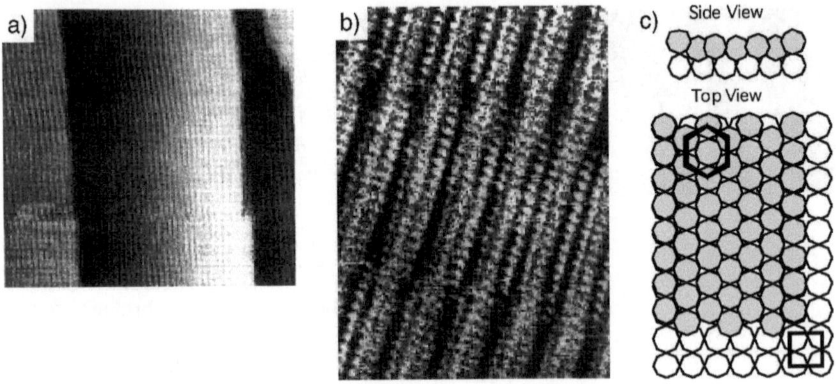

Fig. 13.3. (**a**) A STM image of a reconstructed Au(100) surface. (**b**) An atomic-resolution image. (**c**) A model of the hex-Au(100) structure. (Reprinted from [71])

studied to control surface properties and electron transfer processes, and to stabilize nanoclusters [74–76].

Physisorbed assemblies of organic moleucles at solid/liquid interfaces have been studied on graphite [15–17, 77–81] and gold [82–86] surfaces. Organic molecules seem to adsorb on the surfaces owing to substrate–molecule interactions. Interfacial forces can play a role in SA [2] in some cases when molecules are dissolved in a solvent.

We focus on molecular assemblies on Au(111) and (100) surfaces observed in neat liquids. These systems might give us the simplest picture of the mechanism of SA at solid/liquid interfaces. It is important to study these systems to understand an interface structure itself.

13.4.2 Assembly of Chemisorbed Molecules: Alkanethiols

13.4.2.1 Structures of Thiol SAMs on Au(111)

An alkanethiol SAM formed on the Au(111) surface is the most widely studied system. Since the proposal of the concept of SAM in the early 1980s [87, 88], significant progress has been made in basic characterization and applications of SAMs in the last two decades [74–76, 89–91].

SAMs of alkanethiols are formed in various environments, such as UHV [92,93], gas [94–96] and solutions [88]. There are no significant differences in the structures of monolayers formed in different environments. Formation from solution phases is the most practical and widely used way because large molecules that have small vapor pressure can be easily used. Typically, a several millimolar solution of alkanethiols in organic solvents such as ethanol and hexane is used to form SAMs.

Figure 13.4a,b shows schematic drawings of top and side views of the monolayer, respectively. The basic molecular arrangement is $(\sqrt{3} \times \sqrt{3})$R30° with respect to the

Au(111) surface [97–99]. Closer inspection of the structure revealed the existence of a c(4 × 2) superlattice of $(\sqrt{3} \times \sqrt{3})R30°$ [100, 101], which is explained later. The alkyl chain is tilted from the surface normal about 30° with all-*trans* conformation. This tilt angle comes from the conditions for closed packing of alkyl chains. Close inspection of IR data revealed that the plane defined by an all-trans carbon molecular skeleton alternatively changes its direction [102].

A high-resolution STM image using large tunneling impedance reveals small modulations in height among molecules as shown in Fig. 13.5 [103–105]. The structure considering these modulations is called c(4 × 2) of $(\sqrt{3} \times \sqrt{3})R30°$ and is consistent with the structure predicted from results obtained by using diffraction techniques [103]. Sometimes, two kinds of c(4×2) structures are observed by a STM [88] and a noncontact atomic force microscope [106, 107] as shown in Fig. 13.6. One has an oblique and the other has a square unit cell. Grazing incidence angle X-ray diffraction (GIXD) analysis showed that only an oblique structure is possible [89, 100, 101]. Although it is difficult to deny polymorphism, the convolution effect of the unsymmetrical tip is likely to be the reason for the different structures imaged by the STM.

One of the origins of the c(4 × 2) structure is attributed to the different orientations of the alkyl termination owing to the different twist angles among alkyl chains [103, 104]. In addition, arrangement of the sulfur atom is also considered. GIXD [108], sum frequency generation spectroscopy [109] and X-ray standing waves measurement [110, 111] showed that the sulfur position has a small deviation from the hexagonal symmetry, indicating the existence of two kinds of sulfur positions [89]. The different sulfur adsorption sites can result in variable electronic structure and height in the monolayer. These differences can be the origin of the c(4 × 2) structure.

In fact, the position of the sulfur atom on a gold surface is still under debate. Early studies supposed that the sulfur atom sat on a threefold hollow site of gold atoms [112–114], however, recent theoretical and experimental results have shown that the sulfur stays on atop [115–117], bridge [118–121] and multiple [108, 110, 111, 123–125] sites.

An STM image of a large area revealed various defect structures as shown in Fig. 13.7a. One significant feature is the pitlike structure. The depth of the pits is

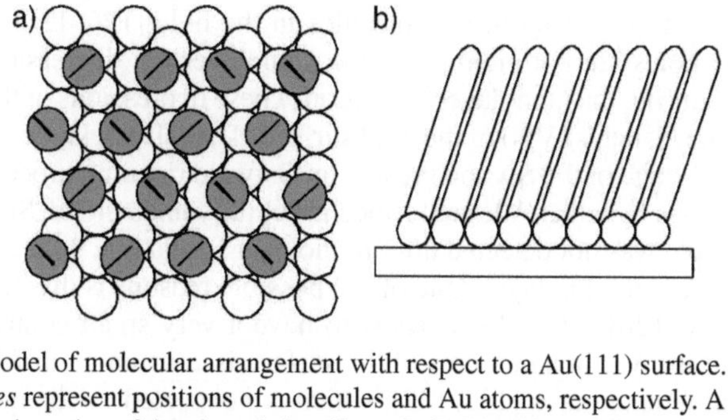

Fig. 13.4. (**a**) Model of molecular arrangement with respect to a Au(111) surface. *Shaded circles* and *open circles* represent positions of molecules and Au atoms, respectively. A *slash* indicates the azimuthal orientation of the plane defined by a C–C–C backbone of an all *trans*-hydrocarbon chain. (**b**) Side view of the monolayer. *Circles* represent sulfur atoms

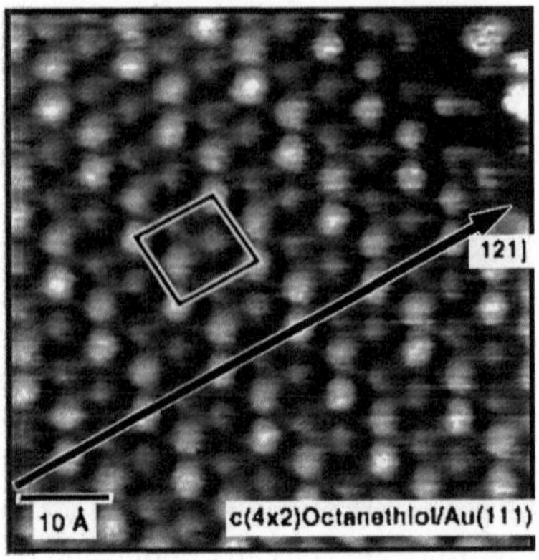

Fig. 13.5. A STM image of a Au(111) surface covered with an octanethiol monolayer. The *rectangular cell* represents the unit cell for c(4 × 2) of ($\sqrt{3} \times \sqrt{3}$)R30°. (Reprinted from [91])

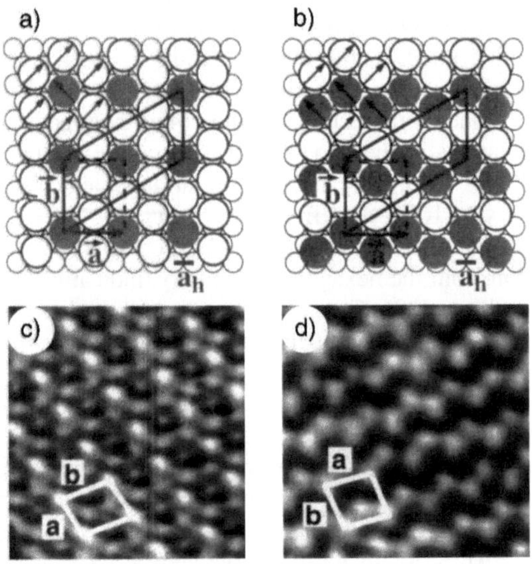

Fig. 13.6. Other kinds of c(4 × 2) of ($\sqrt{3} \times \sqrt{3}$)R30° observed in a STM measurement. (**a**),(**b**) Models of the STM images shown in (**c**) and (**d**), respectively. (Reprinted from [103])

equal to the monoatomic height of the Au(111) surface. Close inspection of the inside of the pits showed the existence of molecules in the hole [126–128]. Considering these facts, the holes are not pinholes in the monolayer but depressions of the Au surface created during the monolayer formation. These depressions of the Au surface are called vacancy islands (VIs) of the gold surface. The VIs are known to be formed at the very initial stage of the SA and grow in an Ostwald ripening process [129, 130].

The mechanism by which VIs are formed is not fully understood. Since shrinking of the gold surface was not detected after monolayer formation, gold atoms seem to be taken away from the top layer. One of the possible reasons is dissolution of gold into the solution [126]. This effect seems to have a very small contribution since the total area of the VIs in a unit area of the surface is constant in different solvents and at different temperatures [131, 132], which are expected to change the etching rate. The extraction of an excess amount of gold atoms might take place when the reconstruction of the gold surface is lifted [133].

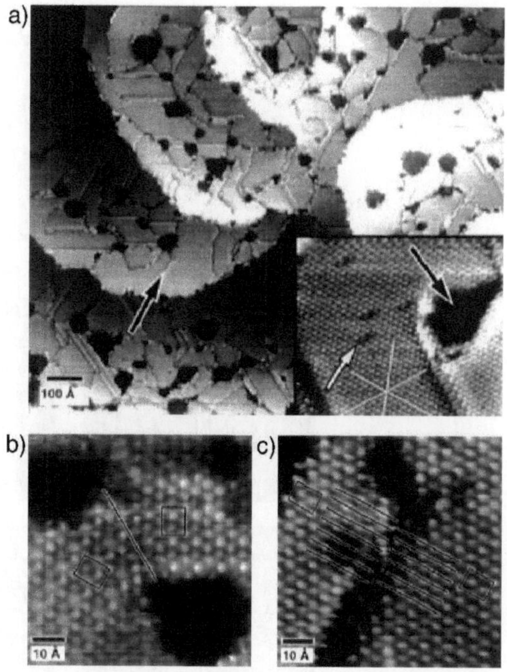

Fig. 13.7. (**a**) A STM image of a Au(111) surface covered with a dodecanethiol monolayer. (**b**),(**c**) Orientational and translational domain boundaries, respectively. (Reprinted from [91])

The other striking feature is the domain boundary. Typically, a domain boundary consists of void lines with a space of single or several molecules. These defects originate from the misfits in tilt angles, staking geometry and rotational direction of c(4 × 2) geometry. Figure 13.7b,c shows typical domain boundaries caused by rotational and stacking misfits, respectively.

The grains are known to become larger with increase in temperature during the modification as shown in Fig. 13.8 [132]. Although the average size of the VIs is larger at a higher temperature, the number of VIs is less. Consequently, the total area of the VIs on the surface was constant regardless of the temperature. The total perimeter of the VIs, representing the total length of the line defects in the monolayer formed at the step of the VIs, becomes shorter at a higher temperature. A similar effect is obtained by annealing of the monolayer [134, 135]. The solvent also has a strong impact on the defect density in the SAMs as shown in Fig. 13.9, though the general tendency is not fully understood [132].

Alkanethiols are known to change their orientations as a function of coverage [7]. Poirier et al. [136] conducted STM investigations to make a phase diagram in UHV conditions as a function of coverage as shown in Fig. 13.10.

When the coverage is very low (Fig. 13.10a), only deformation of the reconstruction of the Au surface is observed, indicating the existence of a highly mobile molecular phase, i.e., 2D gas phase of the molecule. Ordered pinstripe structures were observed at slightly higher coverage (Fig. 13.10b). The period of the stripe is twice as long as the length of the molecules; thus, the head-to-head molecular arrangement shown in Fig. 13.11, panel β is expected. The bright stripe line is formed

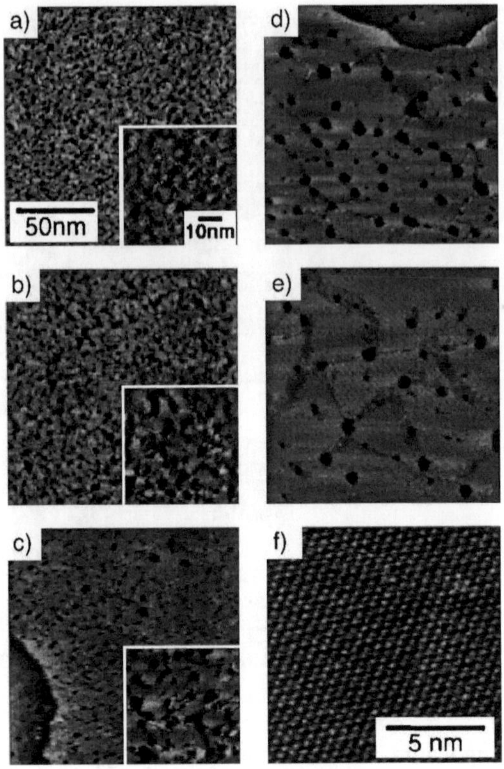

Fig. 13.8. STM images of a Au(111) surface covered with a decanethiol monolayer. Modification was carried out at (**a**) −20 °C, (**b**) 5 °C, (**c**) 25 °C, (**d**) 60 °C and (**e**) 78 °C (boiling point) in 1 mM solution in ethanol for 1 h. (**f**) Typical image of molecular resolution. (Reprinted from [131]

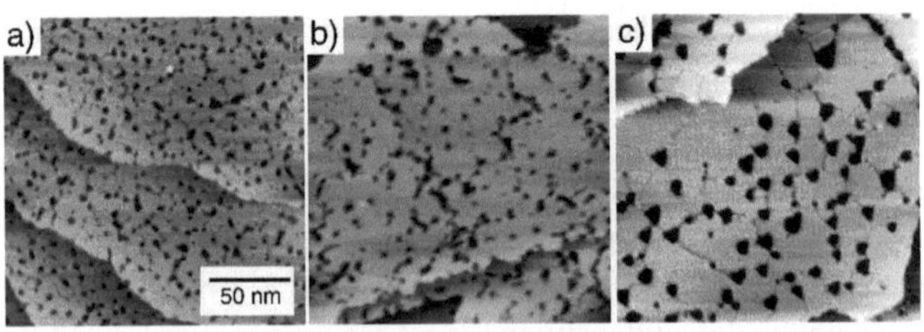

Fig. 13.9. STM images of a Au(111) surface covered with a decanethiol monolayer modified in 1 mM solution in (**a**) ethanol, (**b**) dimethylformamide and (**c**) toluene for 1 h. (Reprinted from [132])

by a thiol group and is directed to the ⟨110⟩ direction or the next-nearest neighbor (NNN) of gold atoms on the (111) surface. Alkyl chains are perpendicular to the stripe row.

As coverage is increased (Fig. 13.10c), other kinds of pinstripe structures are formed. The period of the stripe is shorter than twice the molecular length, although the double bright line in the STM images indicates a couple of thiols form the row in

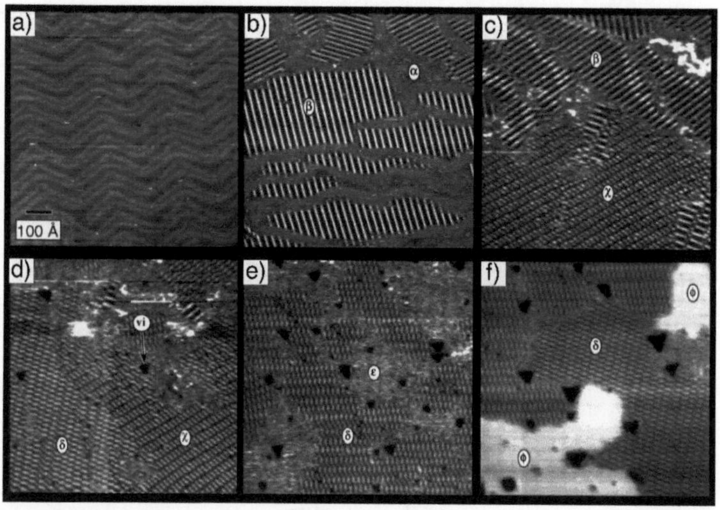

Fig. 13.10. Isothermal growth of decanethiol on Au(111) in ultrahigh vacuum conditions at 22 °C. (**a**) Clean Au(111) surface showing herringbone reconstruction. (**b**) Surface exposed to decanethiol flux. β phase and two-dimensional gas coexist. (**c**) β phase and χ phase observed at higher surface coverage As adsorption proceeded, phases with higher coverage were generated (**d**)–(**f**). (Reprinted from [136])

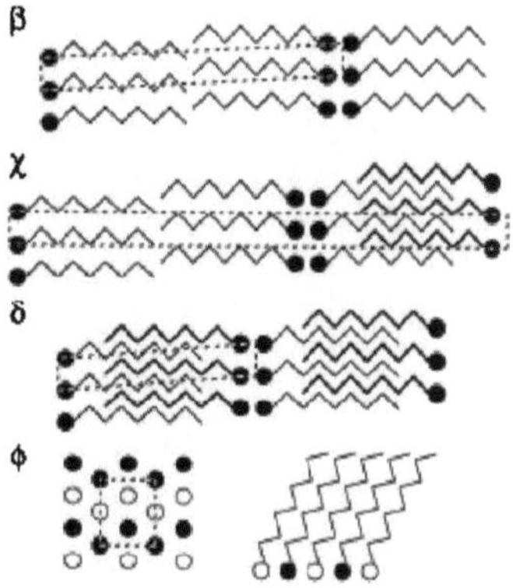

Fig. 13.11. Models of molecular arrangement in various phases of alkanethiols. (Reprinted from [136])

head-to-head configuration. Complex patterns with the same period are also formed at higher coverage as shown in Fig. 13.10d–e. Plausible models for these structures are shown in Fig. 13.11, panels χ, δ and ϕ. In these structures, the alkyl chains are piled up to those in the next rows. These structures are called interdigit structures. The phase behavior of an alkanethiol layer was investigated at various temperatures (Fig. 13.12).

In situ STM observations of the SA process of alkanethiols in solutions were carried out by Yamada and Uosaki [137, 138]. To control the coverage in a stepwise manner, small droplets of a dilute thiol solution were injected into the STM cell that was filled with pure solvent. The STM tip was scanning during the process. Figure 13.13 shows experimental results. When the 0.5 μM solution of decanethiol

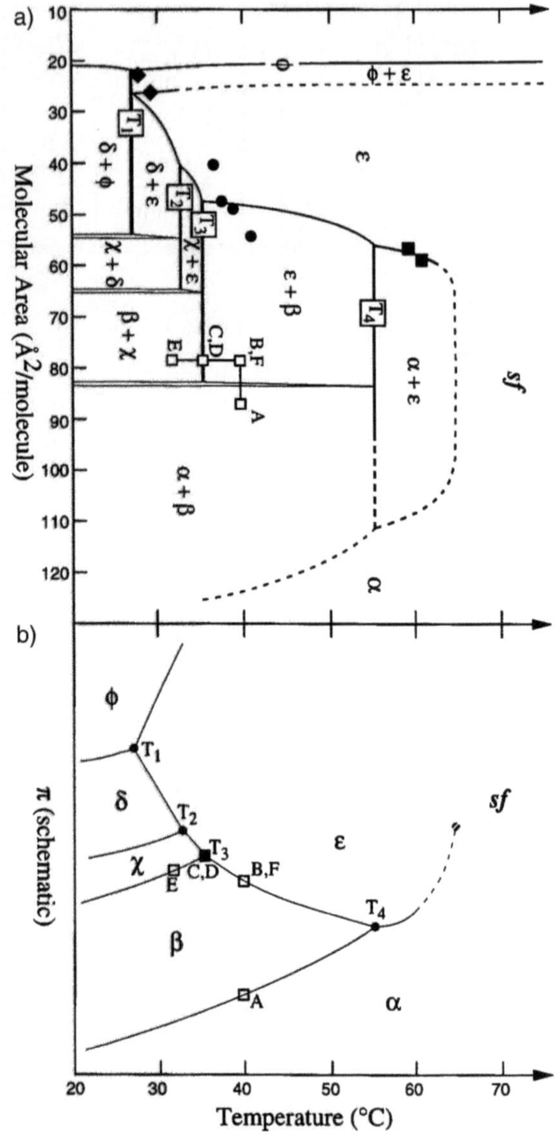

Fig. 13.12. Two-dimensional phase diagram of (**a**) temperature vs molecular area and (**b**) pressure vs temperature. Note that surface pressure was not measured. (Reprinted from [136])

in heptane was injected into the cell, the molecular phase began to be observed as bright thin bands in the image (Fig. 13.13a, A). Close inspection of these regions showed that pinstripe patterns, i.e., the lying-down phase of alkanethiols were formed (Fig. 13.13b, B). The VIs of the gold surface were also observed at the initial stage. As the coverage increased, islands seen as the brightest region began to grow and covered the surface. They were the islands of the upright phase of alkanethiols since the $\sqrt{3} \times \sqrt{3}$ molecular arrangement was observed on these islands. Thus, the formation of the lying-down phase and the following islandlike growth of the upright phase were evident in solutions.

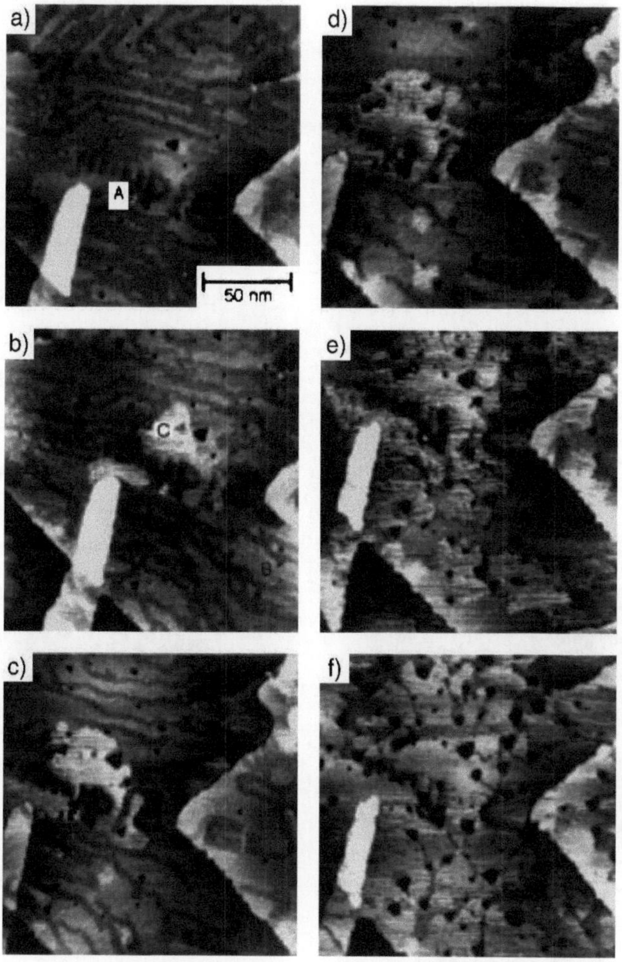

Fig. 13.13. Sequentially obtained STM images of a Au(111) surface in heptane solution. (**a**) Two minutes after the first addition of droplets of 0.5 μM thiol solution. (**b**) Two minutes and (**c**) 7 min after the second addition. (**d**),(**e**) Two minutes and 12 min after the third addition, respectively. (**f**) Thirteen minutes after the fourth addition. (Reprinted from [137])

The pinstripe structures were investigated in detail in dilute solutions. A variety of structures were observed on a surface as shown in Fig. 13.14. Figure 14a shows a stripe pattern with a period of 3.2 nm. This period is close to twice the molecular length. In addition, close inspection of the image reveals the individual molecular structures. This structure corresponds to the β phase in Fig. 13.11. Other pinstripe structures shown in Fig. 13.14b–d have a period of 2.3 nm. These structures correspond to the interdigit-structure or the δ phase shown in Fig. 13.11.

It is interesting that the lying-down phases of alkanethiols are formed in heptane, which is expected to have a strong affinity to the alkyl chains. In fact, there is a strong interaction between alkyl chains and a gold surface as discussed later.

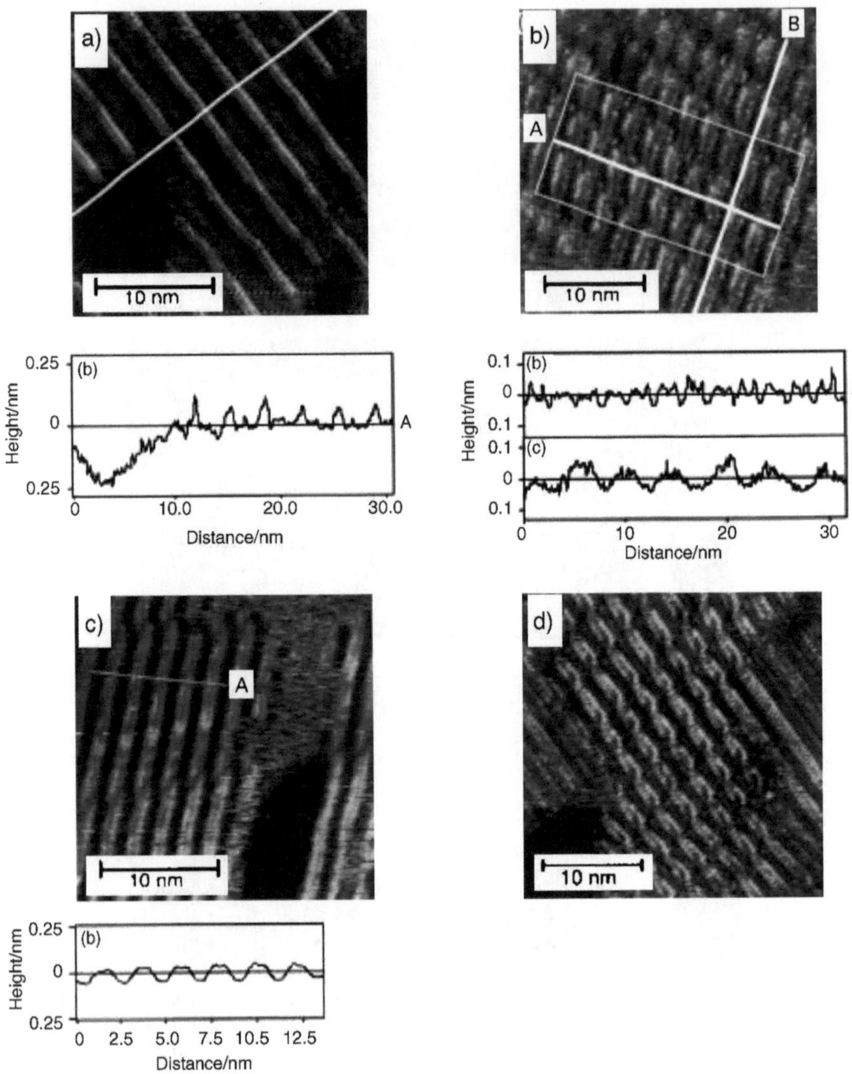

Fig. 13.14. Various pinstripe structures observed in a dilute solution of decanethiol in heptane. (**a**) β phase. (**b**)–(**d**) δ phase. (Reprinted from [137])

13.4.2.2
Structures of Thiol SAMs on Au(100)

The structure of SAMs of alkanethiols on a Au(100) surface has not been studied in detail. A Au(100) surface is known to be reconstructed, resulting in the formation of a hexagonal atomic arrangement as discussed before. The reconstructed Au(100) surface shows a periodic protrusion owing to the dislocations between the subsurface and the topmost layer.

STM investigation of butanethiol SAMs grown in UHV conditions revealed that butanethiol molecules form a c(2 × 8) molecular lattice with a 1 × 4 Au missing row as shown in Fig. 13.15a [139]. A plausible model is shown in Fig. 13.15b,c. This structure is supported by STM measurements carried out in electrolyte solutions [140, 141] and by GIXD measurement [142]. However, another kind of structure, an incommensurate oblique cell, has been also reported [143–145].

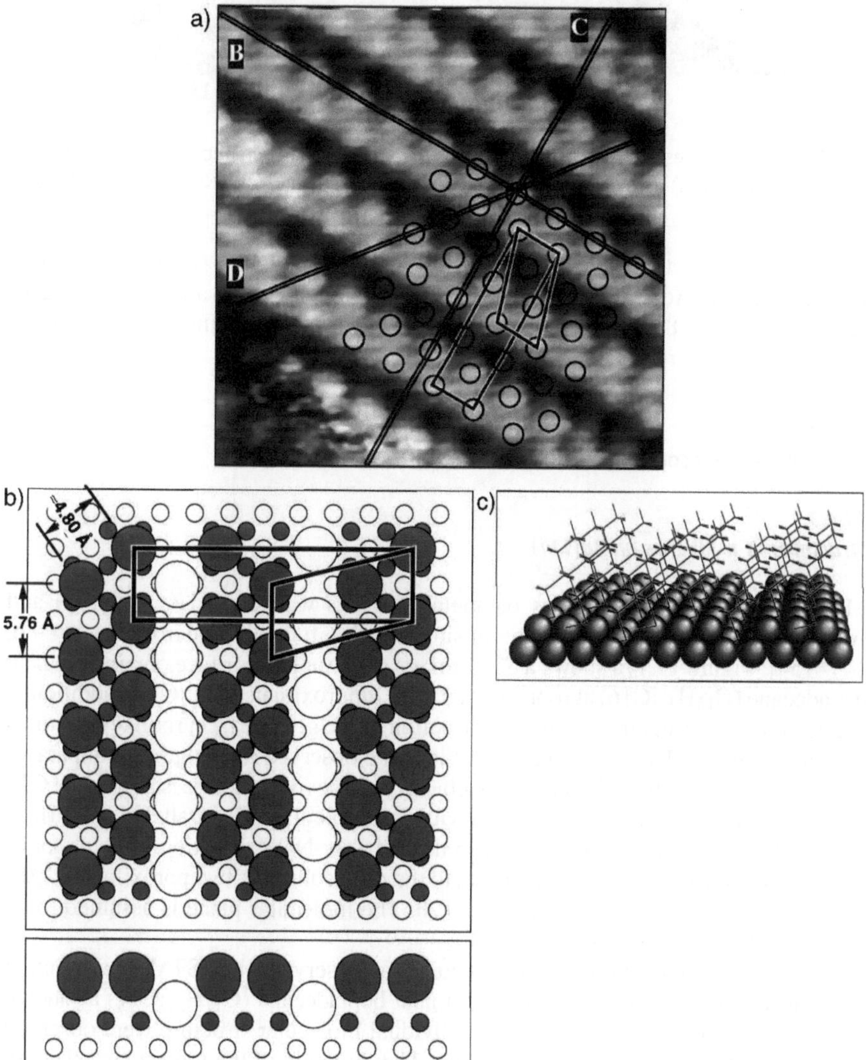

Fig. 13.15. Structure of a Au(100) surface covered with a butanethiol monolayer. (**a**) STM image. (**b**),(**c**) Models of molecular arrangement. The rectangle in (**b**) shows a unit cell for a c(2 × 8) molecular lattice with a 1 × 4 Au missing row. (Reprinted from [139])

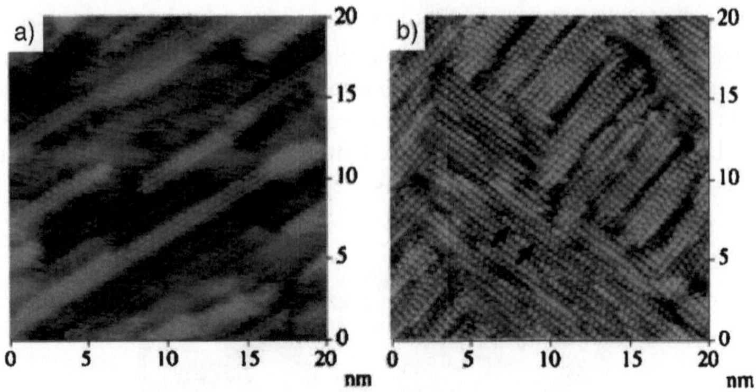

Fig. 13.16. STM images of a Au(100) surface covered with a decanethiol monolayer. Modification was carried out on (**a**) reconstructed and (**b**) (1 × 1) Au(100) surfaces. (Reprinted from [146])

In fact, both of these structures were observed on a Au(100) surface as shown in Fig. 13.16 [146]. Adsorption of the decanethiol on reconstructed and (1 × 1) surfaces resulted in a c(2 × 8) molecular lattice (Fig. 13.16a) and an incommensurate oblique cell (Fig. 13.16b), respectively.

13.4.3 Assembly of Physisorbed Molecules: *n*-Alkanes

13.4.3.1 Structures of n-Alkanes on Au(111)

Molecular assembly of alkanes on metal surfaces was studied on Au(111) and Au(100) surfaces in neat liquids by in situ scanning tunneling microscopy [84–86, 147–154]. Figure 13.17a shows a STM image of a Au(111) surface obtained in neat hexadecane ($C_{16}H_{34}$:C16) at room temperature (approximately 25 °C). In addition to the herringbone structure due to reconstruction of the gold surface, rodlike structures with a length of 2.2 nm were observed. Since the observed rod length is comparable with the length of C16, 2.2 nm, it is concluded that the rod structure is individual C16 molecules in an all-*trans* conformation with the molecular axis parallel to the surface plane. The C16 formed rows running in the ⟨110⟩ or NNN direction of gold atoms. The molecular axis is tilted with respect to the molecular row by approximately 60°. A plausible model is shown in Fig. 13.17b. The molecular plane is assumed to be parallel to the surface plane.

A different molecular packing structure was observed when STM measurement was carried out on a Au(111) surface in neat heptadecane ($C_{17}H_{36}$:C17) as shown in Fig. 13.18a. Ordered lamellae and individual molecular structure were observed by in situ scanning tunneling microscopy. The direction of the molecular axis with respect to the atomic arrangement of the Au(111) surface was the same as that observed in C16. The angle formed by the molecular axis and the lamellae was 90°. A plausible model is shown in Fig. 13.18b. The molecular packing structure is different from that of C16. The difference is explained by the symmetry of

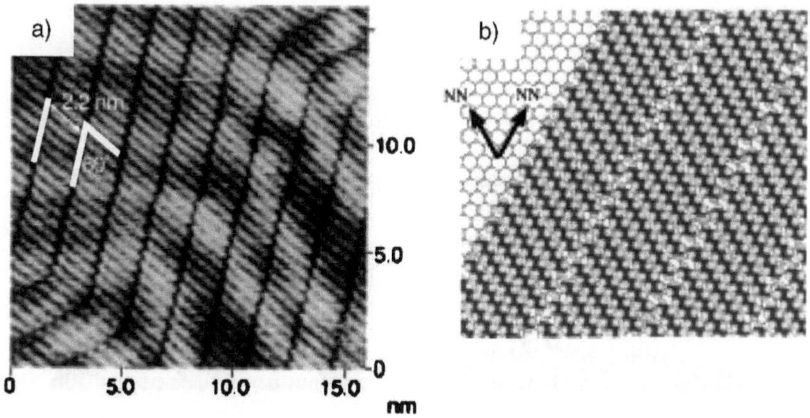

Fig. 13.17. (**a**) A STM image of a Au(111) surface in $C_{16}H_{34}$. (**b**) Plausible model. *NN* nearest neighbor. (Reprinted from [84])

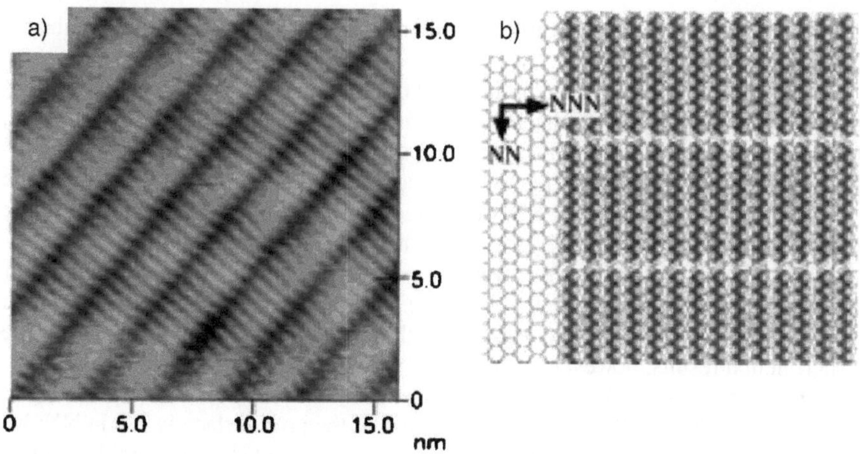

Fig. 13.18. (**a**) A STM image of a Au(111) surface in $C_{15}H_{32}$. (**b**) Plausible model. *NNN* next-nearest neighbor. (Reprinted from [84])

molecules owing to the orientation of the methyl-terminal unit as shown in Fig. 13.19. This observation revealed that molecule–molecule interactions dominated the SA structure of alkanes.

Detailed analysis of the molecular packing structure revealed that the distortion in the atomic arrangement of the Au(111) surface influenced the orientation of the alkane molecules. Reconstruction of the Au(111) surface results in uniaxial compression of the gold lattice in the direction perpendicular to the bridging row as mentioned already. There are, thus, two nearest-neighbor (NN) directions. One is in the most compressed direction as shown in Fig. 13.20a. We call this direction NN(90) since this direction is perpendicular to the stripe pattern formed by the bridging row of the gold surface. The other NN direction is rotated by 60° with respect to NN(90), and we call this direction NN(30) as shown in Fig. 13.20b. The alkanes were always

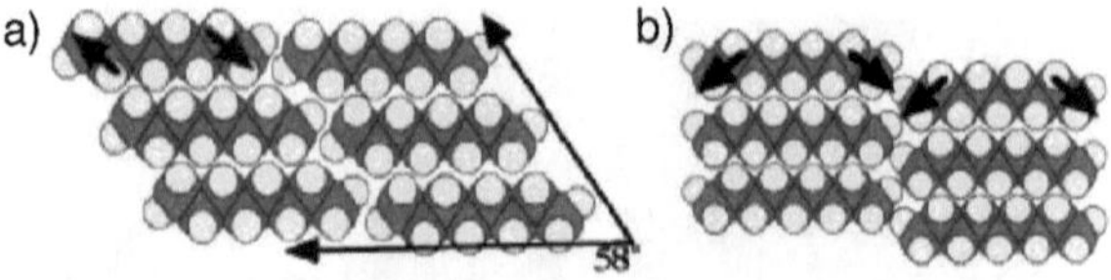

Fig. 13.19. Proposed models of SAMs of (**a**) even-number and (**b**) odd-number alkanes. (Reprinted from [86])

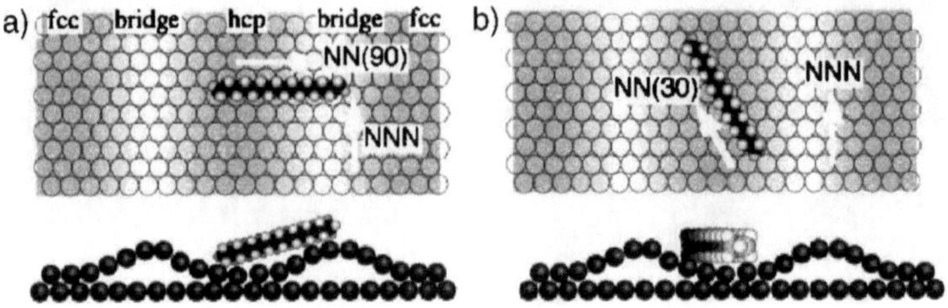

Fig. 13.20. Models of the reconstructed gold surface. An alkane molecule is oriented to (**a**) NN(90) and (**b**) NN(30). (Reprinted from [86])

found to adsorb along the NN(30) direction regardless of the odd or even number of the carbons.

One possible reason for the preferential orientation is the adjustment of the substrate lattice to the carbon chain. For example, the length of the C–C–C zigzag, 0.254 nm, is close to the atomic spacing of graphite, 0 246 nm (space between hollows). The close lattice period allows the carbon skeleton to fit to the lattice of the graphite surface. The NN and NNN atomic distances of the Au(111) surface in the compressed direction are 0.277 and 0.486 nm, respectively, while they were originally 0.288 and 0.499 nm, respectively. The adsorption of alkanes with a molecular axis along NN(90) is expected to be more favorable because the NN(90) atomic distance is closer to the C–C–C zigzag length than the NN(30) atomic distance. The experimental results, however, do not agree with this expectation.

The surface corrugation due to the surface reconstruction explains the preferential alkane orientation in NN(30). The corrugation of the gold surface is less in NN(30) (Fig. 13.20a) than that in NN(90) (Fig. 13.20b), and it should be unfavorable to adapt the molecular skeleton to NN(90).

The importance of the lattice matching between the gold surface and carbon chains is evident in the fact that SAMs of alkanes are not observed when the chain

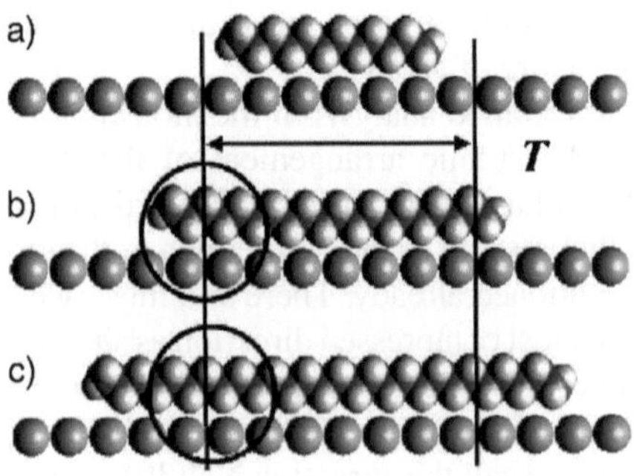

Fig. 13.21. Models showing possible arrangement of *n*-alkanes. (**a**) Misfit between the Au lattice and the carbon position is small ($n < 17$). (**b**) A stacking fault was created at the end of the molecule. (**c**) When the carbon chain is much longer than that in (**b**), the edge of the molecule again sits on the favorable site and the molecule is *locked* on the surface. (Reprinted from [147])

length is longer than 18 [147]. This phenomenon is explained by the formation of stacking faults or discommensurations at the edge of the molecule owing to the large misfit between the gold–gold and carbon–carbon distance as shown in Fig. 13.21. Interestingly, an ordered monolayer of alkanes is again formed when the chain length is longer than 28. This is because both ends of the molecule are positioned between the gold atoms again as shown in Fig. 13.21, model C.

13.4.3.2
Structures of n-Alkanes on Au(100)

Figure 13.22a,b shows STM images of a Au(100) surface obtained in neat C16 and C17, respectively [85]. These images clearly show that C16 and C17 molecules form an ordered molecular layer. The molecular rows run in the [011] direction of the Au(100) surface. The molecular axis is perpendicular to the lamellar structure. There are small height variations among molecules. The distance between the most-elevated molecules is approximately 1.4 nm. As mentioned before, the clean Au(100) surface shows a (5 × 20) or "hex" reconstruction, which has a periodic corrugation owing to the misfit between the first and the second layers as shown in Fig. 13.23a,b. The observed period and direction of the height variation in the molecules were consistent with those of the hex-Au(100) surface; thus, it is concluded that the reconstructed Au(100) surface was maintained.

The schematic models of the alkane adlayer are shown in Fig. 13.23c,d. Alkane molecules are adsorbed on the Au(100) surface with their molecular axis parallel to the stripe pattern formed by the reconstructed Au(100) substrate. This direction corresponds to the NN direction of the top layer of the hex-Au(100) surface and is consistent with the molecular orientation observed on a Au(111) surface. The lattice of the alkane SAMs seems to be incommensurate with that of the gold surface because a shift in the molecular location with respect to the surface corrugation is evident. For example, the most elevated position contains one and two molecules at A and B, respectively, in Fig. 13.22a.

The packing structures of C16 and C17 SAMs have the same symmetry, unlike the case on the Au(111) surface. The absence of the odd–even effect on the

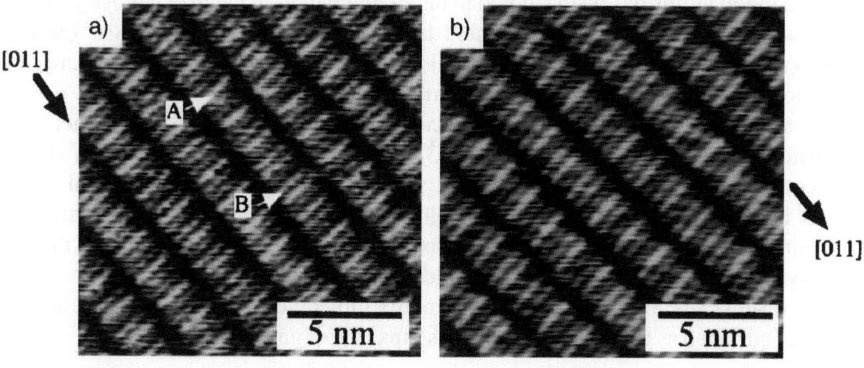

Fig. 13.22. STM images of Au(100) surfaces in (**a**) C16 and (**b**) C17 (Reprinted from [85])

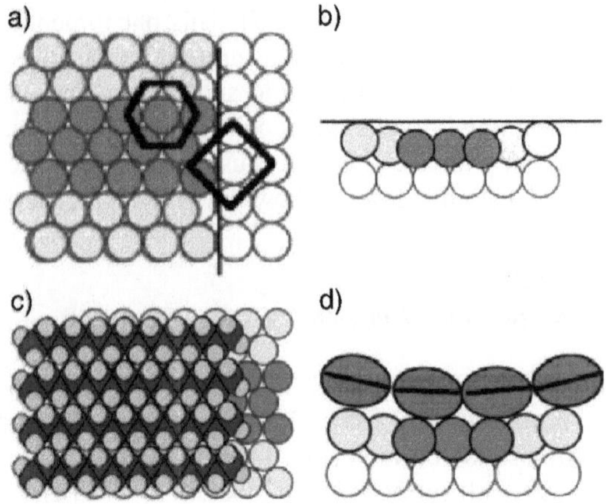

Fig. 13.23. A model of hex-Au(100) ((**a**) *top view*; (**b**) *sectional view*) and the proposed model of the self-assembled monolayer (*SAM*) of C16 ((**c**) *top view*; (**d**) *side view*). The *ellipses* and the *line inside* shown in (**d**) represent the alkanes and the plane formed by the carbon skeleton. (Reprinted from [85])

Au(100) surface is probably because the close packing of alkanes is hindered by the corrugation of the gold surface.

Studies on the SA of *n*-alkanes on metal surfaces have shown that the alkane–surface interaction is fairly strong and can affect the process of the molecular assembly. The formation of the *lying-down* phase of alkanethiols is one of the representative examples. The ensemble of interactions should be carefully taken into account to design the SA structure of the molecules.

13.5 SA of Inorganic Complexes

13.5.1 Introduction

Molecular assemblies of inorganic molecules are not noticed as much as those of organic molecules. In fact, the formation of an ordered monolayer or submonolayer of ions and metal atoms has been studied in detail in the field of electrochemistry [5–8]. The adsorption of small molecules and atoms is usually described as a *specific adsorption* onto a surface because lateral molecule–molecule interactions are not likely to be major driving forces for the order of the monolayer.

Metal-complex monolayers were discovered in the process of electrochemical deposition of precious metals such as Pt [53,155], Rh [156] and Pd [54–56,157–159]. The concept of SA might be applicable to the monolayers of these metal complexes because the geometry of the molecules influences the packing structure. A chemisorbed inorganic molecular assembly would provide a new class of monolayer system, although only a few examples have so far been reported.

We discuss the SA of some metal-complex molecules and metal oxide clusters on metal surfaces as representative SA systems of inorganic molecules.

13.5.2 Assembly of Metal Complexes

13.5.2.1 *Structures of Adsorbed $PdCl_4^{2-}$ on Au(111) and Au(100)*

SA of metal-complex molecules is known to take place when a very small amount of complex molecules exists in electrolyte solutions. Spontaneous adsorption of $PdCl_4^{2-}$ molecules was detected by electrochemical quartz crystal microbalance (QCM) as shown in Fig. 13.24 [54]. The frequency decreased after injection of the $PdCl_4^{2-}$ molecules. Neither electrochemical oxidation nor reduction of the complex molecules occurred because no current was observed during the adsorption. The total mass increase was estimated to be 86.9 ng/cm^2. The number of adsorbed $PdCl_4^{2-}$ molecules was estimated to be 2.1×10^{14}/cm^2, which is equal to 15% of the Au(111) surface atomic density.

Structures of the SAM of $PdCl_4^{2-}$ molecules were investigated by in situ electrochemical scanning tunneling microscopy on a Au(111) surface [54, 157]. The Au(111) surface shows a (1×1) structure, i.e., reconstruction was lifted at 0.95 V as shown in Fig. 13.25a. A totally different structure was observed when a 0.1 mM solution of $PdCl_4^{2-}$ was added as shown in Fig. 13.25b. The potential of the electrode was kept at 0.95 V during the addition of the solution. The newly observed structure has a hexagonal symmetry with a NN distance of 0.78 nm. The adlattice was found to be rotated by 20° with respect to that of the Au(111) surface. Although a bisulfate is known to form an adlayer in sulfuric acid solutions [160, 161], structures assignable to bisulfate were not observed. Thus, the observed structure is attributed to the SAMs of $PdCl_4^{2-}$. A lattice structure was assigned to $(\sqrt{7} \times \sqrt{7})$R19.1°. A slightly different incommensurate structure was also proposed [157].

SA of $PdCl_4^{2-}$ molecules was observed on a Au(100) surface [55, 158]. Figure 13.26a,b shows STM images of the Au(100) surface after the addition of $PdCl_4^{2-}$ molecules to 50 mM H_2SO_4 solution. An ordered structure, which is different from the original Au(100) surface, was observed when $PdCl_4^{2-}$ molecules were added at 1.05 V, at which no electrochemical reactions took place. The newly observed structure is attributed to a molecular layer of $PdCl_4^{2-}$. Figure 13.26c shows a ball model of the $PdCl_4^{2-}$ complex adsorbed on the (1×1)-Au(100) surface.

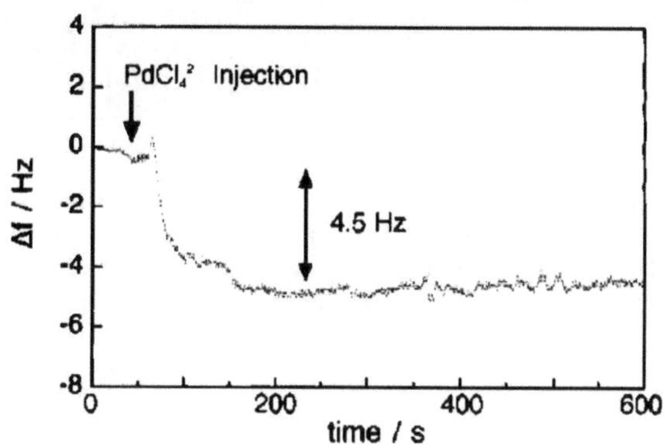

Fig. 13.24. Time dependence of frequency change measured by quartz crystal microbalance when the $PdCl_4^{2-}$ solution was added to the cell at 0.95 V. The cell contained 50 mM H_2SO_4 solution. The final concentration of $PdCl_4^{2-}$ in the cell was 0.1 mM. (Reprinted from [54])

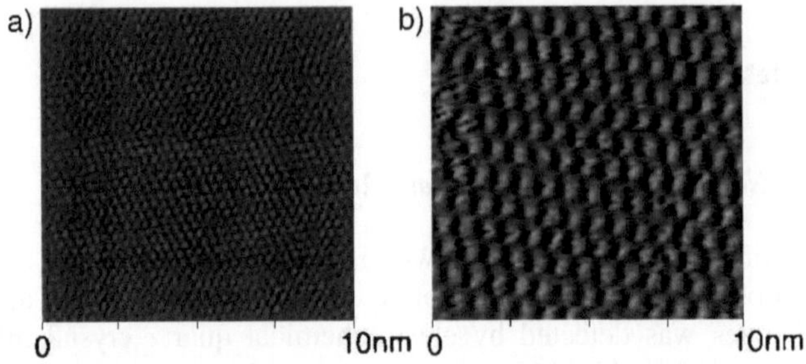

Fig. 13.25. (**a**) A STM image of a (1 × 1)-Au(111) surface obtained at 0.95 V in a 50 mM H_2SO_4 solution. (**b**) The same electrode after the addition of $PdCl_4^{2-}$ solution. The total concentration of $PdCl_4^{2-}$ was 0.1 mM. The potential of the electrode was 0.95 V vs the reversible hydrogen electrode (*RHE*). (Reprinted from [54])

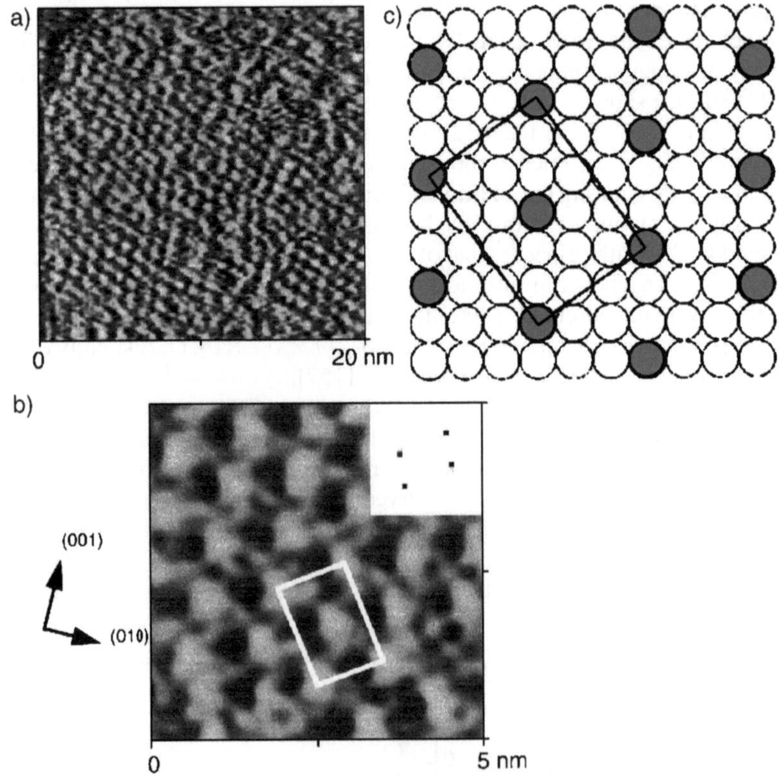

Fig. 13.26. STM images of a Au(100) surface in 50 mM H_2SO_4 solution after the addition of $PdCl_4^{2-}$ in (**a**) 20 nm × 20 nm and (**b**) 5 nm × 5 nm. The concentration of $PdCl_4^{2-}$ was 0.5 mM. The Fourier-transformed 2D spectrum is shown in the *inset* in (**b**). (**c**) A ball model of the $PdCl_4^{2-}$ layer on the (1 × 1)-Au(100) surface. (Reprinted from [55])

Electrochemical formations of a Pd layer in the presence of a $PdCl_4^{2-}$ monolayer are interesting and are discussed here, although they are out of the scope of this section. Figure 13.27 shows a cyclic voltammogram and the QCM response of a Au(111) electrode in 50 mM H_2SO_4 containing 0.1 mM $PdCl_4^{2-}$ [54]. A cathodic current started to flow at a potential below 0.95 V when the potential of the electrode was scanned negatively from 0.95 to 0.35 V. QCM measurements showed an increase in surface mass. Detailed analysis of the relationship between charge and mass revealed that the Pd layer was deposited in the following reaction:

$$PdCl_4^{2-} + 2e^- \rightarrow Pd + 4Cl^- .$$

The anodic current starting from approximately 0.8 V in the positive potential scan was attributed to dissolution of Pd.

In situ scanning tunneling microscopy measurements revealed a layer-by-layer growth of monoatomic Pd layers as shown in Fig. 13.28. $PdCl_4^{2-}$ molecules were found not only on the Au(111) substrate but also on the Pd layer. These observations suggest that the $PdCl_4^{2-}$ layer plays an important role in the formation of Pd layers. In fact, the Pd layer obtained is significantly flatter and has fewer defects than does a Pd layer formed by vacuum deposition [54].

Figure 13.29 shows a plausible mechanism of the Pd deposition in the presence of a $PdCl_4^{2-}$ monolayer. Adsorbed $PdCl_4^{2-}$ molecules on a terrace are expected to be less reactive than those at a step edge of Pd islands, resulting in a 2D growth process of Pd layers. Similar layer-by-layer growth of Pd monolayers was observed on a Au(100) surface.

GIXD measurements were carried out to reveal the position of the Pd atoms with respect to the underlying Au atoms [162]. Pd atoms were shown to follow the stacking

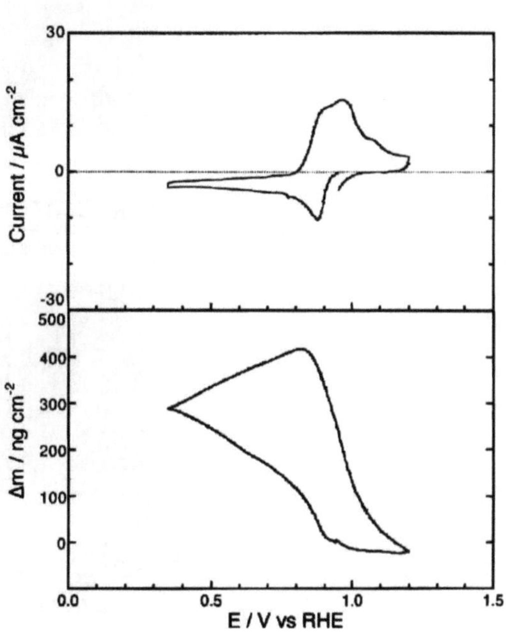

Fig. 13.27. Potential dependence of current (*upper panel*) and mass (*lower panel*) change observed on a Au(111) surface in 50 mM H_2SO_4 + 0.1 mM $PdCl_4^{2-}$. The potential of the electrode was swept at 5 mV/s. (Reprinted from [54])

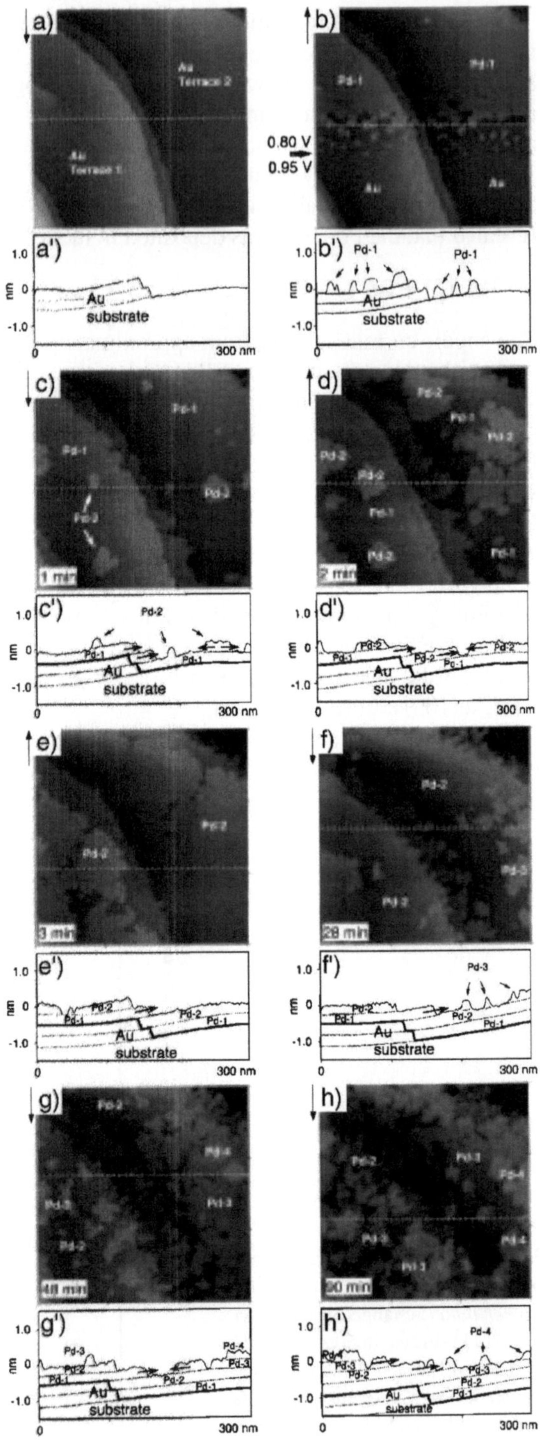

Fig. 13.28. STM images (300 nm × 300 nm) of a Au(111) surface in 50 mM H_2SO_4 + 0.5 mM $PdCl_4^{2-}$ solution. (**a**) The electrode potential was 0.95 V vs the RHE. (**b**) The potential was changed to 0.8 V. (**c**) $t = 1$ min, (**d**) $t = 2$ min, (**e**) $t = 3$ min, (**f**) $t = 28$ min, (**g**) $t = 48$ min and (**h**) $t = 90$ min after (**b**). A cross section along the *white dotted line* in each figure is shown below each image. (Reprinted from [54])

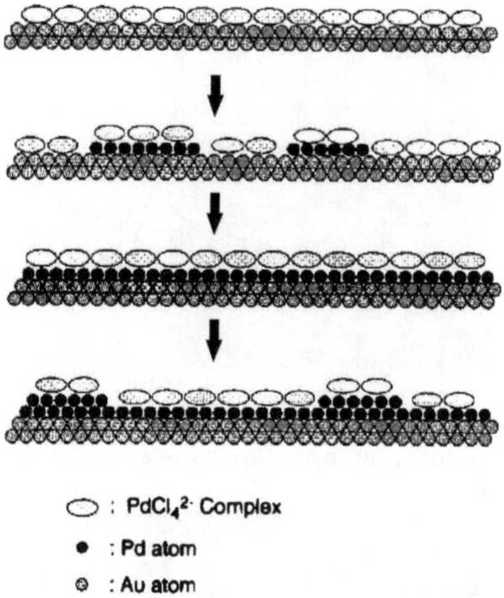

Fig. 13.29. Models for Pd deposition on a Au(111) surface in the presence of the $PdCl_4^{2-}$ SAM. See text for details. (Reprinted from [54])

distance and sequence of a Au(111) substrate, which is a very unique case only seen in an electrochemically grown Pd layer. Notable electrochemical reactivities on an electrochemically formed *pseudomorphic* Pd layer have been reported [163, 164].

13.5.2.2
Structures of Other Metal Complexes Adsorbed on Au(111)

Various metal complexes have been found to form SAMs in electrolyte solutions. Figure 13.30 shows an STM image of a $PtCl_6^{2-}$ monolayer formed on a Au(111) surface in 50 mM $HClO_4$ solution containing 0.6 mM H_2PtCl_6 and a corresponding model [53]. The observed structure was $(\sqrt{7} \times \sqrt{7})R19.1°$. Electrochemical reduction of $PtCl_6^{2-}$ resulted in the formation of a Pt(111) phase on the Au(111) electrode.

Nagahara et al. [165] investigated structures of SAMs formed by a series of haloplatinate complexes by in situ electrochemical scanning tunneling microscopy. SAMs of haloplatinate complexes were formed by immersing a Au(111) substrate in a solution containing 50 μM PtX_4^{2-} (X is Cl and Br) for 1 min. Then the substrate was transferred to a STM cell. Thus, the solution did not contain any complex molecules in the measurement. Figure 13.31a,b shows STM images of Au(111) surfaces modified with $PtCl_4^{2-}$ and $PtBr_4^{2-}$, respectively. The high-resolution images revealed the internal structures of the molecules. Basically, both molecules formed a $(\sqrt{7} \times \sqrt{7})R19.1°$ structure. A different kind of $PtBr_4^{2-}$ adlayer was found as shown in Fig. 13.32. A superstructure containing three small spots was observed. These spots were tentatively attributed to Br atoms or Br^- anions possibly contained as an impurity.

The monolayer of PtI_6^{2-} was also observed with an internal molecular resolution as shown in Fig. 13.33a. The image seems to consist of two kinds of structures, small

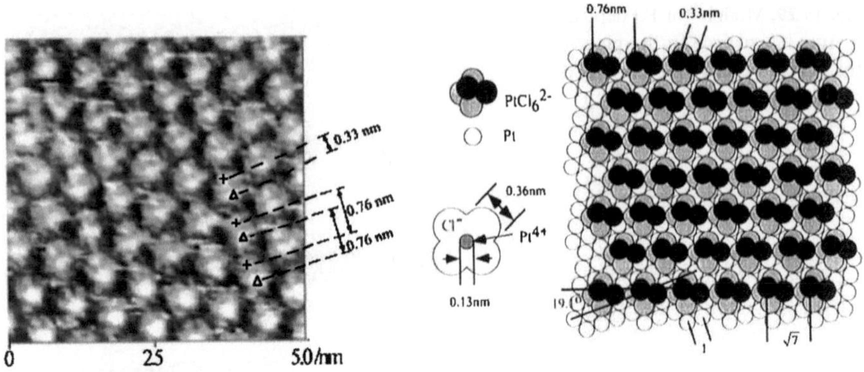

Fig. 13.30. A STM image of a Au(111) electrode in 50 mM $HClO_4$ + 0.6 mM H_2PtCl_6 solution at 0.70 V vs the RHE (*left*). *Right*: A model for the adlayer structure. (Reprinted from [53])

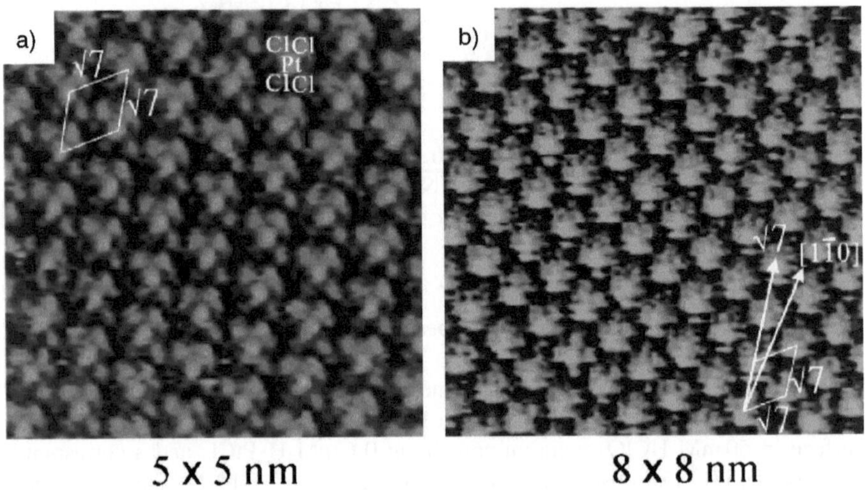

Fig. 13.31. STM images of a Au(111) electrode covered with (**a**) $PtCl_4^{2-}$ and (**b**) $PtBr_4^{2-}$. See text for details. (Reprinted from [165])

spots and large bright spots. The small and large bright spots seem to be attributed to iodine and PtI_6^{2-} molecules, respectively. Interestingly, a pinwheellike structure was observed when the potential was swept to negative as shown in Fig. 13.33b. This structural transition was probably due to the reduction of the PtI_6^{2-} molecule to PtI_4^{2-}.

Adsoptions of Pt complexes with bulky organic ligands, $[Pt(tpy)Cl]^+$ and $Pt(C_5H_7O_2)_2$, on Au(111) surfaces were investigated [166]. Figure 13.34a shows a STM image of a Au(111) electrode obtained in a solution containing 50 mM H_2SO_4 and 0.1 mM $[Pt(tpy)Cl]^+$. The inset in the image shows the molecular structure. An ordered adlayer was observed. Although details of the adlayer structure have not been established yet, the molecules are likely to adsorb via Cl^- on gold surfaces with a 4 × 4 structure as shown in Fig. 13.34b. When the potential of

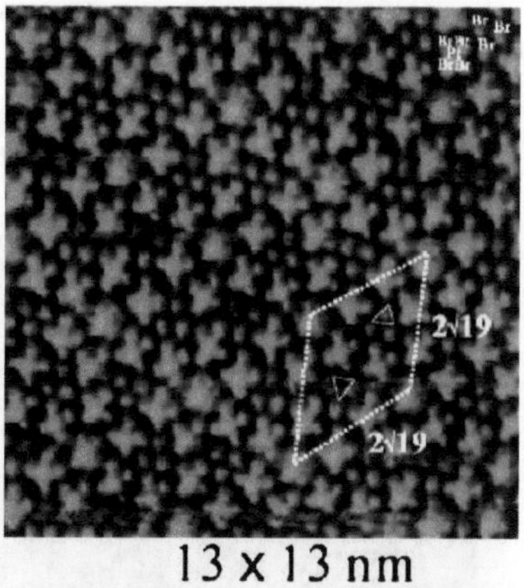

Fig. 13.32. A STM image of a Au(111) electrode covered with $PtBr_4^{2-}$. This structure coexisted with the one shown in Fig. 13.31b. (Reprinted from [165])

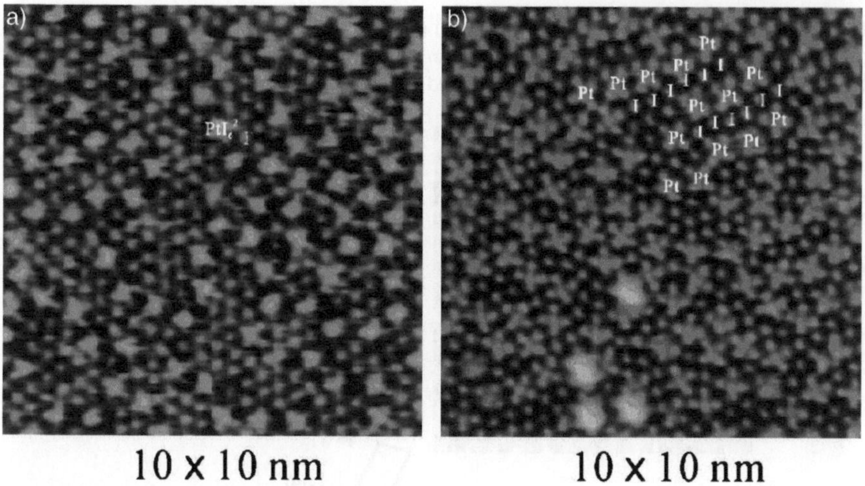

Fig. 13.33. In situ STM images of a Au(111) surface at (**a**) 0.9 V and (**b**) 0.5 V vs the standard calomel electrode. The substrate was immersed in 50 μM K_2PtI_6 solution for 1 min before the measurement. (Reprinted from [165])

the electrode was scanned to positive, the ordered structure disappeared. This is probably because the surface negative charge hindered the adsorption of negatively charged Cl^-.

STM observations of SAMs of $Pt(C_5H_7O_2)_2$ were carried out in a 3:2 mixture of 50 mM $HClO_4$ aqueous and ethanol solution of $Pt(C_5H_7O_2)_2$ since the solubility of $Pt(C_5H_7O_2)_2$ in water is poor. The concentration of $Pt(C_5H_7O_2)_2$ molecules in the mixed solution was 0.2 mM. Figure 13.35a shows a STM image of a Au(111) electrode obtained in the mixture solution. A row structure consisting of propeller-like structures was observed. A plausible model structure is shown in Fig. 13.35b. The bright spot (a shaft of the propeller) is attributed to a Pt atom. The packing structure seems to indicate that molecule–molecule interactions dominate the adlayer structure.

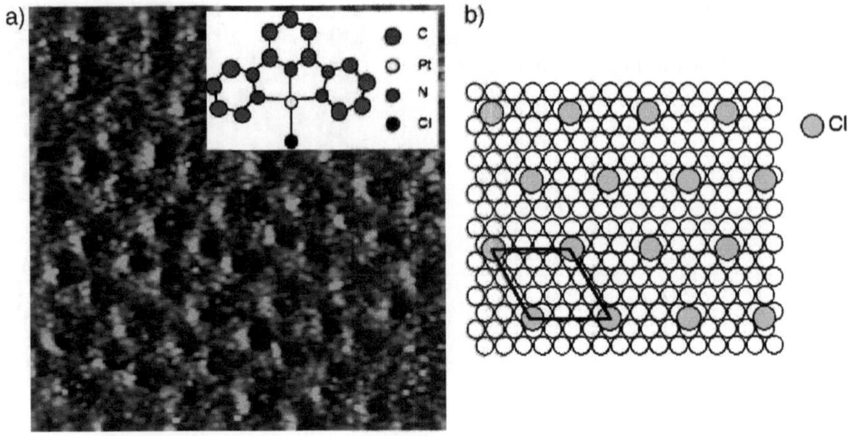

Fig. 13.34. (**a**) A STM image (10 nm × 10 nm) of a Au(111) surface in a solution containing 50 mM H_2SO_4 + 0.1 mM $[Pt(tpy)Cl]^+$. The *inset* shows the molecular structure of $Pt(tpy)Cl^+$. (**b**) Plausible molecular arrangement. (Reprinted from [166])

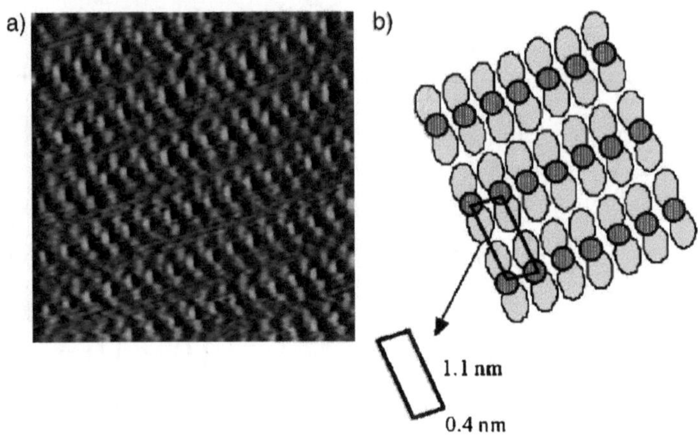

Fig. 13.35. (**a**) A STM image (10 nm × 10 nm) of a Au(111) surface in a 3:2 mixture of 50 mM $HClO_4$ and ethanol solution of $Pt(C_5H_7O_2)_2$ (**b**) Plausible adlayer structure. (Reprinted from [166])

It should be noted that no ordered structures were detected by scanning tunneling microscopy in solutions containing some kinds of Pt complexes such as $PtCN_4^{2-}$, $Pt(SCN)_4^{2-}$ and $[Pt(C_2O_4)_2]^{2-}$, probably because of the immediate decomposition of the molecules.

13.5.3 Assembly of Metal Oxide Clusters: Polyoxometalates

Polyoxometalates (POMs) are metal oxide anionic clusters that exhibit notable properties, including various charge states without change in structure [167–171]. Modification of an electrode with POMs has been extensively studied because of their potential applications such as in catalysis and corrosion inhibition. The adsorption POMs on highly oriented pyrolytic graphite (HOPG) [172–176], Au [168,177–181], Hg [182] and Ag [183–187] electrodes has been studied. Ordered monolayers are known to exist on HOPG, Ag and Au surfaces. Interesting electronic properties such as negative differential resistance have also been reported [174–176].

The monolayer and multilayer formation of α-$H_4SiW_{12}O_{40}$ or silicotungstic on Ag surfaces are quite interesting since they occur in a spontaneous man-

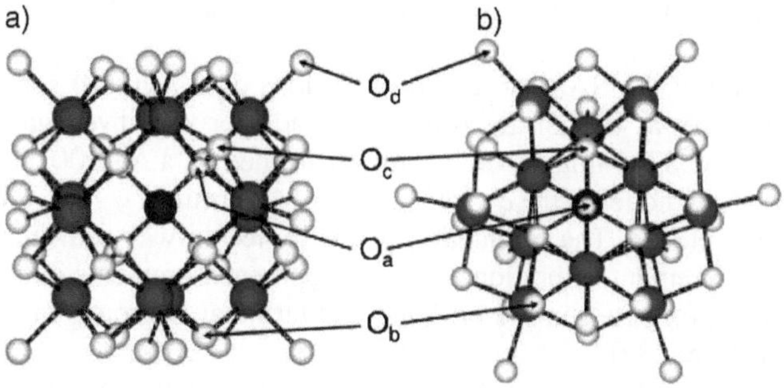

Fig. 13.36. Model structures of α-$SiW_{12}O_{40}^{4-}$ viewed along an S4 axis (**a**) and a C3 axis (**b**). *Larger gray balls* are W, *small light gray balls* are O and the *black central ball* is Si. (Reprinted from [185])

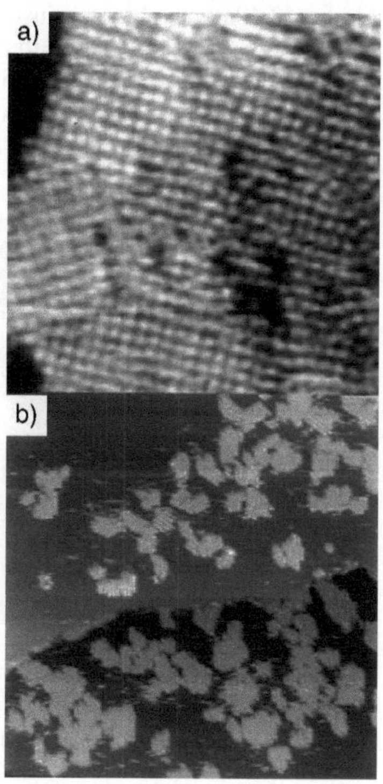

Fig. 13.37. (**a**) A 28 nm × 28 nm STM image of a Ag(111) surface covered with a monolayer of α-$SiW_{12}O_{40}^{4-}$ in 0.1 M $HClO_4$. (**b**) A 110 nm × 110 nm STM image showing submonolayer formation of α-$SiW_{12}O_{40}^{4-}$ on the Ag(111) surface in 0.1 M H_2SO_4. (Reprinted from [183])

ner, i.e., SA, with a strong interaction between POMs and Ag [183–187]. The structure of STA is shown in Fig. 13.36. Figure 13.37a shows a STM image of a Ag(111) surface taken in 0.1 M $HClO_4$ after 3-h immersion of the Ag(111) surface in a solution containing 0.1 M H_2SO_4 and 10^{-4} M α-$SiW_{12}O_{40}^{4-}$ [183]. The image shows a square-lattice structure having a NN distance of 1.02 nm. The observed spacing matches the diameter of the α-$SiW_{12}O_{40}^{4-}$ ion and, hence, each spot is attributed to an individual α-$SiW_{12}O_{40}^{4-}$ ion. The image of a large area shown in Fig. 13.37b reveals island structures that are attributed to α-$SiW_{12}O_{40}^{4-}$. The surface was covered with a full monolayer after 24 h. This result indicates that α-$SiW_{12}O_{40}^{4-}$ ions spontaneously form a monolayer on a Ag(111) surface.

The structure of the SAM of α-$SiW_{12}O_{40}^{4-}$ ions on a Ag(100) surface that has the same symmetry as that of an α-$SiW_{12}O_{40}^{4-}$ adlayer was investigated in detail. Figure 13.38 show STM images of the adlayer formed on a Ag(100) surface [184]. The corresponding model of the α-$SiW_{12}O_{40}^{4-}$ SAM, forming $(\sqrt{13} \times \sqrt{13})R33.69°$, is shown in Fig. 13.39. The orientation of the molecule was confirmed by X-ray reflectivity measurement. Vibrational spectroscopic measurements indicate a strong chemical interaction between the bridging and terminal O groups in the molecule and the Ag surface [187].

Interestingly, a multilayer was formed when the potential of the electrode was set more negative than the first one-electron reduction of STA in the solution [186]. Figure 13.40 shows time-dependent scanning tunneling microscopz images of a Ag(100) surface sequentially obtained in 0.5 mM STA + 0.1 M $HClO_4$ solution at −0.1 V. Figure 13.40a was obtained just after the potential was held at −0.1 V. The dark parts of the image correspond to the bare Ag surface. The gray area, covering most of the surface, represents a STA monolayer. The bright domains indicated by arrows correspond to the newly formed islands on top of the STA monolayer. These islands become larger with time as shown in Fig. 13.40b–f. Multilayer formation of STA was

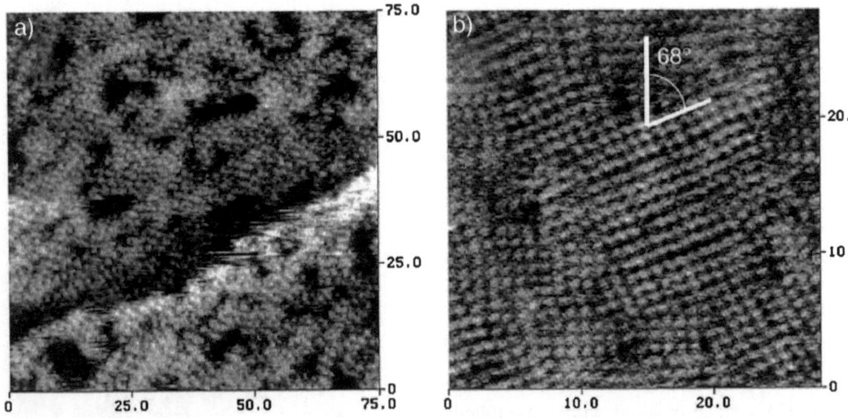

Fig. 13.38. STM images of a Ag(100) surface in solutions containing 0.5 mM α-$SiW_{12}O_{40}^{4-}$ + 0.1 M $HClO_4$ in (**a**) 75 nm × 75 nm at open circuit potential and (**b**) 30 nm × 30 nm at 0.25 V vs the normal hydrogen electrode. (Reprinted from [184])

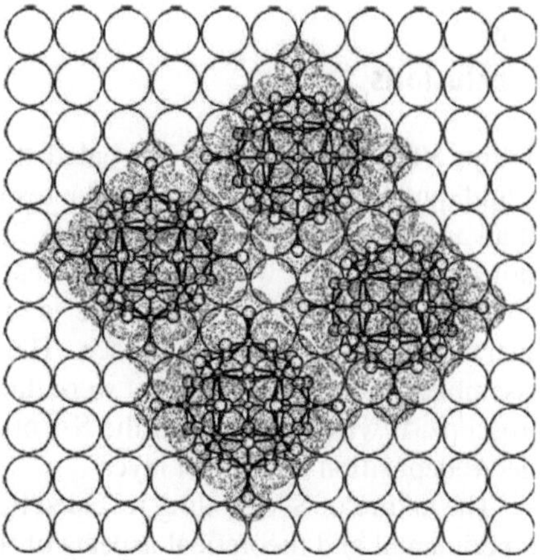

Fig. 13.39. Model of the $(\sqrt{13} \times \sqrt{13})$ structure proposed for the SA layer of α-$SiW_{12}O_{40}^{4-}$. (Reprinted from [184])

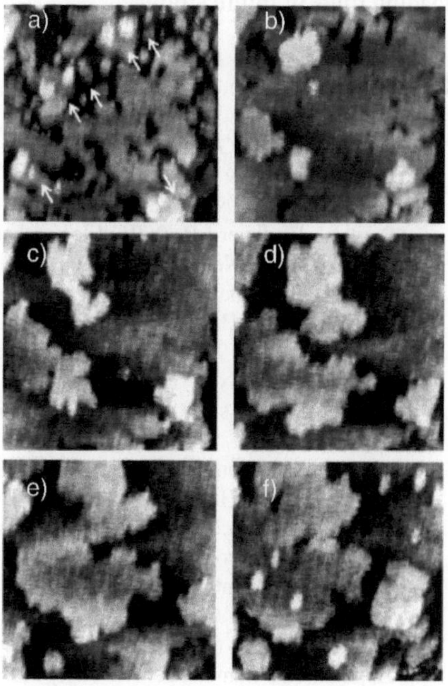

Fig. 13.40. Sequentially obtained STM images of a Ag(100) surface in 0.5 mM α-$SiW_{12}O_{40}^{4-}$ + 0.1 M $HClO_4$ solution at −0.10 V vs Ag/AgCl. (**a**) $t = 0$ min, (**b**) $t =$ 9 min, (**c**) $t = 15$ min, (**d**) $t = 24$ min, (**e**) $t = 30$ min, (**f**) $t = 56$ min. (Reprinted from [186])

evident from other experimental results obtained from cyclic voltammetry, QCM, X-ray photoemission spectroscopy and surface X-ray scattering measurement. A very small number of Ag^+ ions, existing in the solution owing to the finite solubility of Ag, are thought to play an important role in the mechanism of the multilayer formation. The negative potential possibly stabilizes Ag^+, which facilitates holding STA or reduced STA species on top of the STA monolayer through electrostatic interactions.

These results show that STA is a promising material to form inorganic SAMs and multilayered frameworks via the SA process.

13.6
Conclusions

In situ STM measurements revealed the SA process and structures of SAMs at solid/liquid interfaces with molecular resolution. A chemisorbed molecular assembly of alkanethiols proceeds with complex 2D phase transitions. A physisorbed organic molecular assembly was observed in neat liquids of n-alkanes on Au(111) and (100) surfaces. The monolayer structure is strongly related to the molecule–molecule interactions on a Au(111) surface. The chemisorbed and physisorbed molecular assemblies of metal complexes were discovered by in situ scanning tunneling microscopz. It was clarified that the SA of molecules is one of the important steps in electrodeposition of a metal layer.

The nature of SA is being revealed by in situ analytical tools. Further studies on experimental and theoretical aspects of the SA process would make SA techniques much more predictable and reliable. Scanning probe techniques will continue to play leading roles in future studies.

References

1. Adamson AW, Gast AP (1997) Physical Chemistry of Surfaces, 6th edn. Wiley-Interscience, New York
2. Israelachvili J (1991) Intermolecular and Surface Forces, 2nd edn. Academic, San Diego
3. Bard AJ, Faulkner LR (2001) Electrochemical Methods Fundamentals and Applications, 2nd edn. Wiley, New York
4. Bockris JOM, Khan SUM (1993) Surface Electrochemistry. A Molecular Level Approach, Springer, Berlin Heidelberg New York
5. Wieckowski A (1999) (ed) Interfacial Electrochemistry: Theory, Experiment, and Applications. Dekker, New York
6. Lipkowski J, Ross PN (1992) (eds) Adsorption of Molecules at Metal Electrodes. Wiley, New York
7. Gewirth A, Siegenthaler H (1995) (eds) Nanoscale Probes of the Solid/Liquid Interface. Springer, Berlin Heidelberg New York
8. Gewirth AA, Niece BK (1997) Chem Rev 97:1129
9. Itaya K (1998) Prog Surf Sci 58:121
10. Chen CJ (1993) Introduction to Scanning Tunneling Microscopy. Oxford University Press, New York
11. Wiesendanger R (1994) Scanning Probe Microscopy and Spectroscopy: Methods and Applications. Cambridge University Press, Cambridge
12. Bonnell D (2000) (ed) Scanning Probe Microscopy and Spectroscopy: Theory, Techniques, and Applications, 2nd edn. Wiley-VCH, New York
13. Guntherodt HJ, Wiesendanger R (1994) (eds) Scanning Tunneling Microscopy I: General Principles and Applications to Clean and Absorbate-Covered Surfaces, 2nd edn. Springer, Berlin Heidelberg New York
14. Giancarlo LC, Flynn GW (1998) Annu Rev Phys Chem 49:297
15. McGonigal GC, Bernhardt RH, Thomson DJ (1990) Appl Phys Lett 57:28
16. Rabe JP, Buchholz S (1991) Science 253:424
17. Smith DPE, Hörber JKH, Binnig G, Nejoh H (1990) Nature 344:641
18. Sonnenfeld R, Hansma PK (1986) Science 232:211
19. Itaya K, Tomita E (1988) Surf Sci Lett 201:L507

20. Lustenberger P, Rohrer H, Christoph R, Siegenthaler H (1988) J Electroanal Chem 243:225
21. Wiechers J, Twomey T, Kolb DM, Behm RJ (1988) J Electroanal Chem 248:451
22. Lev O, Fan FE, Bard AJ (1988) J Electrochem Soc 135:783
23. Ye S, Ishibashi C, Uosaki K (1999) Langmuir 15:807
24. Vogt MR, Moller FA, Schilz CM, Magnussen OM, Behm RJ (1996) Surf Sci 367:L33
25. Vogt MR, Lachenwitzer A, Magnussen OM, Behm RJ (1998) Surf Sci 399:49
26. Trevor DJ, Chidsey CED, Loiacono DN (1989) Phys Rev Lett 62:929
27. Honbo H, Sugawara S, Itaya K (1990) Anal Chem 62:2424
28. Nichols RJ, Magnussen OM, Hotlos J, Twomey T, Behm RJ, Kolb DM (1990) J Electroanal Chem 290:21
29. Serafin JM, Hsieh SJ, Monahan J, Gewirth AA (1998) J Phys Chem B 102:10027
30. Gewirth AA, Craston DH, Bard AJ (1989) J Electroanal Chem 261:477
31. Kolb DM, Simeone FC (2005) Electrochim Acta 50:2989
32. Xie ZX, Kolb DM (2000) J Electroanal Chem 481:177
33. Azumi K, Araki K, Seo M (1997) J Electroanal Chem 427:15
34. Uosaki K, Ye S, Sekine N (1996) Bull Chem Soc Jpn 69:275
35. Naohara H, Ye S, Uosaki K (1998) Appl Phys A 66:S457
36. Reichelt K, Lutz HO (1971) J Cryst Growth 10:103
37. Chidsey CED, Loiacono DN, Sleator T, Nakahara S (1988) Surf Sci 200:45
38. Lang CA, Dovek MM, Nagami J, Quate CF (1989) Surf Sci 224:L947
39. Holland-Moritz E, Gordon JH, Borges G, Sonnenfeld R (1991) Langmuir 7:301
40. Zei MS, Nakai Y, Lehmpfuhl G, Kolb DM (1983) J Electroanal Chem 150:201
41. Hecht D, Stark D (1994) Thin Solid Films 238:258
42. Uosaki K, Shen Y, Kondo T (1995) J Phys Chem 99:14117
43. Josowicz M, Janata J, Levy M (1988) J Electrochem Soc 135:112
44. Tisone TC, Drovek J (1972) J Vac Sci Technol 9:271
45. Ashwell GWB, Heckingbottom RJ (1981) J Electrochem Soc 128:649
46. Holloway PH (1978) Gold Bull 12:99
47. Ponjee JJ, Nelissen JWA, Verwijlen CJA (1984) Eur Patent Appl EP 111:957
48. Allara DK, Hebard FJ, Paden FJ, Nuzzo RG, Falcone DR (1983) J Vac Sci Technol A 1:376
49. Wasserman SR, Biebuyck H, Whitesides GM (1989) J Mater Res 3:886
50. Goss CA, Charych DH, Majda M (1991) Anal Chem 63:85
51. Clavilier J (1980) J Electroanal Chem 107:211
52. Clavilier J (1980) J Electroanal Chem 107:205
53. Uosaki K, Ye S, Naohara H, Oda Y, Haba T, Kondo T (1997) J Phys Chem B 101:7566
54. Naohara H, Ye S, Uosaki K (1998) J Phys Chem B 102:4366
55. Naohara H, Ye S, Uosaki K (1999) J Electroanal Chem 473:2
56. Kibler LA, El-Aziz AM, Kolb DM (2003) J Mol Catal A199:57
57. Gündel A, Cagnon L, Gomes C, Morrone A, Schmidt J, Allongue P (2001) Phys Chem Chem Phys 3:3330
58. Zangwill A (1988) Physics at Surfaces. Cambridge University Press, New York
59. Somorjai GA (1994) Introduction to Surface Chemistry and Catalysis. Wiley, New York
60. Wöll Ch, Chiang S, Wilson RJ, Lippel PH (1989) Phys Rev B 39:7988
61. Barh JV, Brune H, Ertl G, Behm RJ (1990) Phys Rev B 42:9307
62. Kolb DM (1996) Prog Surf Sci 51:109
63. Titmuss S, Wander A, King DA (1996) Chem Rev 96:1291
64. Dishner MH, Hemminger JC, Feher FJ (1997) Langmuir 13:2318
65. Marchenko VI, Eksp PZ (1981) J Exp Theor Phys Lett 33:397
66. Alerhand OL, Vanderbilt D, Meade RD, Joannopoulos JD (1988) Phys Rev Lett 61:1973
67. Narasimhan S, Vanderbilt D (1992) Phys Rev Lett 69:1564

68. Heine V, Marks LD (1986) Surf Sci 165:65
69. Rhead GE (1973) J Phys F 3:L53
70. Mochrie GJ, Zehner DM, Ocko BM, Gibbs D (1990) Phys Rev Lett 64:2925
71. Magnussen OM, Hotlos J, Behm RJ, Batina N, Kolb DM (1993) Surf Sci 296:310
72. Ulman A (1991) An Introduction to Ultrathin Organic Films From Langmuir-Blodgett to Self-Assembly. Academic, San Diego
73. Ulman A (1996) Chem Rev 96:1533
74. Ulman A (1998) Self-Assembled Monolayers of Thiols. Academic, San Diego
75. Love JC, Estroff LA, Kriebel JK, Nuzzo RG, Whitesides GM (2005) Chem Rev 105:1103
76. Schreiber F (2004) J Phys: Condens Matter 16:R881
77. Yeo YH, McGonigal GC, Thomson DJ (1993) Langmuir 9:649
78. Elbel N, Roth W, Gunther E, Vonseggern H (1994) Surf Sci 303:424
79. Hibino M, Hatta I (1995) Jpn J Appl Phys I 34:610
80. Giancarlo LC, Fang H. Rubin SM, Bront AA, Flynn GW (1998) J Phys Chem B 102:10255
81. Tao NJ, Cardenas G, Cunha F, Shi Z (1995) Langmuir 11:4445
82. Kunitake M, Batina N, Itaya K (1995) Langmuir 11:2337
83. Si SK, Gewirth AA (2000) J Phys Chem B 104:10775
84. Uosaki K, Yamada R (1999) J Am Chem Soc 121:4090
85. Yamada R, Uosaki K (2000) Langmuir 16:4413
86. Yamada R, Uosaki K (2000) J Phys Chem B 104:6021
87. Sagiv J (1980) J Am Chem Soc 102:92
88. Nuzzo RG, Allara DL (1983) J Am Chem Soc 105:4481
89. Schreiber F (2000) Prog Surf Sci 65:151
90. Schreiber F (2001) Annu Rev Phys Chem 52:107
91. Poirier GE (1997) Chem Rev 97:1117
92. Poirier GE, Tarlov MJ (1996) Science 272:1145
93. Kondoh H, Kodama C, Sumida H, Nozoye H (1999) J Chem Phys 111:1175
94. Thomas R, Sun L, Crooks RM (1991) Langmuir 7:620
95. Godin M, Williams PJ, Tabard-Cossa V, Laroche O, Beaulieu LY, Lennox RB, Grütter P (2004) Langmuir 20:7090
96. Deering AL, Van Lue SM, Kandel SA (2005) Langmuir 21:10263
97. Widrig XA, Alves CA, Porter MD (1991) J Am Chem Soc 113:2805
98. Strong L, Whitesides GM (1988) Langmuir 4:546
99. Chidsey CED, Liu G-Y, Rowntree P, Scoles G (1989) J Chem Phys 91:4421
100. Camillone N, Chidsey CED, Liu G-Y, Scoles G (1993) J Chem Phys 98:3503
101. Fetner P, Eisenberger P, Liang KS (1993) Phys Rev Lett 70:2447
102. Nuzzo RG, Korenic EM, Dubois LH (1990) J Chem Phys 93:767
103. Delamarche E, Michel B, Gerber CH, Anselmetti D, Güntherodt HJ, Wolf H, Ringsdorf H (1994) Langmuir 10:2869
104. Poirier GE, Tarlov MJ (1994) Langmuir 10:2853
105. Bucher JP, Santesson L, Kern K (1994) Appl Phys A 59:135
106. Fukuda T, Ichii T, Kobayashi K, Yamada H, Matsushige K (2004) J Appl Phys 95:1222
107. Fukuda T, Ichii T, Kobayashi K, Yamada H, Matsushige K (2005) Appl Phys Lett 86:034103
108. Fenter P, Eberhardt A, Eisenberger P (1994) Science 266:1216
109. Yeganeh MS, Dougal SM, Polizzotti RS, Rabinowitz P (1995) Phys Rev Lett 74:1811
110. Fenter P, Schreiber F, Berman L, Scoles P, Eisenberger P, Bedzyk M (1998) Surf Sci 412/413:213
111. Fenter P, Schreiber F, Berman L, Scoles P, Eisenberger P, Bedzyk M (1999) Surf Sci 425:138
112. Sellers H, Ulman A, Shnidman Y, Eilers JE (1993) J Am Chem Soc 115:9389

113. Gronbeck H, Curioni A, Andreoni W (2000) J Am Chem Soc 122:3839
114. Yourdshahyan Y, Zhang HK, Rappe AM (2001) Phys Rev B 63:081405
115. Roper MG, Skegg MP, Fisher CJ, Lee JJ, Dhanak VR, Woodruff DP, Jones RG (2004) Chem Phys Lett 389:87
116. Kondoh H, Iwasaki M, Shimada T, Amemiya K, Yokohama T, Ohta T, Shimomura M, Kondo S (2003) Phys Rev Lett 90:066102
117. Kitagawa T, Idomoto Y, Matsubara H, Hobara D, Kakiuchi T, Okazki T, Komatsu K (2006) J Org Chem 71:1362
118. Hayashi T, Morikawa Y, Nozoye H (2001) J Chem Phys 114:7615
119. Vargas MC, Giannozzi P, Selloni A, Scoles G (2001) J Phys Chem B 105:9509
120. Gottschalck J, Hammer B (2002) J Chem Phys 116:784
121. Molina ML, Hammer B (2002) Chem Phys Lett 360:264
122. Nara J, Higai S, Morikawa Y, Ohno T (2004) J Chem Phys 120:6205
123. Torrelles X, Barrena E, Munuera C, Rius J, Ferrer S, Ocal C (2004) Langmuir 20:9396
124. Teran Arce F, Vela ME, Salvarezza RC, Arvia AJ (1998) J Chem Phys 109:5703
125. Torrelles X, Vericat C, Vela ME, Fonticelli MH, Millone MAD, Felici R, Lee TL, Zegenhagen J, Munoz G, Martin-Gago JA, Salvarezza RC (2006) J Phys Chem B 110:5586
126. Edinger K, Golzhauser A, Demota K, Woll C, Grunze M (1993) Langmuir 9:4
127. Schönenberger C, Sondag-Huethorest JAM, Jorritsma J, Fokkink LGJ (1994) Langmuir 10:611
128. McDermott CA, McDermott MT, Green JB, Porter MD (1995) J Phys Chem 99:13257
129. Poirier GE, Tarlov MJ (1995) J Phys Chem 99:10966
130. Cavallcri O, Hirstein A, Kern K (1995) Surf Sci 341:L960
131. Yamada R, Wano H, Uosaki K (2000) Langmuir 16:5523
132. Yamada R, Sakai H, Uosaki K (1999) Chem Lett 28:667
133. Poirier GE (1997) Langmuir 13:2019
134. Cavalleri O, Hirstein A, Bucher JP, Kern K (1996) Thin Solid Films 284–285:392
135. Cavalleri O, Gilbert SE, Kern K (1997) Chem Phys Lett 269:479
136. Poirier GE, Fitts WP, White JM (2001) Langmuir 17:1176
137. Yamada R, Uosaki K (1998) Langmuir 14:855
138. Yamada R, Uosaki K (1997) Langmuir 13:5218
139. Poirier GE (1996) J Vac Sci Technol B 14:1453
140. Schweizer M, Hagenström, Kolb DM (2001) Surf Sci 490:L627
141. Loglio F, Schweizer M, Kolb DM (2003) Langmuir 19:830
142. Li J, Liang KS, Camillone N III, Leung TYB, Scoles G (1995) J Chem Phys 102:5012
143. Strong L, Whitesides GM (1988) Langmuir 4:546
144. Dubois LH, Zegarski BR, Nuzzo RG (1993) J Chem Phys 98:678
145. Camillone N III, Chidsery CED, Liu G, Scoles G (1993) J Chem Phys 98:4234
146. Yamada R, Uosaki K (2001) Langmuir 17:4148
147. Marchenko O, Cousty J (2000) Phys Rev Lett 84:5363
148. Cousty J, Marchenko A (2002) Surf Sci 520:128
149. Marchenko A, Cousty J, Van LP (2002) Langmuir 18:1171
150. Zhang HM, Xie ZX, Mao BW, Xu X (2004) Chem Eur J 10:1415
151. Xie ZX, Huang ZF, Xu X (2002) Phys Chem Chem Phys 4:1486
152. Xie ZX, Xu X, Tang J, Mao BW (2000) Chem Phys Lett 323:209
153. Xie ZX, Xu X, Mao BW, Tanaka K (2002) Langmuir 18:3113
154. Xie ZX, Xu X, Tang J, Mao BW (2000) J Phys Chem B 104:11719
155. Waibel HF, Kleinert M, Kibler LA, Kolb DM (2002) Electrochim Acta 47:1461
156. Kibler LA, Kleinert M, Kolb DM (1999) J Electroanal Chem 467:249
157. Kibler LA, Kleinert M, Randler R, Kolb DM (1999) Surf Sci 443:19
158. Kibler LA, Kleinert M, Kolb DM (2000) Surf Sci 461:155

159. Kibler LA, Kleinert M, Lazarescu V, Kolb DM (2002) Surf Sci 498:175
160. Edens GJ, Go X, Weaver MJ (1994) J Electroanal Chem 375:357
161. Magnussen OM, Hageböck J, Hotlos J, Behm RJ (1992) Faraday Discuss 94:329
162. Takahasi M, Hayashi Y, Mizuki J, Tamura K, Kondo T, Naohara H, Uosaki K (2000) Surf Sci 461:213
163. Kibler LA, El-Aziz AM, Kolb DM (2003) J Mol Catal A 199:57
164. Naohara H (1999) Thesis, Hokkaido University
165. Nagahara Y, Hara M, Yoshimoto S, Inukai J, Yau SL, Itaya K (2004) J Phys Chem B 108:3224
166. Uosaki K, unpublished results
167. Pope MT (1983) Heteropoly and Isopoly Oxometalates. Springer, Berlin Heidelberg New York
168. Mizuno N, Misono M (1998) Chem Rev 98:199
169. Sadakane M, Steckhan E (1998) Chem Rev 98:219
170. Coronado E, Gomez-Garcia CJ (1998) Chem Rev 98:273
171. Katsoulis DE (1998) Chem Rev 98:359
172. Keita B, Nadjo L (1993) J Electroanal Chem 354:295
173. Rong CY, Anson FC (1996) Inorg Chim Acta 242:11
174. Kaba MS, Song IK, Barteau MA (1996) J Phys Chem 100:19577
175. Song IK, Kaba SM, Barteau MA (2002) Langmuir 18:2358
176. Kinne M, Barteau MA (2000) Surf Sci 447:105
177. Keita B, Nadjo L, Belanger D, Wilde CP, Hilaire M (1995) J Electroanal Chem 384:155
178. Ge M, Niece BK, Wall CG, Klemperer WG, Gewirth AA (1997) Mater Res Soc Symü Proc 451:99
179. Nicholson KT, Zhang KZ, Banazak Holl MM (1999) J Am Chem Soc 121:3232
180. Schneider KS, Nicholson K, Fosnacht DR, Orr BG, Holl MMB (2002) Langmuir 18:8116
181. Nicholson K, Zhang KZ, Holl MMB, McFeely FR, Pernisz UC (2000) Langmuir 16:8396
182. Rong CY, Anson FC (1994) Anal Chem 66:3124
183. Ge M, Zhong B, Klempere WG, Gewirth AA (1996) J Am Chem Soc 118:5812
184. Lee L, Wang JX, Adzie RR, Robinson IK, Gewirth AA (2001) J Am Chem Soc 123:8838
185. Teague CM, Li Xiao, Biggin E, Lee L, Kim J, Gewirth AA (2004) J Phys Chem B 108:1974
186. Kim J, Lee L, Niece BK, Wang JX, Gewirth AA (2004) J Phys Chem B 108:7927
187. Kim J, Gewirth AA (2003) Langmuir 19:8934

14 Single-Molecule Studies on Cells and Membranes Using the Atomic Force Microscope

Ferry Kienberger · Lilia A. Chtcheglova · Andreas Ebner · Theeraporn Puntheeranurak · Hermann J. Gruber · Peter Hinterdorfer

Abbreviations

AFM	atomic force microscope
DFM	dynamic force microscopy
PEG	poly(ethylene glycol)
BR	bacteriorhodopsin
SPM	scanning probe microscopy
NHS	*N*-hydroxysuccinimide
PSGL-1	P-selectin glycoprotein ligand-1
EM	electron microscopy
TEM	transmission electron microscopy
NPC	nuclear pore complex
NE	nuclear envelope
APTES	aminopropyltriethoxysilane
HSP	heat shock protein
S-layer	surface layer
HRV	human rhinoviruses
HIV	human immunodeficiency virus
LFA-1	leukocyte function-associated antigen-1
ICAM-1	intercellular adhesion molecule-1
WLC	worm-like chain
OspA	outer surface protein of Borrelia burgdorferi
NhaA	sodium-proton antiporters
TREC	simultaneous topography and recognition imaging
MyEnd cells	microvascular endothelial cell line from mouse myocardium
VE-cadherin	vascular endothelial cadherin
EDTA	ethylenediaminetetraacetic acid

14.1 Abstract

Atomic force microscopy (AFM) has been developed into an imaging method that yields fine structural details on biological samples in their physiological environment. Topographic AFM imaging enables a broad range of biomolecular complexes that are bound to the cells and membranes to be investigated. Dynamic processes can be visualized, single biomolecules can be observed at work, and the formation of biomolecular assemblies can be studied. In addition to high-resolution imaging, the measurement of mechanical forces at the molecular level has provided detailed insights into structure–function relationships of many biological systems. The detection of inter- and intramolecular forces yields information of structural parameters of the binding pocket, on the molecular dynamics of the recognition process, and on the energy landscape of the interaction. Single-molecule recognition measurements have also been demonstrated on cells and membranes and open new perspectives in exploring the regulation of cellular processes. By combining topographic imaging with force measurements, receptor binding sites can be localized with nanometer accuracy rendering possible the identification of specific components on the cellular surfaces.

14.2 Introduction

At present, atomic force microscopy (AFM) is extensively used in a wide range of disciplines such as molecular biology, solid-state physics and material science [1]. The major application is imaging of surfaces on scales from micro- to nanometers with the objective to visualize and properly characterize surface textures and shapes. For instance, it has been applied to characterize various biological samples (e.g. proteins, DNA, membranes, cells) at high lateral resolution [2]. Moreover, it is the unique technique to provide subnanometer resolution at a reasonable signal-to-noise ratio under physiological conditions. Due to continuous developments of sample preparation, imaging techniques and instrumentation, AFM has evolved into a companion technique of X-ray crystallography and electron microscopy (EM) for the determination of protein structures [3]. Regarding membranes and cells, AFM has successfully complemented EM studies. Recently, AFM images of native membranes have been obtained at subnanometer resolution for the first time [1–4]. AFM allows visualization of the samples in buffers that preserve their structure over extended periods of time. Most importantly, AFM does not rely on symmetry averaging and crystallization. Thus, AFM can reveal defects and structural anomalies that are not observable in classical ensemble measurements [5]. Unlike EM, AFM yields three-dimensional maps with an exceptionally good vertical resolution (less than a nanometer). When applied to living cells, new cellular surface structures and their physiological functions have been identified [6–8]. For instance, topographical images of living pancreatic acinar cells revealed the presence of fusion pores at the apical plasma membrane and their structure and dynamics [6].

In addition to high-resolution imaging of proteins, nucleotides, membranes and living cells [9], the measurement of mechanical forces at the molecular level has provided detailed insights into the function and structure of biomolecular systems [10]. Inter- and intramolecular interactions can be studied directly at the molecular level, as exemplified by the analysis of polysaccharide elasticity [11, 12], DNA mechanics [13, 14], the function of molecular motors [15, 16], and the binding potentials of receptor–ligand pairs involved in cell adhesion [17, 18]. In the latter case, defined forces are exerted on a receptor–ligand complex and the dissociation process is followed over time. Dynamic aspects of molecular recognition are addressed in force spectroscopy experiments, where distinct force–time profiles are applied to monitor the changes of conformational and states during receptor–ligand dissociation. Consequently, dynamic force spectroscopy allows detection of energy barriers not detectable by conventional near-equilibrium assays and to probe the free-energy surface of proteins and molecular complexes [19]. The capability of AFM to resolve nm-sized details, together with its force-detection sensitivity, has led to the development of molecular recognition imaging [20]. By combining topographical imaging with force measurements, receptor sites are localized with nanometer accuracy. Topography and recognition of target molecules are thereby simultaneously mapped. Thus, the AFM can identify specific components in a complex biological sample and retain its high resolution in imaging. In summary, several examples demonstrate the capability of AFM to resolve characteristic substructures of single individual proteins, to study conformational changes, and to analyze molecular interactions and the assembly of molecular complexes.

14.3 Principles of Atomic Force Microscopy

The AFM, a member of the scanning probe microscopy family, operates by moving a sharp tip on a microcantilever horizontally across a surface, thus rendering it possible to record the topographical map of the scanned surface [21] (Fig. 14.1a). Deflections of the cantilever caused by the tip–sample interaction are usually detected by a four-segment split photodiode. The magnitude of the deflection is registered by the changes in the angle of a laser beam reflected at the end of the cantilever (Fig. 14.1b). The deflection signal of the cantilever is used to keep the applied force constant by moving either the sample or the cantilever up and down via a feedback loop. The surface topography is then reconstructed from the vertical movements of the sample or the cantilever, respectively. The AFM can be operated in various modes. In the contact-force imaging mode, images are created by bringing the tip and sample into contact and scanning the tip across the surface. In dynamic force microscopy (DFM) imaging, the AFM tip is oscillated near its resonant frequency while it scans over the surface, such that the tip does not continuously touch a sample [22–24]. The amplitude reduction upon intermittent tip–surface contact is held constant through the feedback loop. As the tip touches the sample surface at the very end of its downward movement only, the lateral forces are greatly reduced during imaging. This is of advantage when imaging soft biological samples [25].

Both contact and dynamic force microscopy have been employed for high-resolution imaging of biomolecules. In fact, some of the highest resolution images reported in the literature were obtained using the contact mode. In these studies, mostly two-dimensional arrays of membrane proteins were imaged at subnanometer resolution, using conventional AFM cantilevers in aqueous buffered conditions [26]. The high image resolution was made possible by the close packing of the biomolecules, which makes them more resistant to the deformations possibly caused by the forces of the cantilever during imaging. In contrast, DFM imaging has mainly been applied to single, isolated biomolecules on surfaces and to supramolecular complexes [27] because it overcomes the strong lateral imaging forces associated with contact-mode AFM [28]. Moeller et al. [29] have recently demonstrated that DFM allows for imag-

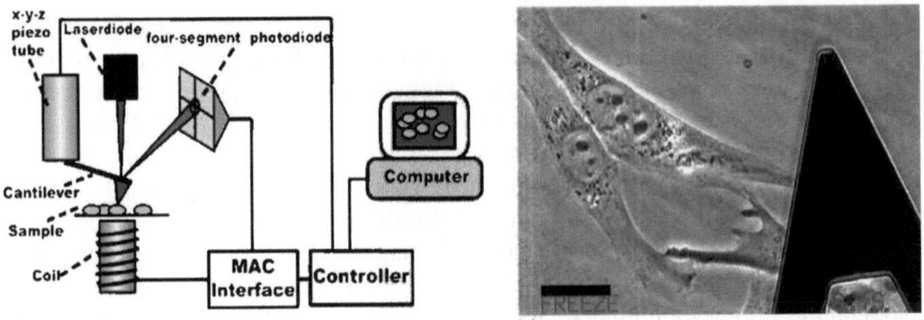

a) b)

Fig. 14.1. Principles of scanning probe microscopy. (**a**) Schematic diagram of an atomic force microscope. A sample is probed by an ultrasharp stylus mounted on a cantilever, which moves in a raster manner over the surface. A reflected laser beam amplifies and reports deflections of the cantilever to a split photodiode. Photoelectric circuitry then converts the deflections into height information recorded as a digital image. Picture taken from [27]. (**b**) Phase-contrast micrograph of a cantilever and living cells. For cell measurements, the can tilever tip is carefully positioned on top of a cell. *Scale bar* = 25 µm. Image taken from [63]

ing of native protein surfaces with similar resolution as in contact-mode AFM. For contact-mode and DFM imaging the samples need to be tightly adhered to the support. In recent years, the art of sample preparations has been optimized in many laboratories with respect to stable sample supports, buffer composition to control the substrate–sample, and tip–sample interactions [30]. The challenge of imaging soft biological material has been met by employing force–distance curves to adjust the forces, the scanning speed and feedback parameters.

14.4 Imaging of Membrane–Protein Complexes

In the following, AFM imaging of biological membranes with a lateral resolution of ∼ 1 nm is discussed and the new insights into biological processes are summarized. In topographical imaging, the heights of membrane-protruding structures can accurately be measured with a vertical resolution of about 1 Å [4]. The high signal-to-noise ratio of the instrument allows for assessing the oligomeric state and the sidedness of membrane proteins directly in raw-data images. Even poorly ordered single particles can be recognized and imaged with high resolution [31]. AFM has developed to a point that now permits detection of flexible and stable extrinsic domains of membrane proteins and the ability to perform time-lapse imaging, thereby monitoring structural changes of molecules and molecular assemblies as a function of time [32].

14.4.1 Membranes of Photosynthetic Bacteria and Bacterial S-Layers

The first application of the AFM in the field of photosynthetic bacteria was an investigation of the peripheral light harvesting complex LH2 that is embedded in a highly organized assembly of transmembrane proteins that are responsible for absorption of light and its transformation into chemical energy [33]. High-resolution images of LH2 (Fig. 14.2a) revealed a nanomeric (i.e. a nine-fold symmetry) organization of the regularly packed cylindrical complexes which were incorporated into the membrane in both orientations. Topographs of LH2 complexes had a lateral resolution of ~8 Å and a vertical resolution of ~1 Å confirming the nanomeric organization of the heterodimers determined by EM. The cylindrical complexes were found to strongly protrude (by ~14 Å) from one side (periplasmic side) and weakly (by ~6 Å) from the other (cytoplasmic) side of the membrane. The heights of the protruding rings provide information about the position of the LH2 cylinder with respect to the lipid bilayer. Such data exemplify the additional benefit of AFM imaging in complementing EM studies, as well as the precision of AFM when analyzing oligomeric states of membrane proteins.

Other native membranes, which have been used for AFM investigations, are bacterial cell surface layers (S-layers) [34]. S-layers were identified as the outermost cell-envelope component in many different species and form highly ordered arrays with various symmetries [35]. S-layer proteins can be functionalized with capture molecules by means of genetic engineering. Such sensing layers have been already exploited for label-free detection methods, as microadsorbents in blood purification, and for the development of novel types of antiallergic vaccines [36]. S-layer

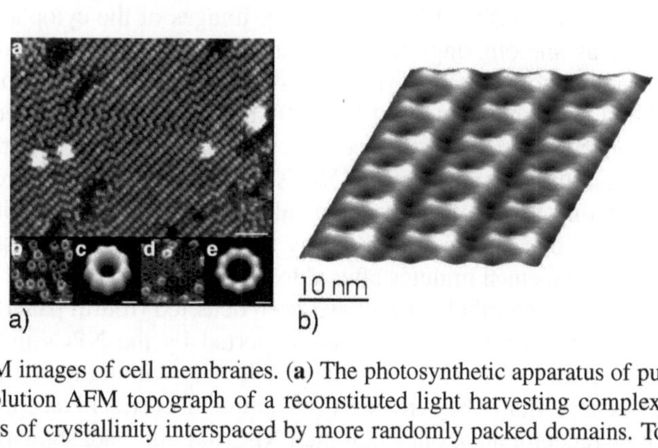

Fig. 14.2. AFM images of cell membranes. (**a**) The photosynthetic apparatus of purple bacteria. Medium-resolution AFM topograph of a reconstituted light harvesting complex LH2 crystal showing areas of crystallinity interspaced by more randomly packed domains. To produce 2D crystals of these complexes, the general method of reconstitution started with the purified proteins and a suitable combination of lipids, both solubilized in detergent. Next, the detergent was removed by dialysis resulting in the formation of lipid bilayers in which the proteins were reconstituted as 2D crystals or densely packed proteins. *Scale bar*: 50 nm. The *lower panel* shows high-resolution topographical analyzes of the periplasmic (*left two images*) and cytoplasmic surface (*right two images*) of the LH2 complex. The *scale bars* are 10 nm and 2 nm for raw-data topographs and for average images, respectively. The protrusion height is 1.6 nm and 0.5 nm for the periplasmic and cytoplasmic complex, respectively. Image taken from [33]. (**b**) Recrystallized bacterial cell surface layer (S-layer) fused with streptavidin. Functional heterotetramers consisting of one molecule fusion protein and three molecules streptavidin were obtained and recrystallized on gold substrates. The two-dimensional protein crystal displays streptavidin in defined repetitive spacing. In the lattice of the fusion protein, streptavidin shows up with periodical heights of 0.5 nm. The unit cell dimensions of the S-layer crystal are 10.2 nm and 8.0 nm with a base angle of 76°. Image taken from [37]

fusion proteins with an easily addressable binding site, such as fused streptavidin, offer the opportunity to specifically bind other proteins tagged with specific ligands. Figure 14.2b shows a high-resolution AFM image of an S-layer functionalized with streptavidin tetramers [37]. Using gentle DFM imaging, details of the S-layer protein lattice with bound streptavidin have been obtained in buffer solution. Streptavidin molecules are evident with periodical heights of 0.5 nm, with unit cell dimensions of 10.2 nm and 8.0 nm, and a base angle $\gamma = 76°$. These values are in good agreement with data from transmission electron microscopy (TEM) of the same system. The streptavidin moiety was observed in the topographical image at a high signal-to-noise ratio (Fig. 14.2b). In contrast to the TEM data that were derived from negatively stained preparations, the AFM images were obtained under physiological conditions.

14.4.2 Nuclear Pore Complexes

Protein transport into the cell nucleus is mediated by nuclear pore complexes (NPCs), supramolecular structures that are embedded in the nuclear envelope (NE) [38]. Usually, AFM studies of NPC are performed on NE spread on mica or glass. Figure 14.3 shows typical topographical images of NPCs in NE, either at high density (Fig. 14.3a) or at low density (Fig. 14.3b) [39]. Images of the cytoplasmic side of NE show NPCs as smooth rings with diameters of about 100 nm and heights of about 10 nm. Time-resolved images of aldosterone-induced conformational changes of individual NPCs are shown in Fig. 14.3c [40]. Before aldosterone injection, NPCs appear as smooth rings at the NE surface (Fig. 14.3c, first panel). Two minutes after hormone injection, the ring periphery of NPCs are decorated with macromolecules, termed flags (second panel). Eight minutes after injection, NPCs again appear as smooth rings (third panel). This is caused by the lack of flags at this stage of hormone stimulation. Nineteen minutes after aldosterone injection, rather large masses (termed plugs) in the central channels have been detected (fourth panel). The plugs have been identified as ribonucleoproteins exported by the NPCs in response to aldosterone stimulation. The flags were detectable only within the first few minutes after aldosterone injection. They are likely to be mineral corticoid receptors, because (i) a competitive receptor antagonist inhibited NPC flagging (not shown) and (ii) the molecular weight of the flags estimated by AFM matches the molecular mass of the mineral corticoid receptor. In summary, hormone-induced protein import across nuclear pores and ribonucleoprotein export from the nucleus to the cytoplasm have been visualized at the molecular level.

14.4.3 Cell Membranes with Attached Viral Particles

Until recently, transmission electron microscopy (TEM) has been virtually the only source of detailed images of virions and virion-infected cells [41]. Cross sections of cells that had been embedded, stained, and visualized with TEM showed the presence of virions near and at cell surfaces in the processes of maturation and

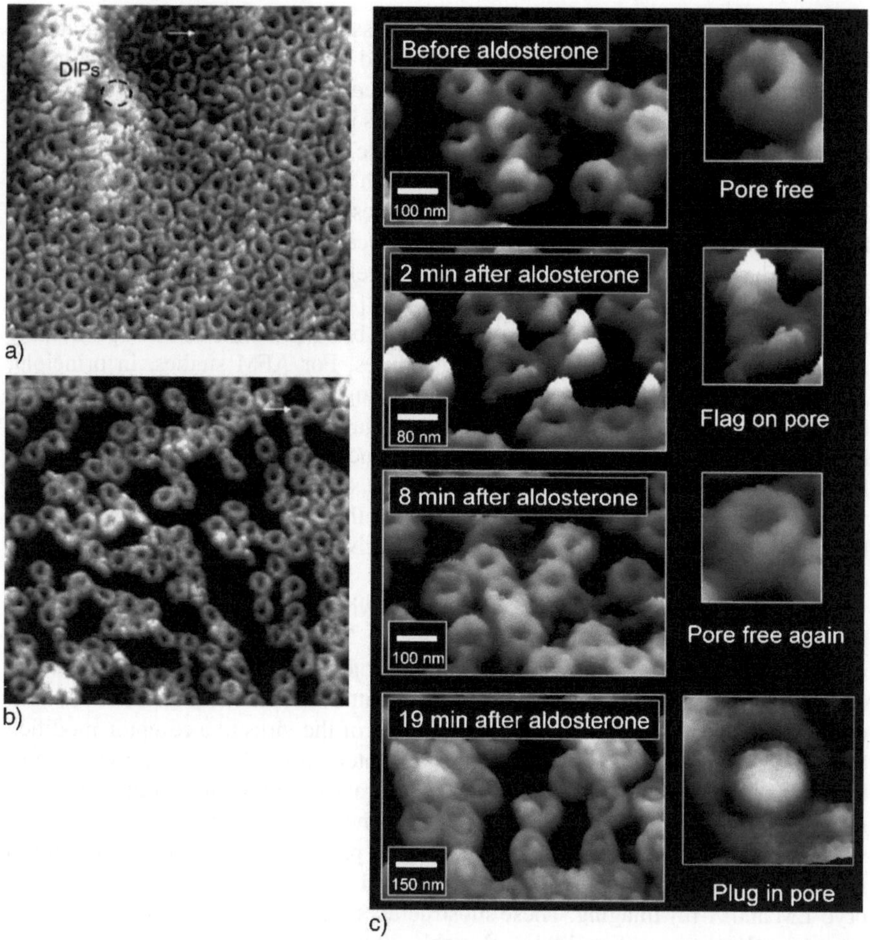

Fig. 14.3. AFM images of nuclear membranes. (**a**) The nuclear envelope of the cell nucleus is perforated by nuclear pore complexes (NPCs). The image shows the cytoplasmic side of the nuclear envelope in Xenopus laevis oocytes. Densely packed NPCs with ~ 100 nm in diameter are observed. Image size is 1.6 μm. The *gray scale* ranges from 0 to 40 nm. Image taken from [39]. (**b**) Topographical image of the cytoplasmic side of the nuclear envelope with a low density of NPCs. Same image size and *gray scale* as before. Image taken from [39]. (**c**) Aldosterone-induced conformational changes of individual NPCs. Images of nuclear envelope cytoplasmic surfaces are shown before and after aldosterone injection. Before aldosterone injection, the NPCs are clearly visible (pore free). Two minutes after injection, most of the NPCs are flagged (flag on pore). Eight minutes after injection, the NPCs are aga in free of flags (pore-free again). Nineteen minutes after injection, the NPCs are plugged (plug in pore). The z-scale ranges from 0 to 20 nm. Image taken from [40]

budding. Although these images are quite impressive, they have certain limitations. First, the cells had been embedded in plastic and subsequently exposed to a heavy metal stain, either or both of which might introduce distortions into the cell and virion structures. While the lateral resolution of TEM is usually high, the determination of height, i.e. of the third dimension is problematic for nonsymmetric objects. More recently, cryo-EM has been used to visualize both immature and mature retroviruses isolated from culture medium [41]. These studies are technically more demanding but they are free of many of the artefacts introduced by staining and dehydration.

AFM has its virtues and its drawbacks as well, but fortunately they tend to be complementary to those of TEM and cryo-EM [42]. With AFM, samples may be imaged in fluids, including culture medium or buffer, in situ, or after processing according to established histological procedures. For AFM studies, in principle, a single cell or virion is all that is required, though several hundred of either may be present on the substrate. In previous investigations of viruses with AFM it has been shown that the technique is sufficiently incisive and reproducible that even individual capsomeres can be visualized on the surfaces of both plant and animal viruses. Furthermore, it was demonstrated that the structures of viruses observed by AFM are entirely consistent with models derived by X-ray crystallography and cryo-EM [42].

Figure 14.4a shows an image of human rhinoviruses (HRV) on an artificial membrane interface that mimics the cell surface [43]. The specific and site-directed immobilization of viral particles via cell receptor proteins resulted in a two-dimensional crystal with hexagonal symmetry; the average lattice constant obtained was similar to the diameter of the virus, i.e. 35 nm. Binding of the virus to a receptor-modified bilayer closely mimics the binding to its receptor on the host-cell plasma membrane and is thus ideally suited for studies of viral dynamics and recognition under near physiological conditions. In addition to the hexagonal arrangement, a regularly spaced pattern was observed on almost all viral particles (Fig. 14.4b). This pattern reflects the protrusions on the surface of the viral capsid, which is well known from cryo-EM and X-ray imaging. These substructures are more clearly discernable upon contrast-enhancement (Fig. 14.4c). Roughly 20 blobs or protrusions with diameters of 2–4 nm and heights of ~ 0.3 nm were counted on each viral particle; they were arranged in lines with ~ 5 protrusions per line (see labeling in Fig. 14.4c). Virus capsids did not appear strictly globular in shape, but rather the projection of the particles exhibited some straight borders akin to their polygonal surface features, which were also observed by cryo-EM. The center-to-center distance of single protrusions is about 7 nm (Fig. 14.4c). This is in the same range as the distance between the plateau at the three-fold axis and the plateau at the five-fold axis as seen in the structures resolved by cryo-EM. In summary, high-resolution imaging of the rhinovirus capsid revealed a regular arrangement of 3-nm-sized protrusions. The achieved resolution was comparable to the resolution of EM, but the latter method required the sample to be frozen in amorphous ice, whereas AFM measurements were carried out under near physiological conditions.

In another study, human immunodeficiency virus (HIV)-infected lymphocytes have been imaged in buffer solution [44]. Figure 14.4d–f show images of HIV virions as they emerge from, or are attached to, the surfaces of infected cells. Presumably, they represent a mixture of immature and mature particles and possibly

Fig. 14.4. AFM images of virus–membrane complexes. (**a**) Crystalline arrangement of human rhinovirus (HRV) on a cell mimicking membrane. Specific and site-directed immobilization of HRV on a receptor modified bilayer results in a hexagonal symmetry and 35 nm spacing. *Inset (upper right)*: The Fourier spectrum exhibits hexagonally arranged spots (*arrows*). *Inset (lower right)*: Averaged three-dimensional representation of the virus arrangement. Image taken from [43]. (**b**) Three-dimensional representation of rhinovirus particles with 30 nm diameters. Roughly 20 protrusions are counted on each single virus particle. Image size is 80 nm. The z-scale ranges from 0 to 10 nm. Image taken from [43]. (**c**) Contrast-enhanced image for a better visibility of the protrusions. Singly resolved structures (labeled one to five) with diameters of ~3 nm and heights of ~0.3 nm can be clearly obtained. Image size is 80 nm. *Inset*: External surface of HRV as derived from X-ray coordinates. Image taken from [43]. (**d**) Human immunodeficiency virus (HIV)-infected lymphocytes imaged under physiological conditions. On the surface of cultured lymphocytes, emerging HIV particles can be seen. The particles are not present on uninfected cells (not shown). The image size is 5 × 5 µm. Image taken from [44]. (**e**) Small scan size image of the surface of an infected human lymphocyte. The image size is 1.5 × 1.5 µm. Image taken from [44]. (**f**) The average diameter of single HIV particles emerging from or attached to the cells is ~127 nm. The image size is 500 × 500 nm. Image taken from [44]

virions in intermediate states. The particles have an average diameter of 127 nm, with a variation of 30 nm. A similar range of size diversity of particle diameters has been observed by cryo-EM as well. Some viral particles can be observed that are considerably smaller than 127 nm that had the gross appearance of virions, indicating that these particles might lack their cores. Also frequently present were some anomalously large particles that otherwise had the appearance of virions. They had diameters in the range of 160 to 240 nm and probably represent virions that contained multiple cores. The external appearances of the virions having diameters of about 127 nm, the vast majority, are similar to one another and are the same as those of the free virions. In either case, the surfaces of HIV particles exhibit protrusions, presumably gp120, which do not physically resemble spikes. The protrusions, which number about 100 per virus particle, have average diameters of about 20 nm, but with a large variance. They likely consist of arbitrary associations of small numbers of gp120 monomers on the surface of HIV particles.

14.5 Single-Molecule Recognition on Cells and Membranes

A number of techniques are presently available to investigate intermolecular forces acting between single biomolecules and cellular surfaces. The most prominent tools are the atomic force microscope, optical tweezers [45], and the biomembrane force probe [46]. These techniques span a measurable force window ranging from entropic forces at several femto-Newtons (1 fN $= 10^{-15}$ N) up to the rupture of covalent bonds at several nano-Newtons (1 nN $= 10^{-9}$ N) [47]. Using AFM, many different types of interactions have been studied either on isolated proteins in vitro or on cellular surfaces in vivo. For studying specific binding between biomolecules, an increasing force is exerted onto the molecular complex, and the dissociation process is followed over time. Dynamic aspects of recognition are addressed in force spectroscopy experiments, in which the timescale is systematically varied in order to evaluate the changes in conformations and states during receptor–ligand dissociation [48]. Such experiments allow the estimation of affinity, rate constants, and structural data of the binding pocket.

14.5.1 Principles of Recognition Force Measurements

In molecular recognition force spectroscopy experiments, the binding of ligands immobilized on AFM tips towards surface-bound receptors (or vice versa) is studied by applying a force to the receptor–ligand complex until the bond breaks at a measurable unbinding force [49]. Such experiments require that one or several ligand molecules are permanently tethered to the apex of the AFM tip, usually by covalent bonding via a flexible linker molecule [50]. As an example, the covalent coupling of ligands to gold-coated tip surfaces via a free SH-group of the ligand guarantees a sufficiently stable attachment, because these bonds are about ten times stronger than typical ligand–receptor interactions [51]. This chemistry has been used for various types of receptor–ligand studies, but it requires a prior deposition of a gold layer onto the silicon nitride or silicon oxide tip surface. Another frequently used method of functionalization is the direct tethering of ligand molecules to the surface of a silicon or silicon nitride tip (Fig. 14.5a). In the first step, amino groups are generated on the tip surface, either with ethanolamine hydrochloride or with silane derivatives such as aminopropyltriethoxysilane (APTES) [52,53]. These two procedures do not cause stickiness and yield a rather low number of amino groups on the apex of the tip [54], as is desired for the single-molecule experiments. In the second step of the anchoring protocol, a distensible and flexible linker is often used to space the ligand molecule from the amino-functionalized tip surface by several nanometers. The ligand on the spacer molecule can freely orient and diffuse within a certain volume

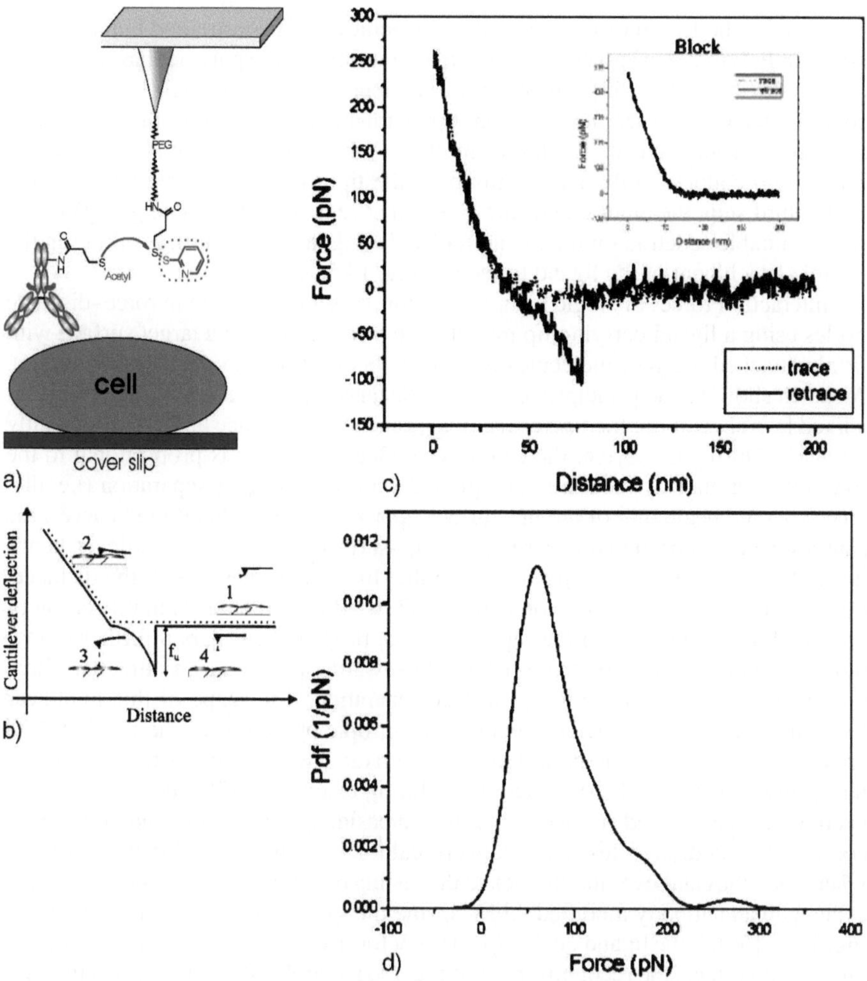

Fig. 14.5. Specific immobilization of biomolecules to the AFM tip for studying single-molecular interaction forces on cells. (**a**) A distensible and flexible polymer linker (polyethylene-glycol, PEG) was used to bind a single antibody covalently to the tip, enabling the ligand to freely reorient and to achieve unconstrained binding to its cognate receptor at the cell surface. (**b**) In force–distance cycles, an AFM tip with ligands is brought into contact with a cell expressing the corresponding receptor at high density, so that a ligand–receptor bond is formed. During a subsequent retraction of the tip, an increasing force is applied to the ligand–receptor bond until the interaction breaks at a critical force, the unbinding force f_u. (**c**) A force–distance cycle on the surface of a living cell with an antibody-coated AFM-tip shows specific interaction in the retrace (jump at 75 nm). The interaction is blocked using free antibody in solution (*inset*), proving the specificity of the detected force jump. (**d**) The probability density function (pdf) is constructed from an ensemble of forces and gives the distribution of unbinding forces f_u, similar to a histogram. The maximum of the pdf depicts the most probable unbinding force at a given loading rate. Data taken from [58]

provided by the length of the tether, thereby achieving unconstrained binding to its receptor. Poly(ethylene glycol) (PEG), an inert water soluble polymer has been often used as flexible linker [55]. The heterobifunctional crosslinkers used for tip–ligand coupling carry two different reactive groups at their ends [56]. One of these is usually a carboxyl group, activated in the form of an *N*-hydroxysuccinimide (NHS) ester. This group couples to the amino group on the tip, yielding a stable amide bond. In the third step, the outer end of the crosslinker (e.g. a thiol-reactive group) reacts with a suitable function on the ligand molecule (e.g. a thiol group), resulting in the flexible attachment of the ligand to the tip (Fig. 14.5a).

Interaction forces of single ligand–receptor pairs are measured in force–distance cycles using a ligand-carrying tip mounted on a cantilever and a target surface with firmly attached receptor molecules (e.g. receptors anchored in the cell membrane). A sketch showing the principles of a force–distance cycle is given in Fig. 14.5b. At a fixed lateral position, the tip vertically approaches the surface and is subsequently retracted. During this cycle, the cantilever deflection (which is proportional to the force) is continuously measured and plotted versus tip–surface separation (i.e. distance). At the beginning of the tip–surface approach (Fig. 14.5b, dotted curve), the cantilever deflection remains zero. Upon tip–surface contact, the cantilever bends upward, consistent with a repulsive force that linearly increases with the distance. Subsequent tip–surface retraction (Fig. 14.5b, solid line) first leads to relaxation of the cantilever bending until the repulsive force drops to zero. Upon further retraction, the cantilever progressively bends downwards, reflecting an attractive force that increases with increasing tip–surface separation. The shape of this nonlinear force–distance profile is determined by the entropic properties of the flexible PEG crosslinker and shows parabolic-like characteristics, which mirrors the increase of the spring constant of the polymer chain during extension [57]. The physical connection between tip and surface exerts the increasing force until the ligand–receptor complex finally dissociates at a certain critical force, termed the unbinding force f_u, whereupon the cantilever jumps back to the resting position. A typical force–distance cycle with an antibody modified AFM-tip and the surface of living cells is shown in Fig. 14.5c [58]. If the ligand on the tip does not form a specific bond with the receptor on the cell surface, the recognition event (i.e. the parabolic-shaped curve) is missing and the retrace looks like the trace (inset in Fig. 14.5c). In addition, the specificity of ligand–receptor binding is usually demonstrated by blocking experiments with free ligands, which are injected into the solution in order to block the receptor sites on the cell surface. As a consequence, almost all specific recognition signals completely disappear and only occasional adhesion events are observed.

After acquiring a few hundred force–distance cycles, empirical probability density functions (pdf) from the detected unbinding forces f_u can be constructed (Fig. 14.5d) [58]. The maximum of the distribution (Fig. 14.5d) reflects the most probable force upon which a single antibody-cell surface bond dissociates under the force ramp used. An overall binding probability, which is the probability to record an unbinding event in a force–distance cycle of 15.4% was obtained. Blocking experiments, performed by injection of free antibody in solution reduced this probability to 5.8% [58]. No binding was found at all when a bare tip was used instead of an antibody-coated tip. These results strongly support the specificity of the binding events observed.

Viewing ligand–receptor binding on the single-molecule level, the average lifetime of a ligand–receptor bond, $\tau(0)$, is given by the inverse of the kinetic off-rate constant, $\tau(0) = 1/k_{\text{off}}$. In the thermal activation model, the lifetime of a complex in solution is described by a Boltzmann equation, $\tau(0) = \tau_{\text{osc}} \exp(E_{\text{b}}/k_{\text{B}}T)$ [59], where τ_{osc} is the inverse of the natural oscillation frequency and E_{b} the energy barrier for dissociation. Hence, due to the thermal energy, there is a finite probability of overcoming the energy barrier E_{b}, which leads to the separation of the ligand–receptor complex. The force acting on a binding complex deforms the interaction energy landscape and lowers the activation energy barrier. The lifetime $\tau(f)$ of a bond loaded with a constant force f is given by $\tau(f) = \tau_{\text{osc}} \exp(E_{\text{b}} - fx/k_{\text{B}}T)$ [60], x being interpreted as the distance of the energy barrier from the energy minimum. Using AFM, an effective force increase or loading rate r can be deduced from $r = \mathrm{d}f/\mathrm{d}t$, equal to pulling velocity times effective spring constant [61]. The combination of the Boltzmann equation with the stochastic description of the unbinding process predicts different unbinding force distributions at different loading rates r [62]. The maximum of each force distribution, $f^*(r)$, reflects the most probable unbinding force for the respective loading rate r. f^* is related to r through $f^*(r) = k_{\text{B}}T/x \ln(r.x/k_{\text{B}}T.k_{\text{off}})$. Apparently, the unbinding force f^* scales linearly with the logarithm of the loading rate. For a single barrier, this would give rise to a simple, linear dependence of the force on the logarithm of the loading rate. In cases where more barriers are involved along the escape path, the curve will follow a sequence of linear regimes, each of which marks a particular barrier. The predicted logarithmic dependence of the unbinding force on the loading rate in the thermally activated regime was confirmed by a large number of unbinding experiments. A beautiful demonstration revealing hidden barriers between biotin and avidin was provided by Merkel et al. [63]. Most notably, the unbinding kinetics was dominated by different activation energy barriers positioned along the force-driven unbinding pathway and were consistent with the location of prominent transition states revealed by molecular dynamics simulations.

14.5.2 Force-Spectroscopy Measurements on Living Cells

Recently, recognition force spectroscopy has also been applied to cells. In one of the early studies, Lehenkari and Horton [17] measured the unbinding forces between integrin receptors present on the surface of intact cells and various ligands bound to the tip. The unbinding forces measured were found to be cell- and amino acid sequence-specific and sensitive to the pH and divalent cationic composition of the cellular culture medium. In another study, Chen and Moy [64] used AFM to measure the adhesive strength between concanavalin coupled to the tip and receptors on the surface of fibroblasts. Crosslinking of receptors on the cell surface led to an increase in adhesion that was attributed to enhanced cooperativity among adhesion complexes. Pfister et al. [58] investigated the surface localization of heat shock proteins (HSP) on stressed and unstressed endothelial cells. By detecting specific single-molecule binding events between the monoclonal antibody tethered to the tip and HSPs on cells, clear evidence was found for the occurrence of HSPs on the surface of stressed cells, but not on unstressed cells.

Figure 14.6 shows force-spectroscopy measurements on living cells, where the interactions between leukocyte function-associated antigen-1 (LFA-1) with its cognate ligand, intercellular adhesion molecule-1 (ICAM-1) was probed [65]. Force–distance cycles between cells expressing LFA-1 and ICAM-1 have been carried out under different buffer conditions (Fig. 14.6a). A typical measurement involved the formation of multiple LFA-1/ICAM-1 complexes that did not necessarily rupture simultaneously during retraction. The "saw-tooth" profile observed in the force–distance cycles suggests that these complexes often ruptured sequentially before the final separation, with each sharp transition in the retraction trace being attributed to a breakage of one or more LFA-1/ICAM-1 complexes (Fig. 14.6a). The unbinding forces required to detach the cell are quantitative measurements of cell adhesion. In terms of the unbinding force, cells stimulated with Mg^{2+} (i.e. activated cells) adhered more tightly to ICAM-1 than resting cells. As shown in Fig. 14.6b, the average unbinding force of the LFA-1/ICAM-1 complex increases over three orders of magnitude change in loading rate. Moreover, two loading regimes in the LFA-1/ICAM-1 interactions were evident in the force-spectroscopy plot (Fig. 14.6b). There was a gradual increase in unbinding force with increasing loading rate up to 10,000 pN/s. Beyond this point, there was a second loading regime that exhibited a faster increase in unbinding force. Activated cells resulted in higher unbinding forces that were pronounced in the slow-loading regime (Fig. 14.6b). Interestingly, there was no significant difference in the unbinding forces of the low and high-affinity complexes in the fast loading regime (at loading rates $> 10{,}000\,\mathrm{pN/s}$). The force-spectroscopy measurements revealed details of the complex process of ligand–receptor unbinding in cell adhesion. Two activation barriers were found in the dissociation of the LFA-1/ICAM-1 complex (Fig. 14.6c), as deduced from the two slopes in the force-spectroscopy plot (Fig. 14.6b). Each of the activation barriers (TS_1 and TS_2) is characterized by two parameters: a dissociation rate constant, k_i, and the position of the transition state, x_i, whereby $i = 1$ refers to the inner activation barrier and $i = 2$ to the outer activation barrier of the complex. Both k_i and x_i can be extracted from the force-spectroscopy plot (Fig. 14.6b). The values obtained are comparable to the dissociation rate constants measured by conventional methods. Furthermore, the dissociation rate constants can be used to estimate the energy differences between transition-state energies of high- and low-affinity complexes (Fig. 14.6c), showing that the outer activation barrier of the high-affinity complex is $3.2k_{\mathrm{B}}T$ higher than that of the low-affinity complex (Fig. 14.6c).

In another study, single-molecule dynamic force spectroscopy has been applied to study the interaction of antibodies and different sugars with the Na^+-glucose cotransporter SGLT1 in living cells (Fig. 14.7) [66]. The transporter drives the translocation of glucose into cells by use of the electrochemical gradient for Na^+ across the membrane and it is a prime example of the large family of biologically important ion-gradient-driven cotransport systems. Several methods have been employed to investigate the topology of the SGLT1, indicating that the SGLT1 contains 14 transmembrane helices [67]. The N-terminal half of the protein contains the Na^+-binding sites, whereas the sugar pathway is located in the C-terminal domain. Phlorizin, an aromatic glucoside, is known to be a potent competitive inhibitor of the SGLT1 transporter. It is postulated that

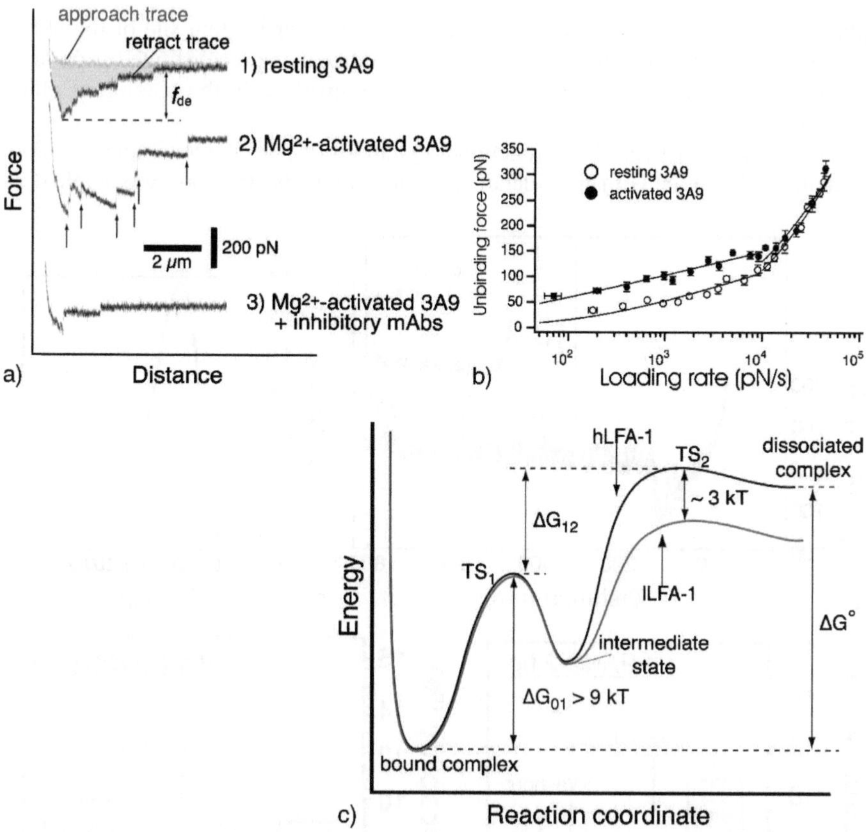

Fig. 14.6. AFM force measurements of cell adhesion. (**a**) A series of force–distance cycles between a living cell and intercellular adhesion molecule-1 (ICAM-1) acquired under different conditions. For traces 2 and 3, the cell was treated with Mg^{++}. F_{de} is the detachment force supported by the adhesive bonds formed between the cell and the substrate. The *arrows* in trace 2 point to positions where a cell/ICAM-1 complex ruptured. (**b**) Measurement of cell/ICAM-1 unbinding forces as a function of the loading rate. Plotting the most probable unbinding force against the loading rate results in the force-spectroscopy plot for ICAM-1 and resting cells (*open circles*) or activated cells (*filled circles*). The slopes of the linear regimes in the force-spectroscopy plot mark distinct activation barriers along the direction of force and yield the energy landscape along the reaction coordinate. The fitted curves (*solid lines*) were derived from fitting Eq. 2 to the data. (**c**) Intermolecular potentials of the cell/ICAM-1 interaction calculated from the force-spectroscopy plot. The dissociation of ICAM-1 from the cell surface occurs via two transition states, TS_1 and TS_2. The inner activation barriers of the high-affinity (hLFA-1) and the low-affinity (lLFA-1) complexes are $> 9k_BT$. Differences between the high- and low-affinity complexes ($\sim 3k_BT$) stemmed from the energetics of the outer barriers. Estimates of the energies of transition states were obtained from the dissociation rate constants. Data taken from [65]

phlorizin binds with a two-step mechanism to the sugar binding site of the transporter. Site-directed mutagenesis studies and studies using reconstituted peptide in vitro have shown that the C-terminal loop 13 is involved in the binding of phlorizin [68].

To study the structural and functional dynamics of plasma membrane transport proteins in living cells in finer detail, dynamic force spectroscopy was applied at

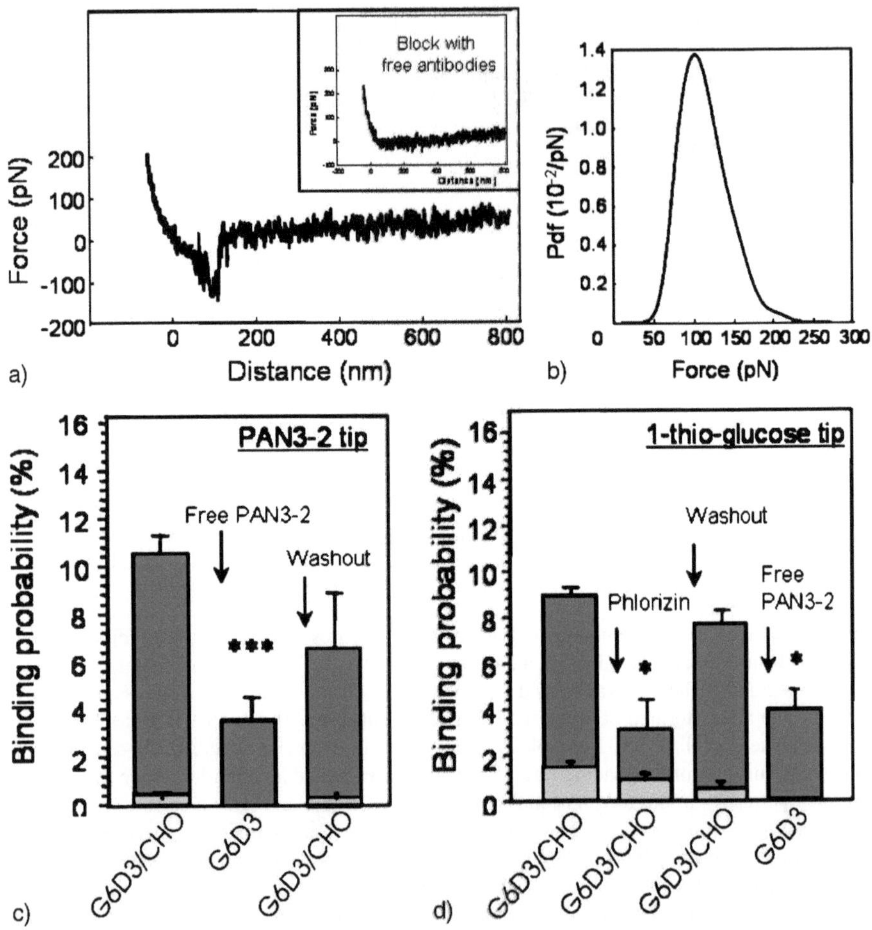

Fig. 14.7. Force measurements of antibodies and substrates with the Na^+-glucose cotransporter SGLT1 in living cells. (**a**) Recognition of SGLT1 on the surface of intact cells by an AFM tip carrying an epitope specific antibody (PAN3-2). Force curve showing specific interaction between the antibody and SGLT1 upon tip–surface retraction. The specific interaction is blocked by injecting free PAN3-2 antibodies in the solution (*inset*). (**b**) Probability density function giving the distribution of the unbinding force (f_u) of PAN3-2 to SGLT1. (**c**) Quantitative comparison of binding probabilities of PAN3-2-coated tips on G6D3 cells (*dark gray*) and CHO cells (*light gray*) in the absence or presence of free PAN 3–2 in the medium. (**d**) Recognition of SGLT1 on the surface of intact cells by a D-glucose primed AFM tip and the effect of phlorizin and the antibody PAN3-2 on the binding probabilities on G6D3 and CHO cells. Data taken from [66]

the single-molecule level [66]. The molecular recognition of SGLT1 on the cell membrane surfaces of living cells in different buffer conditions was probed with single antibody and with single D-glucose molecules that had been coupled to AFM tips. Scanning SGLT1-transfected cells with AFM tips carrying an epitope-specific antibody directed against the C-terminal loop 13 resulted in specific recognition events detected in force–distance cycles. Figure 14.7a shows a typical force curve (retraction) for a single-molecule antibody–antigen recognition event. For confirmation of the specificity of recognition events, blocking experiments were performed by injecting free antibodies. As illustrated in Fig. 14.7a (inset), no binding events were observed in the presence of free antibodies since no free epitopes were present on the cell surface. By constructing an empirical probability density function (pdf) of the unbinding forces (Fig. 14.7b), the maximum of the distribution was found to be 100 ± 30 pN. In addition to unbinding forces, binding probabilities from several experiments were quantified (Fig. 14.7c). For G6D3 cells, a binding probability (the probability to find an unbinding event in one force distance cycle) of 10.6% was observed. Another indication for specificity was that a very low binding probability was found for cells that contained no functional SGLT1. When free PAN3-2 antibodies were present in solution, the binding probability on G6D3 cells was significantly decreased to 3.6% (Fig. 14.7c). After subsequent washout of free PAN3-2 antibodies a high binding probability was recovered. These results demonstrate that SGLT1 on the surface of living cells can be specifically detected at the molecular level by using epitope specific antibodies. In addition, it was concluded that the epitope that is part of the loop 13 of SGLT1 is localized on the extracellular part of the cell membrane. In order to study the initial event of the glucose transport cycle, i.e. the binding of the sugar to SGLT1 on the surface of intact cells, glucose was coupled to the AFM tip. With this tip, force–distance cycles were performed on live cells, either on G6D3 cells expressing SGLT1 or on CHO control cells lacking SGLT1. Distinct glucose-recognition events were only observed in G6D3 cells, whereas in CHO cells the recognition signals were almost absent (Fig. 14.7d). These results demonstrate that the glucose binding site of SGLT1 is accessible on living cells for AFM force spectroscopy investigations.

14.6 Unfolding and Refolding of Single-Membrane Proteins

Dynamic force spectroscopy not only allows investigation of the interaction between receptors and ligands but also measurement of intramolecular force profiles of single molecules. The application of mechanical force to biological polymers like proteins, polysaccharides and DNA produces conformations that are different from those previously investigated by chemical or thermal denaturation [69]. The force-induced domain unfolding in proteins [70], length transitions caused by conformational changes in the sugar rings [71] or in the secondary structure of polysaccharides [72], and alterations of the secondary structure of DNA molecules were studied in great detail [73], elucidating the molecular determinants of mechanical stability and the role of force-induced conformational changes in the regulation of physiological function. In these experiments, the molecule is held between the tip and the support and its viscoelastic properties are studied in force–distance cycles (Fig. 14.8). Similarly as in molecular recognition force-spectroscopy experiments, a detailed picture of the complex mechanical unfolding pathway through a rough energy landscape can be gained by varying the dynamics of pulling.

As a recent example, OspA, an outer-surface protein of Borrelia burgdorferi containing a unique single-layer β-sheet, was unfolded with AFM (Fig. 14.8a) [74]. In this study, a monomeric unit was stretched by pulling it from its N- and C-terminal residues. Two peaks in the force–extension cycle were detected before the final rupture of a fully extended polypeptide occurred (Fig. 14.8b). The peaks in the unfolding curves are consistent with the results of previous thermodynamic studies showing two cooperative units in OspA. Mutations near the boundary of the cooperative units reduced the height of the first unfolding peak to undetectable levels

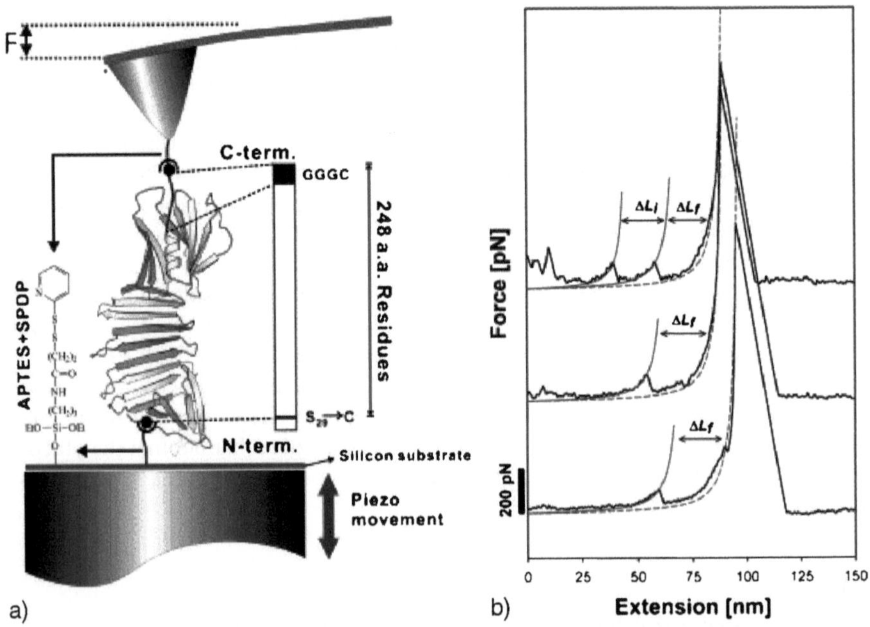

Fig. 14.8. Unfolding mechanics of an isolated membrane protein using AFM. (**a**) Experimental scheme for mechanical unfolding of a single, monomeric form of the outer surface protein A (OspA), a Lyme disease antigen containing a unique single-layer β-sheet. The engineered protein was covalently sandwiched through the formation of disulfide bonds between the substrate and the AFM tip. The sandwiched protein was then mechanically stretched by lowering the sample stage, leading to a characteristic and reproducible unfolding force pattern. (**b**) Force–extension curves of wild-type OspA (*upper panel*) and modified OspA (*lower panels*). Two peaks are detected in the force-extension profile before the final rupture of the fully extended protein occurs, which are interpreted as unfolding of multiple substructures in OspA. The contour length difference between two intermediate force peaks (ΔL_{i}) and that between the rupture peak and its adjacent peak (ΔL_{f}) are given. The persistence length of 0.4 nm was used to fit the worm-like-chain (WLC) model to all force peaks (*dashed lines*). Data taken from [74]

and marginally affected the second one (Fig. 14.8b). Based on a worm-like chain (WLC) analysis of the data, a model for mechanical unfolding of OspA was proposed. The analysis of these data revealed both similarities and differences between the mechanical unfolding process in the AFM and the thermal unfolding process at elevated temperature. This work demonstrates that the dynamic force spectroscopy studies of monomeric proteins can elucidate details of the intramolecular mechanics of protein substructures.

In another study, controlled unfolding and refolding of a sodium-proton antiporter has been analyzed with AFM (Fig. 14.9) [75, 76]. Single-molecule force spectroscopy was employed to unfold and refold single sodium-proton antiporters (NhaA) from membrane patches. For this purpose, the AFM tip was pressed onto the membrane surface with a contact force of about 1 nN for 1 s, and then withdrawn while recording the cantilever deflection versus tip–sample distance. In about 3% of the cases, one terminal domain of the protein actually adhered to the AFM tip, which resulted in a force spectrum over a distance of more than 95 nm (Fig. 14.9a, upper panel). Such force curves reflect the stretching and unfolding of the entire polypeptide AFM stylus (Fig. 14.9a, lower panel). The force spectra contained detailed information on the unfolding process, each peak representing an internal potential barrier that was built up by molecular interactions within the protein. Many of the potential barriers were detected either from the C-terminal or the N-terminal end. In this way, it was found that some molecular interactions stabilizing secondary structural elements were directional while others were not.

The above-described unfolding experiments of membrane proteins were further refined to allow for controlled refolding of individual secondary structures (Fig. 14.9c) [76]. Once distinct helices and loops had been unfolded, the whole system was relaxed for a given period of time, allowing the polypeptide to refold (Fig. 14.9b). Repeated recording of a force spectrum detected whether the polypeptide refolded to its native secondary structure. An example of such experiments is shown in Fig. 14.9b. First, NhaA was unfolded until only two helices remained anchored in the membrane (Fig. 14.9b, upper panel). The force peaks of the unfolding spectra indicated that the protein was unfolded from its C-terminal end. The AFM stylus was then lowered towards the membrane (Fig. 14.9b, lower panels). After the system had been allowed to relax for 5 s, the AFM stylus was separated from the membrane to restretch the polypeptide chain. The force spectrum then indeed contained the record of the repeated unfolding process (Fig. 14.9b, lower panels). All major peaks (potential barriers) observed during the initial unfolding were detected. These data indicate that several secondary structures observed for native NhaA had actually been refolded, and they support the hypothesis that unfolding and refolding of transmembrane helices may be fully reversible, including reinsertion of the transmembrane segments into the lipid bilayer.

14.7 Simultaneous Topography and Recognition Imaging on Cells (TREC)

The combination of high-resolution topography imaging with single-molecule force spectroscopy provides unique possibilities for the detection of specific molecular-

Fig. 14.9. Unfolding and refolding of single-membrane proteins. (**a**) Mechanical unfolding pathways of single sodium-proton antiporters (NhaA) from membrane patches. A characteristic pattern of force peaks corresponding to the unfolding of single NhaA molecules was observed. *Upper panel*: Superimposition of dozens of force curves highlights common features of the unfolding events and reduces deviations. The *gray scale* corresponds to the number of force curves superimposed. *Red lines* represent WLC fits with their numbers indicating the contour length of the stretched polypeptides. *Medium panel*: Detailed analysis of the positions and the probability of main peaks and side peaks. *Lower panel*: To compare the polypeptide length derived from the WLC fits with the NhaA structure, the secondary structure model has been chosen. Data taken from [76]. (**b**) Single-molecule unfolding and refolding of NhaA. *Upper panel*: Mechanical unfolding of NhaA yields force–distance curves containing characteristic peaks (*upper red trace*). The scheme shows the major unfolding pathway of NhaA. WLC fits of major force peaks and side peaks are shown by *continuous* and *broken black lines*, respectively. *Lower panel*: Scheme of a refolding experiment. Initially, the intact NhaA molecule was unfolded. The tip was then retracted from the membrane up to about 80 nm, so that the last helical pair remained anchored in the membrane. Then the stretched polypeptide was relaxed as the tip was brought into proximity (about 10 nm) to the surface (*upper panel, blue trace*). After a specified delay time (ranging from 0.5 s to 15 s; *upper panel*), the polypeptide was pulled repeatedly while recording a force–distance trace (*red traces* in the *upper panel*). After certain refolding times, the force peaks, which were detected of the refolded peptide, occur at positions identical with those measured upon initial unfolding of NhaA. Since this spectrum exhibits the characteristic unfolding peaks of native NhaA, it follows that the peptide folded into the native structure. Data taken from [75]

recognition events [77, 78]. The identification and localization of specific receptor binding sites on biosurfaces such as cells and membranes are of particular interest in this context. For this purpose, high-resolution imaging should be combined with the detection of specific ligand–receptor binding in the way that binding sites can be easily assigned to biomolecular structures. By combining dynamic force microscopy imaging with molecular recognition force spectroscopy, an imaging method for the detection and localization of binding sites has recently been developed [79–81]. This new imaging mode allows separation of the topography signal from the simultaneously recorded recognition signal. Topography and recognition images are thereby acquired using tips, which are magnetically oscillated during scanning and contain ligand molecules directed against the cognate receptor on the surface. For cantilevers driven at frequencies below resonance, the surface contact only affects the downward deflections (i.e. the minima) of the oscillations, whereas binding of the ligand on the tip to the receptor molecule on the surface affects only the upward deflections (i.e. the maxima) of the oscillations. Accordingly, the oscillation minima are used to record the topography image, and the oscillation maxima provide the data for the recognition image.

Figure 14.10 shows the application of the simultaneous imaging of topography and recognition sites on cells [82]. Real-time visualization and quantification of receptor binding sites on cell surfaces is a fundamental challenging task in cell biology. This is normally achieved by using immunofluorescence techniques such as immunostaining. However, the observed lateral resolution is not better than 200 nm. Recently, the simultaneous imaging of topography and recognition sites has been applied to vascular endothelial (VE) cells in order to localize VE-cadherin bind-

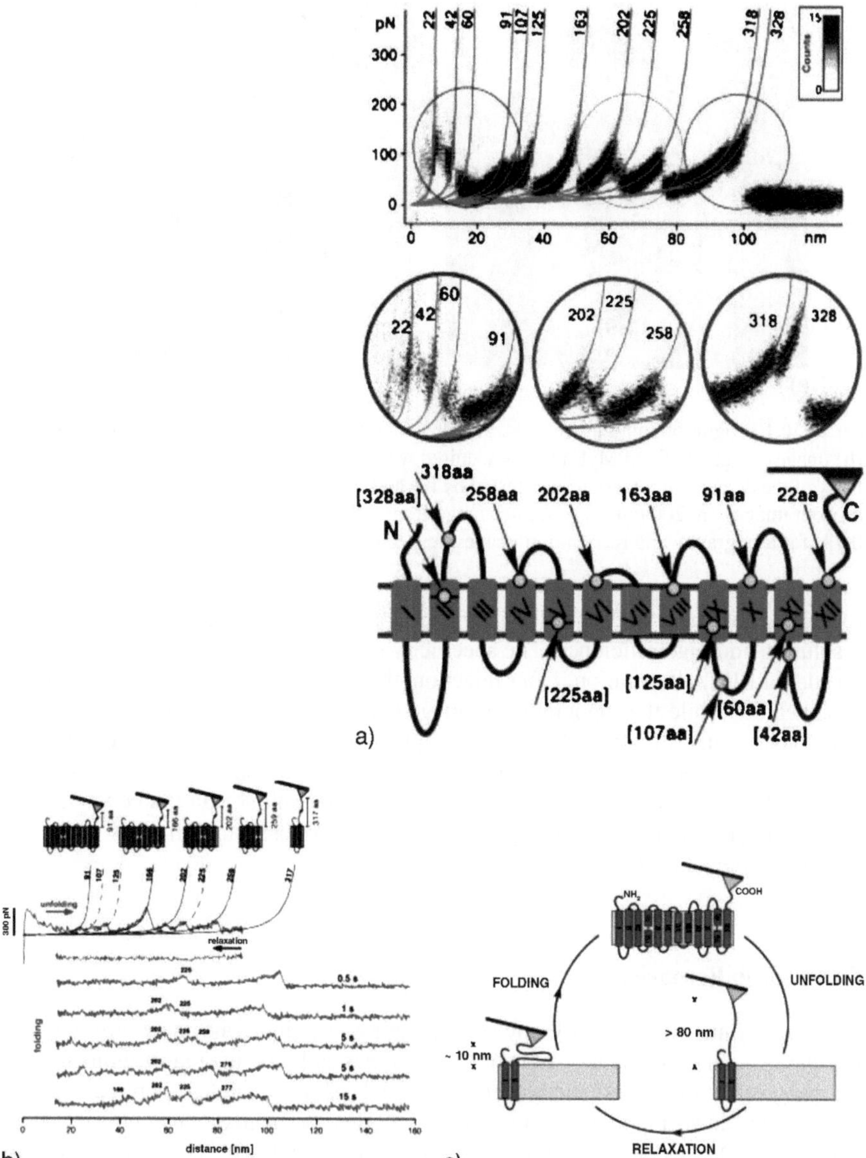

ing sites on the cell membrane [82]. In this study, topography (Fig. 14.10a) and recognition (Fig. 14.10b) images were recorded on gently fixed cells using AFM tips, which were chemically functionalized with fibrinogen. Besides the prominent role of fibrinogen in haemostasis, fibrinogen molecules can bind specifically to VE-cadherins on the cell membrane. Many specific interactions were observed in the recognition image (Fig. 14.10b). The recognition dark spots in the recognition image

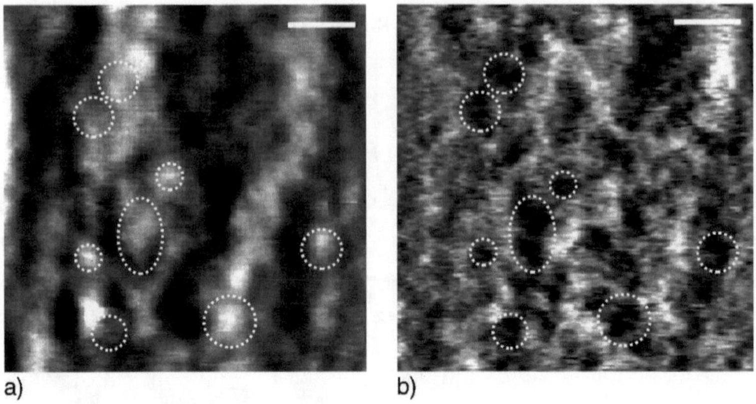

Fig. 14.10. Recognition imaging on cells. Simultaneous recorded topography (**a**) and recognition (**b**) images on gently fixed MyEnd cells acquired with a fibrinogen-coated AFM tip. The specificity of the detected interactions was tested by the addition of EDTA in the solution. *Scale bars* on both images are 200 nm. The z-scales range from 0 to 25 nm (from *black* to *white*) and 0 to 0.7 V for topography and recognition images, respectively. Data taken from [82]

reveal the locations of VE-cadherins in the topographical image with high lateral resolution and high efficiency. The specificity of the recognition process was tested by adding EDTA in solution. After injection, the dark spots in the recognition image disappeared, while the topographical image did not change (not shown). Thus, the simultaneous investigation of both topography and recognition open up a wide field of applications for investigating biological structure–function relationships in native environments on the nanometer scale, because the technique can map composition on top of a topographical image and can detect compositional changes occurring during biological processes.

14.8 Concluding Remarks

By using atomic force microscopy as an imaging technique one can at present gain the fine structural information of native membranes and protein organizations such as membranes of photosynthetic bacteria, bacterial S-layers, nuclear pore complexes, or even of viral arrangement on an artificial cell membrane under near physiological conditions. Due to continuous progress in the technical aspects of the AFM and in tip functionalization procedures, the investigation of receptor–ligand interactions on living cell has become possible. Moreover, the AFM as a force tool enables the dynamic pathways of the mechanical unfolding and subsequent refolding of single proteins in cell membranes to be followed. Thus, details in the intramolecular mechanics of protein substructures and of molecular interactions that apparently stabilize secondary structural elements are elucidated. The recent development of the TREC technique demonstrates a way, alternative to fluorescence studies, to visualize, identify, and quantify local receptor binding sites and to assign their locations to the topographical features of the cell surface with nm lateral resolution.

Thus, with improvements of the sensitivity and acquisition speed the AFM will open the way to fundamental challenging tasks in cell molecular biology.

Acknowledgements. This work was supported by Austrian Science Foundation projects P14549/P15295, the GEN-AU initiative of the Austrian Ministry of Education, Science and Culture, and EU projects Tips4Cells, NASSAP, HFSP RGP 0053/2004C and Force Tools. We thank Dr R. Zhu for the careful reading of the manuscript.

References

1. Binnig G, Quate CF, Gerber C (1986) Phys Rev Lett 56:930
2. Horber JK, Miles MJ (2003) Science 302:1002
3. Mueller DJ, Janovjak H, Lehto T, Kuerschner L, Anderson K (2002) Prog Biophys Mol Biol 79:1
4. Scheuring S, Levy D, Rigaud JL (2005) Biochim Biophys Acta 1712:109
5. Malkin A, Plomp J, McPherson A (2005) Meth Mol Biol 292:85
6. Jeremic A, Kelly M, Cho SJ, Stromer MH, Jena BP (2003) Biophys J 85:2035
7. Franz CM, Muller DJ (2005) J Cell Sci 118:5315
8. Zaman MS, Goyal A, Dubey GP, Gupta PK, Chandra SL, Das TK, Ganguli M, Singh Y (2005) Microsc Res Tech 66:307
9. Engel A, Muller DJ (2000) Nat Struct Biol 7:715
10. Allison DP, Hinterdorfer P, Han W (2002) Curr Opin Biotechnol 13:47
11. Rief M, Oesterhelt F, Heymann B, Gaub HE (1997) Science 275:1295
12. Marzsalek PE, Oberhauser AF, Pang YP, Fernandez JM (1998) Nature 396:661
13. Smith S, Cui Y, Bustamante C (1996) Science 271:795
14. Strick TR, Allemend JF, Bensimon D, Bensimon A, Croquette V (1996) Biophys J 271:1835
15. Veigel C, Coluccio LM, Jontes JD, Sparrow JC, Milligan RA, Molloy JE (1999) Nature 398:530
16. Tanaka H, Homma K, Iwane AH, Katayama E, Ikebe R, Saito T, Yanagida T, Ikebe M (2002) Nature 415:129
17. Lehenkari PP, Horton MA (1999) Biochem Biophys Res Com 259:645
18. Benoit M, Gabriel D, Gerisch G, Gaub HE (2000) Nat Cell Biol 2:313
19. Evans E (2001) Annu Rev Biophys Biomol Struct 30:105
20. Kienberger F, Ebner A, Gruber HJ, Hinterdorfer P (2006) Acc Chem Res 39:29
21. Sarid D (1991) Scanning Force Microscopy, Oxford University Press, New York
22. Putman CAJ, Vanderwerf KO, de Grooth BG, Vanhulst NF, Greve J (1994) Appl Phys Lett 64:2454
23. Hansma PK, Cleveland JP, Radmacher M, Walters DA, Hillner PE, Bezanilla M, Fritz M, Vie D, Hansma HG, Prater CB, Massie J, Fukunaga L, Gurley J, Elings V (1994) Appl Phys Lett 64:1738
24. Han W, Lindsay SM, Jing T (1996) Appl Phys Lett 69:1
25. Kienberger F, Stroh C, Kada G, Moser R, Baumgartner W, Pastushenko V, Rankl C, Schmidt U, Muller H, Orlova E, LeGrimellec C, Drenckhahn D, Blaas D, Hinterdorfer P (2003) Ultramicroscopy 97:229
26. Muller DJ, Schabert FA, Buldt G, Engel A (1995) Biophys J 68:1681
27. Kienberger F, Zhu R, Moser R, Blaas D, Hinterdorfer P (2004) J Virol 78:3203
28. Lantz M, Liu YZ, Cui XD, Tokumoto H, Lindsay SM (1999) Interface Anal 27:354
29. Moeller C, Allen M, Elings V, Engel A, Muller DJ (1999) Biophys J 77:1050
30. Muller DJ, Amrein M, Engel A (1997) J Struct Biol 119:172
31. Scheuring S, Sturgis JN (2005) Science 309:484

32. Ando T, Kodera N, Takai E, Maruyama D, Saito K, Toda A (2001) Proc Natl Acad Sci USA 98:12468
33. Scheuring S, Reiss-Husson F, Engel A, Rigaud JL, Ranck JL (2001) The EMBO Journal 20:3029–3035
34. Pum D, Neubauer A, Gyoervary E, Sara M, Sleytr UB (2000) Nanotechnol 11:100
35. Sara M, Sleytr UB (2000) J Bacteriol 182:859
36. Sara M, Pum D, Schuster B, Sleytr UB (2005) J Nanosci Nanotechnol 5:1939
37. Ebner A, Kienberger F, Huber C, Kamruzzahan ASM, Pastushenko V, Tang J, Kada G, Gruber HJ, Sleytr UB, Sara M, Hinterdorfer P (2006) Chem Bio Chem 7:588
38. Pante N, Aebi U (1993) J Cell Biol 122:977
39. Shahin V, Albermann L, Schillers H, Kastrup L, Schaefer C, Ludwig Y, Stock C, Oberleithner H (2005) J Cell Phys 202:591
40. Schaefer C, Shahin V, Albermann L, Hug MJ, Reinhardt J, Schillers H, Schneider SW, Oberleithner H (2002) Proc Natl Acad Sci USA 99:7154
41. Chiu W, Baker ML, Jiang W, Dougherty M, Schmid MF (2005) Structure 13:363
42. Kuznetsov YG, Malkin AJ, Lucas RW, Plomp M, McPherson A (2001) J Gen Virol 82:2025
43. Kienberger F, Rankl C, Pastushenko V, Zhu R, Blaas D, Hinterdorfer P (2005) Structure 13:1247
44. Kuznetsov YG, Victoria JG, Robinson W, McPherson A (2003) J. Virol. 77:11896
45. Svoboda K, Schmidt CF, Schnapp BJ, Block SM (1993) Nature 365:721
46. Evans E, Ritchie K, Merkel R (1995) Biophys J 68:2580
47. Clausen-Schaumann H, Seitz M, Krautbauer R, Gaub HE (2000) Curr Opin Chem Biol 4:524
48. Hinterdorfer P, Baumgartner W, Gruber HJ, Schilcher K, Schindler H (1996) Proc Natl Acad Sci USA 93:3477
49. Hinterdorfer P (2002) Meth Cell Biol 68:115
50. Hinterdorfer P, Gruber HJ, Kienberger F, Kada G, Riener C, Borken C, Schindler H (2002) Coll Surf B 23:115
51. Grandbois M, Dettmann W, Benoit M, Gaub HE (1999) Science 283:1727
52. Hinterdorfer P, Schilcher K, Baumgartner W, Gruber HJ, Schindler H (1998) Nanobiology 4:177
53. Lyubchenko YL, Gall AA, Shlyakhtenko LS (2001) Meth Mol Biol 148:569
54. Riener CK, Stroh CM, Ebner A, Klampfl C, Gall AA, Romanin C, Lyubchenko YL, Hinterdorfer P, Gruber HJ (2003) Anal Chim Acta 479:59
55. Hinterdorfer P, Kienberger F, Raab A, Gruber HJ, Baumgartner W, Kada G, Riener C, Wielert-Badt S, Borken C, Schindler H (2000) Single Mol 1:99
56. Riener CK, Kienberger F, Hahn CD, Buchinger GM, Egwim IOC, Haselgruebler T, Ebner A, Romanin C, Klampfl C, Lackner B, Prinz H, Blaas D, Hinterdorfer P, Gruber HJ (2003) Anal Chim Acta 497:101
57. Kienberger F, Pastushenko V, Kada G, Gruber HJ, Riener C, Schindler H, Hinterdorfer P (2000) Single Mol 1:123
58. Pfister G, Stroh CM, Perschinka H, Kind M, Knoflach M, Hinterdorfer P, Wick G (2005) J Cell Sci 118:1587
59. Bell GI (1978) Science 200:618
60. Evans E, Ritchie K (1997) Biophys J 72:1541
61. Evans E, Ritchie K (1999) Biophys J 76:2439
62. Strunz T, Oroszlan K, Schumakovitch I, Guntherodt H, Hegner M (2000) Biophys J 79:1206
63. Merkel R, Nassoy P, Leung A, Ritchie K, Evans E (1999) Nature 397:50
64. Chen A, Moy VT (2000) Biophys J 78:2814
65. Zhang X, Wojcikiewicz E, Moy VT (2002) Biophys J 83:2270
66. Puntheeranurak T, Wildling L, Gruber HJ, Kinne KH, Hinterdorfer P (2006) submitted

67. Xia X, Lin JT, Kinne RK (2003) Biochemistry 42:6115
68. Raja MM, Tyagi NK, Kinne RK (2003) J Biol Chem 278:49154
69. Carrion-Vazquez M, Oberhauser AF, Fowler SB, Marszalek PE, Broedel SE, Clarke J, Fernandez JM (1999) Proc Natl Acad Sci USA 96:3694
70. Rief M, Gautel M, Oesterhelt F, Fernandez JM, Gaub HE (1997) Science 276:1109
71. Marszalek PE, Pang YP, Li H, El Yazal J, Oberhauser AF, Fernandez JM (1999) Proc Natl Acad Sci USA 96:7894
72. Li H, Rief M, Oesterhelt F, Gaub HE (1998) Adv Mater 3:316
73. Rief M, Clausen-Schaumann H, Gaub HE (1999) Nat Struct Biol 6:346
74. Hertadi R, Gruswitz F, Silver L, Koide A, Koide S, Arakawe H, Ikai A (2003) J Mol Biol 333:993
75. Kedrov A, Ziegler C, Janovjak H, Kuehlbrandt W, Muller DJ (2004) J Mol Biol 340:1143
76. Kedrov A, Janovjak H, Ziegler C, Kuehlbrandt W, Muller DJ (2006) J Mol Biol 355:2
77. Willemsen OH, Snel MME, van der Werf KO, de Grooth BG, Greve J, Hinterdorfer P, Gruber HJ, Schindler H, van Kyook Y, Figdor CG (1998) Biophys J 57:2220
78. Raab A, Han W, Badt D, Smith-Gill SJ, Lindsay SM, Schindler H, Hinterdorfer P (1999) Nat Biotech 17:1999
79. Stroh CM, Ebner A, Geretschlaeger M, Freudenthaler G, Kienberger F, Kamruzzahan ASM, Smith-Gill SJ, Gruber HJ, Hinterdorfer P (2004) Biophys J 87:1981
80. Stroh CM, Wang H, Bash R, Ashcroft B, Nelson J, Gruber H, Lohr D, Lindsay SM, Hinterdorfer P (2004) Proc Natl Acad Sci USA 101:12503
81. Ebner A, Kienberger F, Kada G, Stroh CM, Geretschlager M, Kamruzzahan ASM, Wildling L, Johnson WT, Ashcroft B, Nelson J, Lindsay SM, Gruber HJ, Hinterdorfer P (2005) Chem Phys Chem 6:897
82. Chtcheglova LA, Waschke J, Kada G, Drenckhahn D, Hinterdorfer P (2006) (manuscript in preparation)

15 Atomic Force Microscopy of DNA Structure and Interactions

Neil H. Thomson

Abbreviations

SPM	scanning probe microscope/microscopy
STM	scanning tunneling microscope/microscopy
AFM	atomic force microscope/microscopy
FM	frequency modulation
AM	amplitude modulation
ShFM/TDFM	shear force microscopy/tranverse dynamic force microscopy
SEM	scanning electron microscopy
UHV	ultrahigh vacuum
TIRFM	total internal reflection fluorescence microsopy
DNA	deoxyribose nucleic acid
RNA	ribonucleic acid
bp	base-pairs
dsDNA	double-stranded DNA
ssDNA	single-stranded DNA
APTES	aminopropyltriethoxysilane
APS	aminopropylsilatrane
ATP	Adenosine triphosphate
ADPNP	5′-adenylyl β,γ-imidodiphosphate
E. coli	*Escherichia coli*

15.1 Introduction: The Single-Molecule, Bottom-Up Approach

Scanning probe microscopes (SPMs) are perceived as tools central to the goals of the burgeoning field known as nanotechnology. In the late 1950s, the physicist Richard Feynman, envisaged and encouraged a new field of manipulating matter down to the atomic scale [1]. In his visionary talk "There's plenty of room at the bottom" he suggested that electron microscopes needed to be made more powerful and introduced the idea of top-down fabrication to reach out to the nanoworld. One of the many ideas that arose naturally from his presentation, is that we should only have one machine or device interfacing between us and the nanoworld, because errors in scaling down will be too large, i.e. in building a machine to make a smaller machine that makes a smaller machine and so on, unacceptable tolerances would be needed. He used the example of smaller and smaller pantographs joined together, but perhaps did not envisage just one device that could not only image directly in the nanoworld but manipulate it also. Indeed, he said "Perhaps you can figure a better way than that to get down to small scale more rapidly." This way has now been achieved (at least in one sense) and the family of scanning probe microscopes that measure and manipulate matter at the nanoscale were possibly of what he was dreaming. Electron microscopes, of course, were well on their way already, and since that time they have been made more powerful and are now important imaging devices for nanotechnology, while electron beams are essential tools for top-down nanofabrication. In contrast, SPMs offer flexibility in addressing a probe directly to a surface, particularly the versatile atomic force microscope (AFM), which can be a nanopen, nanoscribe or nanofinger, allowing the human user to write with molecules [2], scratch or draw lines on surfaces smaller than the wavelength of light [3] or simply feel atoms and molecules. While SPMs can also be used as devices for top-down fabrication in lithographic modes of operation, they are also suited to bottom-up fabrication because they can, in principle, pick up and move atoms and molecules and construct new devices atom-by-atom or molecule-by-molecule.

Another important aspect of bottom-up fabrication in nanotechnology is the property of certain molecules to self-assemble in a spontaneous fashion [4]. Evolutionary theory tells us that biological organisms themselves have been self-assembled through myriad generations of Earth's history and incorporated ever-increasing levels of complexity. To understand biomolecular interactions and molecular mechanisms on a fundamental level, one major approach is to isolate pure samples of the molecules of interest so that only one or two, or a handful of species are interacting within an in vitro model environment. By adding more and more species sequentially as understanding of the fundamental system is gained, the complexity of biology may be grasped in a bottom-up approach. The counterpoint to this approach is also analogous to the top-down approach mentioned in nanofabrication, whereby markers (e.g. fluorescent probes) and imaging devices (e.g. confocal microscopy) are used on cells in an attempt to understand biological processes by seeing what happens in vivo, whilst all the complexity is present.

While the AFM is capable of imaging atomic periodicity, and in certain cases true atomic resolution [5] and even subatomic resolution [6], it has a large dynamic range that enables it to image molecules up to larger structures, such as cells [7]. The resolution on soft, large structures such as cells is compromised to a certain degree because the tip indents the surface, increasing its effective size. Additionally, as a surface-profiling technique, AFM is unable to "look" or should one say feel inside a cell without causing irreversible damage. For these reasons, the AFM is perhaps less suited to the in vivo approach for biology and could be claimed to be better suited to studying in vitro systems. Having said that there are many interesting questions that can be asked about cells using AFM because much of the communication in biology occurs at the cell interfaces, however, in this chapter we will focus on the use of AFM to study biomolecular systems in vitro, from the single-species case (e.g. DNA alone), increasing the complexity molecule-by-molecule and species-by-species (e.g. DNA plus one protein type) in a bottom-up approach to understanding molecular biology, which may eventually lead to spin-off applications in nanotechnology. This review of AFM research in the last twenty years is centered on DNA: the molecule central to the propagation of living organisms, the molecule that preserves the genetic code.

We will see that we have not come far on this bottom-up journey into biological complexity, and that introducing more complexity than DNA alone has only fairly recently been realized. However, it is hoped that this review of the field today will demonstrate the increasing importance of this approach and illustrate that new gains in the knowledge of DNA and its interactions with other molecules will continue to be made using the AFM in this fashion for many years to come.

15.2 DNA Structure and Function

Deoxyribonucleic acid: the name does not exactly trip off the tongue, but its abbreviation DNA does and it is arguably the most well-known of biomolecules outside scientific circles. Its key functions are storage and retrieval of genetic information. It acts both as the blue-print for maintaining the survival and propagation of an organism: it is the "library" of the cell allowing portions of the genetic information to be used, for example, for the synthesis of enzymes and other proteins. This chapter will outline the basic structure and interactions of double-stranded DNA (dsDNA) in the context of AFM studies but for more details on DNA structure and function see for example [8, 9].

Its double-stranded structure is the key to its function, as first recognized by Watson and Crick [10]. The hydrogen bonding between the base-pairs makes the structure very stable allowing the propagation of genetic information through generations upon generations of organisms with low error rates and little genetic drift. The base-pairs (bp) that encode the information are hidden "inside" the helix, protected by the sugar phosphate backbone (with negatively charged phosphate groups) meaning that the structure has to be actively distorted for the genetic code to be read or changed.

The base-pairing in DNA occurs between the two classes of chemical bases, known as purines and pyrimidines. Both are based on carbon-based rings with nitrogen atom substitutions, the pyrimidines being a single ring and the purines being a double ring of 5 and 6 atoms fused, where two carbon atoms are shared between the rings (see Fig. 15.1). The most stable base-pairings are always between purines and pyrimidines giving rise to the standard Watson–Crick pairings: adenine and thymine (A–T) and cytosine and guanine (C–G). There are two hydrogen bonds between A–T base-pairs and three between C–G, giving rise to a binding energy difference

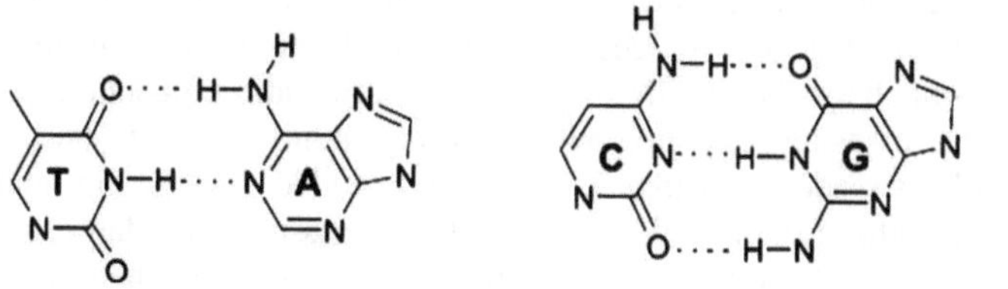

Fig. 15.1. The chemical structures of base-pairing between adenine and thymine (A–T) and between cytosine and guanine (C–G) through 2 and 3 hydrogen bonds respectively. A and G are purine bases while T and C are pyrimidine bases. Courtesy of Neal Crampton

between the two sets of genetic letters of around $2kT$ [11], allowing discrimination in energetic terms as well as structural between the letters in the genetic code. Other base-pairings can occur, i.e. base mismatches, where two purines or two pyrimidines are paired together, but these situations cause distortions in the overall DNA structure because the pyrimidines cannot bridge the gap between strands very easily and purines are too large to be accommodated by the usual helix structure.

The bases of DNA are arranged into a helical structure through the presence of the scaffold of the sugar moieties, deoxyribose, linked together to by phosphate groups via phosphodiester bonds. The chemical moiety consisting of the base and sugar is known as a nucleoside. The bases are attached to the C1′ carbon of the sugar ring, while the alternating phosphates between these nucleosides are attached at the hydroxyl groups on the C3′ and C5′ positions. (Note: a prime (′) is used to distinguish the carbons on the sugar from those in the base.) This staggering of the phosphodiester bonds gives rise to a polarity in each single-strand of DNA such that one is running in the 3′ to 5′ direction, while the other runs from the 5′ to 3′ end in an antiparallel arrangement. The two strands are intertwined into a double-helix that takes two basic structural forms: known as A and B. Both these structures were originally identified from X-ray diffraction of DNA fibers, in low (75%) and high (92%) humidities respectively. The "fully" hydrated B-form is thought to be the most common conformation found in vivo [8]. In B-DNA, the base-pairs lie essentially perpendicular to the helix axis, while in A-DNA the base-pairs are tilted with respect to the helix axis and lie a longer way off axis (Fig. 15.2). While the base-pairs in B-DNA span the helix, the helix in A-DNA opens up to form a channel down the center. The consequences of this structural transition are that the number of bp per helix turn goes from 10.5 to 11.0, while the helix diameter goes from about 2 to 2.6 nm, as the DNA is dehydrated. In essence, the pitch of the helix becomes less steep, which causes the molecule to become broader, and shortens the length of a linear dsDNA molecule. In B-DNA the axial rise per base-pair is 0.33 nm whereas in A-DNA this falls to only 0.23 nm. It is unclear as yet whether A-DNA even exists in vivo but has been proposed to occur at specific sites for protein binding [8]. There are also more unusual DNA structures, such as Z-DNA, where the helix turns inside out and becomes left-handed. Again, the prevalence of such structures within organisms is still under debate [8].

The genomes of organisms are organized into compressed structural architectures, chromosomes, which serve the dual purposes of conserving space and regulating gene expression. In cells containing a nucleus, the DNA is wound around histone protein complexes, like cotton on reels, to form chromatin, which is then further hierarchically condensed into chromosomes. For the survival and maintenance of cells a large number of genetic processes are required, which fall into the categories: RNA and protein synthesis; DNA repair; DNA replication; genetic recombination. All these processes are mediated by enzymes (proteins) that carry out work on the DNA templates made accessible as a result of decondensing the chromatin through the action of other proteins. When genes are expressed, a fundamental molecular process, known as DNA transcription is required to copy the genetic code into messenger RNA, which is subsequently used by the cell as a code to manufacture proteins in the process known as translation. Transcription is mediated (or catalyzed) by the enzyme RNA polymerase and later, we will see that AFM has been used to visualize

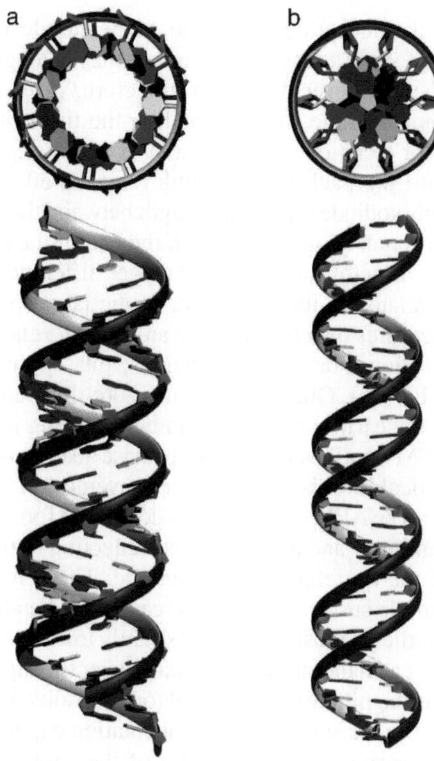

Fig. 15.2. The structure of double-stranded DNA in the A-form (*left*) and the B-form (*right*). The *upper panels* show top down views of the two structures demonstrating that dehydration of the helix leads to a opening up of the two strands to form a channel down the middle in the A-form. These model helices are formed from repeating sequences of GC bases for 30 bp. Note that the top view of the B-form is slightly magnified compared to the other images for clarity. Courtesy of Sarah A. Harris

the process of transcription in situ at single-molecule resolution [12]. This was the first visualization of an enzyme in action by AFM and heralded the beginning of AFM single-molecule enzymology. Since that time, still only a handful of real-time visualizations by AFM have been achieved (see Sect. 15.5.3.5). This highlights the difficulty of the technique and this chapter attempts to explain the reasons behind this slow progress. For more details on the molecular mechanisms of genetic expression and maintenance, see for example [13].

15.3 The Atomic Force Microscope

The atomic force microscope (AFM) [14] can be considered as the most versatile of all scanning probe microscopes (SPM). Since forces arise from the properties of matter, they are ubiquitous, and therefore the AFM can in principle be applied to any solid-state material. As with all SPMs, the roughness of the sample is a limiting factor, and curvature and topographical gradients on the sample should ideally not be comparable to or larger than the radius of curvature and opening angle of the AFM tip respectively. The "atomic forces" that arise between the apex of the AFM tip and the sample are detected because the tip is on the end of a flexible cantilever spring. The cantilever obeys Hooke's law ($F = kx$), so that the force detected is proportional to the deflection of the end of the cantilever beam (Fig. 15.3). The deflections are very small compared to the cantilever length and hence the deformation is elastic. The cantilever deflection, therefore, gives a measurement of the resistance of the sample to the force exerted by the tip. It is commonly measured using an optical lever technique whereby a laser beam (e.g. 5 mW red solid-state laser) is focused onto the back of the cantilever and reflected onto a split photodetector (usually a photodiode). The path length between the cantilever end and the detector amplifies the small angular motion of the end of the cantilever into large linear motion across the detector interface. The power difference on the two halves of the photodetector is related to the movement of the laser spot. In this sense, this technique actually measures the change in the angle of the end of the cantilever rather than its absolute position. For small deflections, this approximation of the cantilever deflection is valid [15]. Other ways to measure the cantilever deflection are interferometry and in the earliest cases a scanning tunneling microscope (STM), piggy-backing on the AFM cantilever [14]. Both these techniques are more difficult to set up and the optical lever is preferred for convenience.

The AFM can be used to determine forces by calibrating the spring constant of the cantilever and measuring the deflection of the cantilever as a function of the position of the tip relative to the surface. These forces are of course acting normal to the sample surface because the cantilever bending occurs along this axis. In reality, the cantilever must be inclined slightly toward the surface so that the cantilever itself does not catch the surface. This cantilever tilt angle can be 10 to 15°, which slightly affects the magnitude of the normal forces resolved. This tilt also means that the tip will travel across the surface during indentation experiments. Lateral forces can be inferred by measuring torsional motion of the cantilever using a quadrant photodetector and taking the power difference between the left and right sides of the detector.

As well as measuring forces along a single axis, the AFM can be used as an imaging tool to gain three-dimensional topographical information about samples by using a scanning system with a feedback loop to manipulate the tip across the sample. Figure 15.4 shows a schematic representation of the AFM and its major components. The scanning is usually generated through the use of piezoceramic materials that change

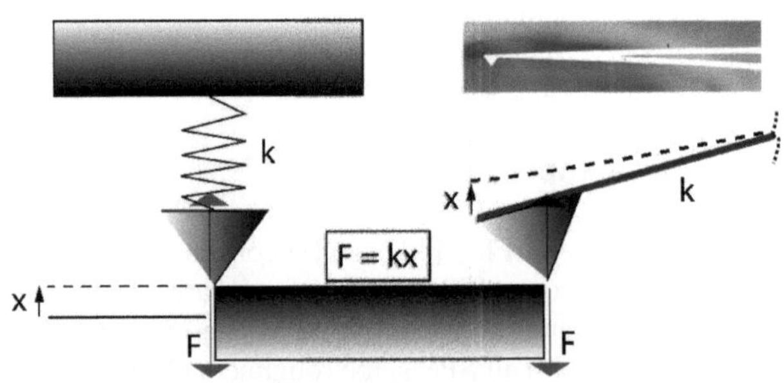

Fig. 15.3. RHS: The AFM cantilever (RHS) obeys Hooke's law for the small nanometer scale deflections it makes compared to the length of the cantilever, typically 100 or 200 μm. LHS: The AFM tip can represented as being on the end of a spring with stiffness k, known as the spring constant of the cantilever. The *top right panel* shows an SEM micrograph of a 100 μm-long silicon nitride AFM cantilever, with an integral tip, which has an open triangular shape

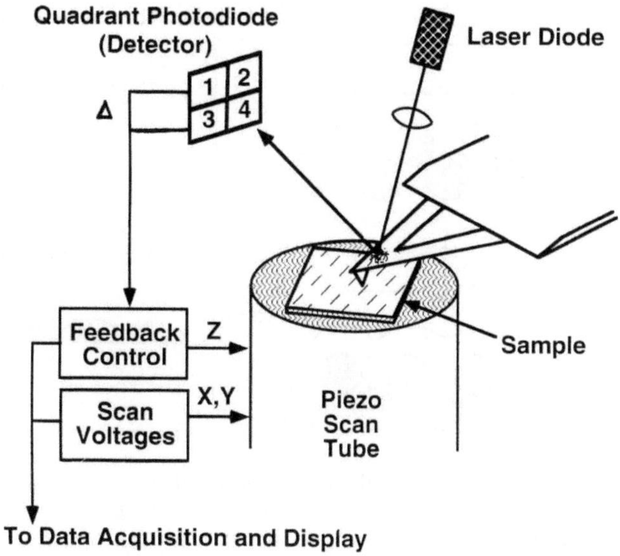

Fig. 15.4. Schematic representation of the major components of an atomic force microscope. Either the sample or tip assembly is scanned on a piezoceramic transducer that can create motion in all three axes. The microfabricated cantilever is addressed to the surface and the deflection of the end of the cantilever measured using an optical lever technique. The relative size of the cantilever in this schematic is greatly exaggerated. The deflection signal of the cantilever is used to determine how the feedback circuit and scanners drive the tip across the three-dimensional surface

shape on application of high voltages, such as lead zirconium titanate (PZT). Minute changes in the equilibrium position of the atoms in the crystals allow subnanometer positioning accuracy, although they do exhibit nonlinear behavior and hysteresis. Newer generations of commercial AFMs overcome these limitations by accurately measuring the position of the scanner and using an additional feedback loop to hold the tip in the required position. These are known as closed loop scan systems and eliminate image distortion from piezo nonlinearity and thermomechanical drift. In principal, any scan pattern can be generated by the scanners, but typical images are square and built up in a raster pattern in a series of lines, so that there are fast and slow scan axes. Typical scan rates of current commercial AFMs are between 1 and 4 Hz (i.e. scan lines per second), although the tip velocity across the surface is the limit for imaging speed, and this is typically up to 10 μm/s. Images are typically sampled at 512 × 512 pixels, which means that a 1 μm^2 scan will take nearly one minute. AFM scan technologies are developing all the time, however, and some groups are working on making systems that perform at video rates or better (see Sect. 15.6).

There are two imaging strategies for AFM, (or indeed any SPM): to have the feedback enabled or disabled. Figure 15.5 shows schematically how these strategies work and what information is obtained. Without feedback, the tip is simply scanned across the surface and the deflection measured to build up an image of the surface. In this case, the force applied to the sample is proportional to the topographical changes across the sample, since this is related to the cantilever deflection. This approach

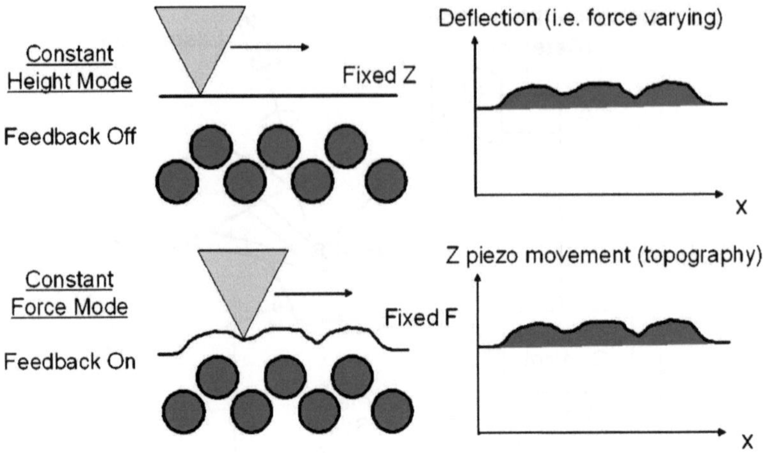

Fig. 15.5. Schematic representation of the two fundamental ways of operating a scanning probe microscope. For the atomic force microscope, constant-height mode implies information on the surface topography resides in the cantilever deflection (**a**), whereas in constant-force mode with feedback on the movement of the vertical Z-piezo gives topographical information also (**b**)

is almost always unsuitable for soft samples, such as polymers and biomaterials because the forces that build up as the deflection increases when a sample feature is encountered cause sample damage. To eliminate this problem a feedback loop can be used to sample the deflection as a function of time and move the scanner up or down to keep the cantilever deflection and therefore the contact force constant and ideally minimized. With feedback on, the imaging method is known as constant force, while with feedback off the imaging method is known as constant height. In practice, there are user-controlled gains for the amplifiers in the feedback loop that allow essentially a continuous spectrum between the feedback off and the fully on situations. The two fundamental signals that are typically monitored are the cantilever deflection and the voltage sent to the vertical scanner element, denoted as the Z-piezo. The distribution of information between these two images depends on the response time of the microscope. If the feedback loop could be driven infinitely quickly then the deflection signal would be perfectly flat. In practice, the cantilever usually has the quickest response time in the whole system (being a small mechanical device) and therefore there is always residual information in the deflection signal. The inertial response of the piezos means that the topographical signal (i.e. Z-piezo position) contains the lowest frequency information about the surface, while the deflection signal contains the higher-frequency information up to the frequency of the cantilever. Altering the gains in the feedback loop essentially allows the user to partition information between these two signals, but for biomolecular samples such as DNA, gains are almost always kept at the maximum possible (above which scanner resonance would occur).

Whether operation is performed in constant height or constant force imaging, there are two basic imaging modes that the AFM can be operated in. In analogy with electronics, these are termed DC or AC modes, which relate to whether the cantilever is used statically or dynamically (Fig. 15.6). In the typical and traditional DC mode, known as contact mode, the AFM tip is addressed mechanically towards the sample surface and the quasi-static deflection of the cantilever is monitored as the tip is scanned back and forth across the sample surface [14]. This mode creates large shear forces between tip and sample because the tip is in constant contact with the surface. Contact mode is generally unsuitable for soft materials such as polymers and biomaterials because they are distorted and/or damaged quite readily. To overcome these limitations, AC techniques in which the cantilever is vibrated around its resonant frequency were developed. In these methods, the tip is intermittently addressed to the sample surface, once per cycle of cantilever vibration and importantly contacts the surface in essentially the surface normal direction (i.e. vertically). It appears that viscoelastic material can support compressive forces more readily than the shear forces present in contact mode. AC imaging techniques can be divided into those that use the cantilever amplitude as the detection signal and those that use the cantilever frequency. In analogy with radio electronics, these are known as amplitude modulation (AM) and frequency modulation (FM) techniques, respectively [16]. Historically, AM AFM has been used in ambient and liquid environments on soft samples (e.g. polymers and biomaterials) while FM AFM is used under UHV conditions, usually to try and achieve atomic resolution on solid-state systems. The divergence in the use of AM or FM detection in AFM AC techniques for different applications arose because FM detection usually requires cantilevers with a very high quality factor (Q) where environmental damping in UHV is negligible. By contrast, in AM AFM a significant change in the cantilever amplitude is required when the tip interacts with the sample, therefore cantilevers with a lower Q and higher damping environments (air and liquids) are acceptable. The AM imaging mode is often known as tapping mode, since the AFM tip repeatedly taps the sample surface as it is scanned [17].

Having established that constant-height imaging is not suitable for imaging biomolecular samples and realizing that the feedback is almost always set to the quickest response time possible, of the two imaging modes the AC (AM) approach is almost always more successful for biological specimens. Having said this, as we will learn later, the imaging environment plays an influential role on the ability of the AFM tip to image molecular structure reproducibly. The AFM can be operated under the full range of environments, from vacuum through gaseous or ambient

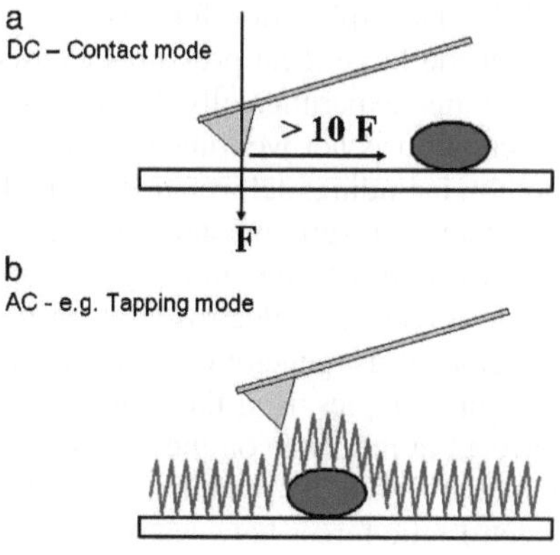

Fig. 15.6. Schematic representation of the two fundamental imaging modes of an atomic force microscope. (**a**) In contact mode the static DC deflection is used as the detection of the surface, but large lateral shear forces can build up between tip and sample damaging or sweeping away weakly bound molecules. (**b**) In amplitude-modulation techniques, such as the tapping mode, the cantilever amplitude is the detected signal. In this mode the tip contacts the surface in the normal direction eliminating damaging shear forces

(air) to liquid, including aqueous media. The environment affects the magnitude of different forces operating between the tip and the sample. For instance, in ambient air conditions there is usually a capillary neck between the tip and surface formed through condensation of water vapor on the contacting surfaces. This produces a large attractive force that sucks the tip onto the surface and limits the lowest tracking force that can be achieved. This effect can be eliminated by going to vacuum, using dry environments or submersing the tip under liquid. The cantilever is chosen to have a suitable stiffness (spring constant) depending upon the imaging environment and mode. For example, tapping mode in air requires relatively stiff cantilevers (e.g. $k > 10\,\mathrm{N/m}$), to prevent the tip from adhering irreversibly to the sample surface through the capillary interaction. Contact mode requires relatively soft cantilevers (e.g. $k < 1\,\mathrm{N/m}$) to give a measurable deflection at loading forces that are sufficiently low as to not cause sample (or tip) damage. Tapping mode under liquid also uses these softer levers to prevent sample damage during mechanical drifting of the microscope.

The ultimate resolution in the AFM is determined by the size and shape of the apex of the scanning tip. The cantilevers are microfabricated from silicon and silicon nitride and usually include integrated tips that have slightly different shapes and are often sharpened through oxidation to produce the finest apexes. Typically, the radius of curvature of the tips is around 10 nm, which means that in AFM images, DNA will also appear broadened by approximately twice the radius of the tip [18], limiting the lateral resolution obtained. Resolution along the backbone could be higher than this because the contact area between tip and sample is at its lowest when the sample is locally convex.

It should be realized that the AFM only produces a true topographical image of a surface if the mechanical response under the AFM tip is homogeneous across the surface. For soft biological molecules adhered to a relatively hard support surfaces, it is unlikely that the AFM will produce an exact topograph. For instance, biological cells are readily indented by the AFM tip and the image produced is usually representative of the internal cytoskeleton, i.e. the stiffest components of the cell. For single biomolecules, the situation is more complex but it is interesting to note that the heights of DNA, particularly when imaged in air are almost always lower than would be expected. Typically, the maximum height of the DNA backbone above the support surface obtained by AFM is around 1 nm or so, sometimes less, whereas we might expect 2 to 2.6 nm depending on the local conformation of the DNA. One explanation for this effect has been to say that the DNA is compressed under the large local pressures exerted by the AFM tip, however, it is difficult to verify this experimentally. In tapping mode at least, the image-contrast-generation mechanism is not well understood. Other factors may also influence the measured height, including: deformation of the DNA molecule due to surface-binding forces; erroneous background support levels due to surface deposition of salt layers; complex dynamics of the vibrating cantilever in AC imaging modes; influence of longer range forces (e.g. electrostatic) on topographical imaging. The full picture of the image-contrast-generation mechanisms for AFM will require much more work. For imaging samples in air the capillary neck present between tip and sample is likely to have a big influence on the imaging mechanism, both in contact mode and tapping mode [19]. On 2D protein crystals under aqueous fluid it has been shown that the electrostatic forces between tip and sample also play an important role [20].

15.4 Binding of DNA to Support Surfaces

15.4.1 Properties of Support Surfaces for Biological AFM

Since the AFM, like all SPMs, is a surface-profiling instrument, study of molecular systems requires that the molecules of interest are deposited irreversibly onto a suitable support surface (sometimes known as the substrate[1]) that can be scanned by the AFM tip. This support surface should fulfil a number of criteria.

First, the roughness of the surface should be small compared to the height of the molecules in question so that the AFM tip can distinguish them from the background in the topographical (Z-piezo) signal. Secondly, the interaction between the molecules and the surface should be strong enough such that the influence of the scanning tip does not distort, move or damage the molecules during repeated scanning. Finally, the surfaces must be easy to prepare, reproducible and free of contamination. These different aspects of the supporting surface will now be discussed in more detail.

The criterion that the surface should be extremely flat, at least on local scales (nm to μm), tends to be more important for globular biomolecules, such as proteins, compared to long polymer molecules, such as DNA. Even if the resolution is not high enough to pick up a globular feature in a rough background the correlation of many points along an elongated structure (even if it is only one pixel wide in an image) allows the structure to be discerned from the noise (background surface features). Development of high-resolution chemical force microscopy (CFM) [21] or molecular recognition force microscopy (MRFM) [22], where the tip is functionalized with a chemical species or biological receptor respectively, would generate better material contrast and may, to some extent, obviate the need for very flat surfaces (see Sect. 15.6 for a brief discussion of CFM and MRFM techniques).

There are a number of ways that molecules can be adhered to a surface such that they are immobile under the scanning AFM tip. Essentially we can think of these as either a chemical strategy where molecules are covalently linked to the surface, or a physisorption process, where the natural forces involved attract and immobilize the molecules. The most straightforward sample preparation uses physisorption processes, whereby the chosen surface and the environment (air, vacuum or liquid) determine if and how molecules bind. Chemical immobilization requires either a derivatization of the surface or a chemical crosslinker to be introduced between the molecule and surface or a combination of both approaches. Molecular immobilization strategies, with particular reference to DNA, are discussed in more detail in Sect. 15.4.2 below.

Ideally, the support surface should be very easy to prepare, having as few preparation steps as possible to prevent introduction of any contamination. This means that physisorption approaches are preferred over chemical strategies if and only

[1] The term substrate is also used in biochemistry to mean the entity, e.g. molecular species, upon which an enzyme acts during its catalytic cycle. Consequently, we will refer to the underlying surface in the AFM sample as the support surface.

if they achieve the required molecular surface binding. Introduction of chemical functionalization may also make reproducible formation of surfaces more difficult to achieve. The preferred support surface for many biomolecular AFM studies that traditionally has been used is the mineral, mica. It is a layered crystal in which the silicate layers are ionically bonded together comparatively weakly. The anisotropy in the crystal, means it forms flat flake-like structures, and allows the top surface of the flakes to be simply cleaved away using sticky tape (the best kinds to use are those with low tackiness, usually semitransparent and white in appearance). This produces a clean surface every time molecules are required for deposition. The layers are atomically flat over many hundreds of micrometers to millimeters, allowing easy distinction of biomolecules such as DNA. There are other such layered materials available to use in SPM studies of molecular systems, such as highly oriented pyrolytic graphite (HOPG), crystalline graphite in which the planes are regularly ordered over large dimensions. They are usually a couple of centimeters across and can also be cleaved with sticky tape because of the weak bonding between the 2D covalently bonded carbon planes. It is, however, a very hydrophobic surface (due to its non-polar nature) to which hydrophilic molecules, such as DNA, do not bind very strongly and is therefore inappropriate for most AFM studies. Mica, by contrast, is a hydrophilic surface that is well-suited to biomolecular deposition for all the reasons given above.

15.4.2 DNA Binding to Surfaces

As we have discussed, mica is the most common surface on which to deposit DNA and its complexes for study by AFM. Mica is in fact a group of minerals with over 30 members belonging to the silicates class. Silicates are thought to make up 90% of the Earth's crust and are based on tetrahedral SiO_4 motifs. These tetrahedra can be arranged singularly or doubly and in chains, sheets, rings and other frameworks. Mica is a phyllosilicate with a sheet-like structure and is related to clay minerals, all of which have a plate-like or layered structure. Clays hold water in between the silicate sheets, whereas micas are anhydrous with no water present. The mica used in AFM is one of the most common members of the group, called muscovite mica, being potassium aluminum silicate hydroxide $KAl_2(AlSi_3O_{10})(OH)_2$. It consists of silica tetrahedra and alumina octahedra covalently bonded together into sheets 1 nm thick that are weakly bonded together ionically with potassium ions (K^+) (Fig. 15.7). This gives rise to its good basal cleavage, revealing a surface terminated with oxygen atoms. Potassium ions sit in a regular array of binding sites at the surface. When immersed in aqueous solution the K^+ ions diffuse away from the surface to be replaced by a counterion cloud. This renders the mica surface negatively charged with a diffuse double layer of cations above it.

Since DNA is a negatively charged semiflexible rod [23] it will not bind to the mica unless the electrostatic repulsion between itself and mica is overcome. This is usually achieved through the introduction of divalent metal cations into the aqueous medium. These replace the K^+ ions in the mica binding sites and effectively render the mica positively charged. Different cations have a different efficacy of inducing DNA binding to mica, which appears to be related to their

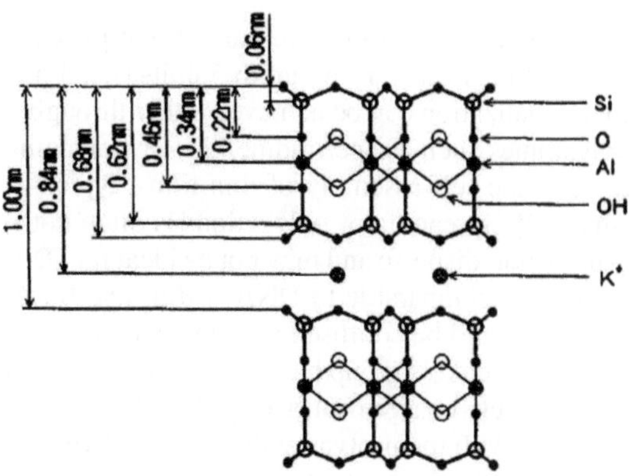

Fig. 15.7. The atomic structure of mica. Reused with permission from Shojiro Miyake, Applied Physics Letters, 67:2925 (1995). Copyright 1995, American Institute of Physics

enthalpy of hydration [24]. The transition-metal ions, Ni(II), Co(II) and Zn(II) have been shown to be the most effective. A simplistic picture is the idea that smaller ions are the only ones that physically fit into the binding sites, however, along with these ions, Mg(II), despite being a group II element, has a similar ionic radius and yet does not bind DNA well enough for it to be imaged in tapping mode, at least not under aqueous fluid. Larger transition-metal cations, like Mn(II) bind DNA weakly, while Cd(II) and Hg(II) do not appear to bind DNA at all to mica. Those cations, however, that have a larger enthalpy of hydration bind DNA more efficiently (e.g. Ni(II)), while Mg(II) and Mn(II) having intermediate values. The enthalpy of hydration of an ion is related to the strength of the coordination bonds it makes with ligands. Those ions that make strong bonds with water can also make strong bonds with the recessed hydroxyl groups in the mica surface once the coordinated water molecules are removed. It appears that the immobilization of the divalent cations into the mica binding sites is the most important factor for immobilization of DNA. The real situation is somewhat more complex, however, since in Mg(II) buffers DNA binds sufficiently strongly to mica that it can be imaged using tapping-mode AFM in air, after the sample has been rinsed with excess water and dried with a stream of nitrogen gas. There are two ideas used to explain the mechanism of binding of DNA to the mica; one is a static model based on salt bridges and the other is a dynamic picture invoking counterion correlation forces. In the static picture, the divalent cation simply acts as an electrostatic bridge between the negatively charged DNA and mica [25]. In the dynamic model, the ions are in constant motion, and an overall attractive force arises through sharing of divalent counterions in a correlated way [26]. This picture seems realistic since the attraction is ion-concentration dependent, where higher concentrations of divalent cations electrostatically screen these correlated interactions. Pretreating the mica surface with transition-metal ions that bind irreversibly into the K^+ sites on the mica (e.g. Ni(II)) increases the correlation force and lowers the net negative surface charge, encouraging DNA to bind more strongly.

In essence, immobilization of the DNA requires a positively charged surface since the DNA is highly negatively charged. Another way to create positively charged mica surfaces is to chemically functionalize the surface with positively charged groups, such as amines (NH_2). These can accept a proton to give a $+1$ charge around neutral pH, although it should be remarked that pKa values between solution and surface environments can vary by up to 5 units (i.e. 5 orders of proton concentration). This functionalization can be achieved either through the spreading of polycations, using polyamines such as spermidine [27] or polyaminoacids such as polylysine [28], or through the self-assembly of thin films, typically a monolayer, using silane chemistry [29]. Spreading of polycations is often not very reproducible because they are usually polydisperse and may not be ideal for DNA applications because the polymer widths are comparable to DNA widths and linear surface features arising from the polymer could be confused with DNA molecules.

Lyubchenko's group has pioneered the formation and use of aminosilanes to create positively charged surfaces [29]. Silanes are chemically unstable molecules that crosslink with themselves and/or surface hydroxyl groups, using a condensation reaction, eliminating water to form Si–O bonds. They react strongly with the hydroxyl groups on silicon surfaces and much of the characterization of the resultant layers has been performed on silicon. The preferred silane for mica functionalization is 3-aminopropyltriethoxysilane (APTES) partly because of its commercial availability. A trifunctional silane such as this has the ability to crosslink to a surface through one alkoxyl group, while the other two can be involved in lateral crosslinking of the molecules into a monolayer. The amine group on the end of the alkane is then expected to be pointing upwards on the surface creating an array of positively charged sites under aqueous fluid. Figure 15.8 shows the chemical structure of APTES and how it self-assembles into monolayers on surfaces. There is increasing evidence that the stability of APTES layers on mica comes solely from lateral crosslinking rather than any direct surface crosslinking, which is unlikely because the hydroxyl groups on mica are recessed 0.22 nm below the surface, whereas the Si–O bond length is 0.16 nm [30]. It is possible that plasma treatment of mica disrupts the surface enough to allow covalent interactions between silane and the surface [31]. The formation of these layers is strongly affected by the formation conditions, e.g. from vapor versus liquid phase, and the instability of the silane can easily lead to 3D polymerization in the bottle before the surface is made. If the silane "goes

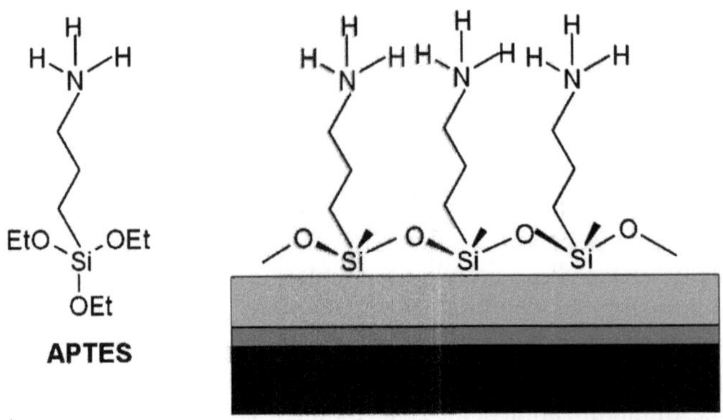

Fig. 15.8. The chemical structure of aminopropyltriethoxysilane (APTES) and its formation of monolayers on silicon and related surfaces through covalent crosslinking. The crosslinks form through a condensation reaction eliminating alcohol. Courtesy of Neal Crampton

off' in this way, reproducible surfaces cannot be made. Liquid-phase preparation also tends to produce uncontrolled polymerization at the surface into 3D globular structures also unsuitable for molecular imaging. Lyubchenko and coworkers [32] have got round these problems by making a cyclic derivative of APTES, aminopropylsilatrane (APS), that is stable in aqueous solution and can only crosslink in the presence of a surface. Figure 15.9 shows schematically how the interactions of DNA in solution are mediated with mica and aminosilane treated mica to promote surface binding.

Vapor-phase deposition is the preferred method of formation of APTES-mica, although the humidity of the surrounding environment can strongly affect the structure of the silane layers formed [30]. The water layers that are present on hydrophilic mica under ambient conditions influence the extent and structure of the layer formation. Below 25% relative humidity (RH) a monolayer of APTES is formed that can bind DNA without the need for divalent cations in the buffer. Above 25% RH a bulk water layer starts forming on the mica and the APTES forms a bilayer with mobile silanes present that induce DNA condensation. This condensation effect can be removed upon baking the surfaces to drive off the mobile silanes, but DNA does then not bind without the presence of divalent cations, suggesting a gross structural rearrangement of the structure. Generally, APTES-mica for DNA applications is therefore produced under environments as dry as possible [32].

DNA that is bound to surfaces adopts conformations that can be influenced by the binding forces present. The two extremes of behavior are known as equilibrated and kinetically trapped [33]. Figure 15.10 shows examples of AFM images taken in these two conditions. In the first situation, the DNA is bound to the surface but free to diffuse sideways in two-dimensions and therefore adopts a minimum energy conformation in the 2D plane of the surface (Fig. 15.10a). In the second case, as part of the DNA molecule touches the surface it is irreversibly pinned at that point and the DNA molecule effectively collapses on top of itself and can no longer move sideways (Fig. 15.10b), although local motions of regions of the DNA can be allowed [34]. The equilibration of DNA occurs on mica surfaces where the interaction is mediated via the Mg(II) cations. This is allowed because the Mg(II) ions are mobile and not bound into the cavities on the mica surfaces, enabling the DNA to move in the plane of the surface. For the divalent cations that are irreversibly bound into the cavities such

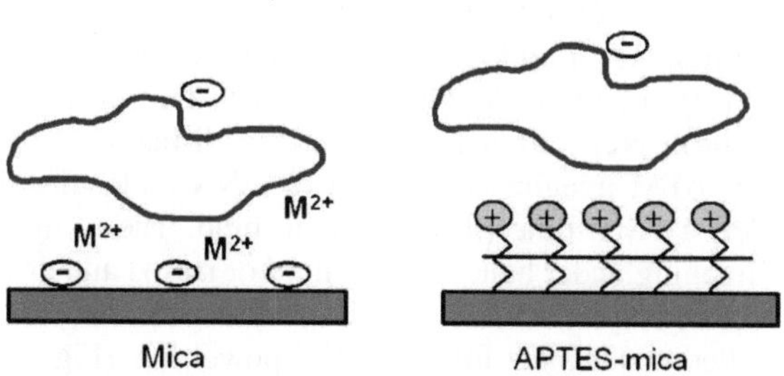

Fig. 15.9. Schematic representation of the binding mechanisms of dsDNA to mica and APTES-mica. The binding to mica is mediated by divalent metal cations, while the negatively charged phosphate backbone of the DNA interacts directly with positively charged amino groups on the APTES surface

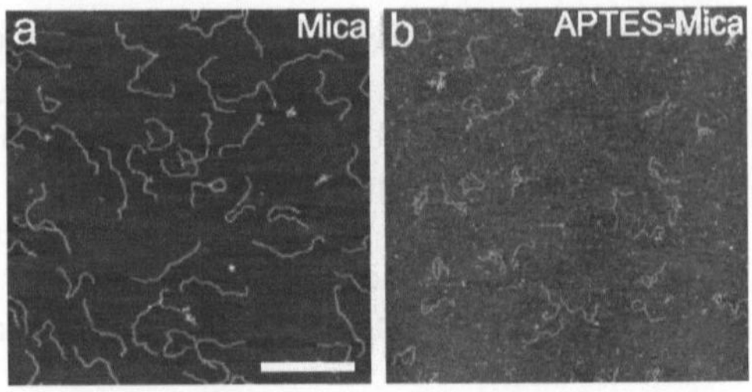

Fig. 15.10. Tapping-mode AFM images of a 1074 bp linear dsDNA fragment on dried surfaces imaged in ambient laboratory conditions. (**a**) The mica surface confers open conformations in 2D equilibrated state. (**b**) APTES-mica binds the DNA molecules in more compact conformations. *Scale bar* = 500 nm. Courtesy of Neal Crampton

as Ni(II) the DNA becomes trapped [35]. Generally, for an electrostatic interaction with the DNA where the positive charges are immobile the DNA will be kinetically trapped at least to some extent. Increasing the ionic strength of the buffer above the Ni(II)-treated mica can alleviate the trapped state and leads to molecules closer to the 2D surface equilibrated conformation [26]. Surfaces that induce trapping include amine-functionalized and plasma-treated mica, essentially those that retain a net positive charge.

Measurement of average end-to-end distance (by the shortest path) of a distribution of imaged DNA molecules indicates whether the DNA is equilibrated or trapped. The expected distance can be calculated using polymer-chain statistics models; the equilibrated case being equivalent to a random walk along the DNA backbone. For strong trapping of molecules, the average end-to-end distance tends towards a limit that demonstrates that the conformation of the DNA on the surface is a projection of the 3D solution conformation (random coil) onto the 2D surface [33]. It is important to know in which conformation the DNA molecules have bound to the surface when considering structural interactions between proteins and DNA, particularly measurement of DNA wrapping and DNA bend angles (see Sect. 15.5.3.4). As a full understanding of the mechanisms by which DNA binds to mica it then becomes possible to control the DNA binding through altering the buffer conditions [35, 36] and/or surface chemistry for real-time in situ experiments (see Sect. 15.5.3.5).

15.4.3 DNA Transport to Surfaces

Before DNA binds to the surface it has to arrive at the surface from the bulk solution. Counting the number of molecules present on the surface as time progresses indicates that for very dilute DNA solutions at least the transport of DNA to the surface in sessile-drop preparation is governed by diffusion [33]. Practically, to make a sample for AFM imaging, a solution of DNA molecules is pipetted onto the surface and left a given time (incubation) to bind. The sample is then rinsed and kept wet for imaging under bulk aqueous fluid or rinsed and dried for structural studies. The surface density of DNA molecules per unit time that are binding through a diffusion-controlled mechanism follows a $t^{1/2}$ power law (Fig. 15.11). Recently, we have demonstrated that there can be some competition in binding from heterogeneous solutions (Crampton, Bourbigot, Thomson; unpublished). This illustrates that for complex multicomponent systems, what is on the surface may not be a true representation of the distribution of structures in solution.

Fig. 15.11. Surface density of 1074 bp linear dsDNA molecules on mica deposited from a Mg(II) containing buffer as a function of time. The fit is to a $t^{1/2}$ dependence. Courtesy of Neal Crampton

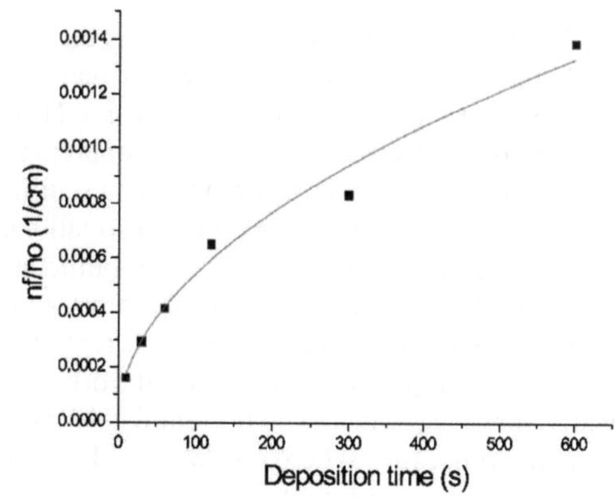

15.5 AFM of DNA Systems

This chapter concentrates on the use of AFM imaging to study DNA structure, interactions and dynamics. While the AFM has the ability to mechanically stretch DNA and thereby study DNA interactions through force measurements, the force-range capabilities of AFM are not always low enough. This is particularly true for protein molecular motors that operate using forces of about 50 pN or less, where optical or magnetic tweezers are usually the preferred biophysical tool. Many more imaging studies have been carried out to date with the AFM and this next section reviews current progress. We begin by discussing the relative merits of undertaking structural or dynamics investigations. Then we relay the history of the efforts made to make DNA imaging by AFM reproducible and to push the boundaries of resolution. In the final sections, we review the types of interactions with DNA that have been studied, from small organic molecules, such as drugs, to larger polymer systems that invoke DNA condensation, to biomolecular protein interactions. We conclude by highlighting the advances made in studying the dynamics of DNA systems in situ, using what is known as real-time or time-lapse imaging.

15.5.1 Static Imaging versus Dynamic Studies

Since the AFM can be operated in air and liquid there are two basic ways that biomolecular systems can be studied: imaging of static structures in air or imaging of dynamic processes under aqueous fluids. In practice, of course, there is nothing to stop one performing studies of static systems under bulk liquid, but biomolecules in buffers will be moving unless rigidly attached to the support surface. We have already discussed protocols for immobilization of molecules with varying degrees of complexity involved. The most straightforward way to immobilize DNA on mica is to deposit it from a Mg(II)-containing buffer, but imaging under Mg(II) buffer with typical conventional scanners (where imaging speeds are restricted to below about 4 Hz line rates) DNA molecules cannot be resolved. The interpretation is that the DNA is diffusing too quickly across the surface for the tip to be able to trace the backbone of the polymer. However, if one rinses the sample with excess water (to avoid salt precipitation) and dries it (using N_2 gas, for example) the DNA molecules are clearly resolved. The relative ease of preparing dried samples means that imaging of biomolecular systems in air make up a significant proportion of AFM experiments, and is a useful starting point for investigations of new systems, e.g. DNA-protein. More importantly, since mica is hydrophilic and a very thin (1 or 2 nm) water layer exists on the surface at typical laboratory humidities (RH – 30 to 40%) [37–41], biomolecules such as DNA appear to remain hydrated [42] and it is not clear that going to bulk liquid for pure structural work is necessarily always an advantage. Some water, however, needs to be present for dsDNA to maintain its correct structure. For dynamic processes, where one would like to study the interactions of molecules in situ, imaging under liquid is required, allowing the imaging buffer to be exchanged online to introduce other components, such as chemical fuel (e.g. ATP) or proteins or other cofactors.

15.5.2 The Race for Reproducible Imaging of Static DNA

Applications of AFM to biology began a few years after its conception and DNA was a particular favorite, no doubt because of its central importance to life processes. It can also be regarded as a model system for the following reasons. Large enough DNA fragments are thermodynamically stable, giving it a well-defined unchanging structure (in the same environment) allowing the morphology (i.e. semiflexible polymer) to be easily distinguished in topographical images. Also, as discussed already it has well-defined physical properties that allow a good understanding of how it binds to model surfaces and what influence those surfaces exhibit on its structure. As mentioned in Sect. 15.2, genetic maintenance, expression and regulation are controlled by DNA-dependent proteins, and proteins generally have globular morphology that is easily recognizable from DNA in micrographs, therefore standard AFM techniques are well suited to studying DNA-protein systems (see Sect. 15.5.3.4 below).

The first images of DNA alone using an SPM were obtained using the scanning tunneling microscope (STM) [43], in essence because the STM was developed before the AFM. It soon became obvious however, that the STM was not, in general, a particularly convenient tool to image DNA. STM is the original SPM developed that uses the quantum-mechanical tunneling current between a conducting tip and sample to map surface electronic states. The DNA, therefore, had to be deposited on a conducting substrate, often chosen to be HOPG because of its atomically flat terraces. However, as mentioned already, DNA does not bind very strongly to such a hydrophobic surface and early claims that atomic resolution of the double-helix was achieved under ultrahigh vacuum conditions [44] were hard to justify in the longer term, especially after steps on the HOPG surface were shown to be realistic mimics of a helical structure of similar dimensions [45]. The low hit rate of finding DNA on these surfaces, indicated that it either did not bind well enough and/or was swept away by the scanning tip without being imaged, supporting the idea that most, if not all, images were artefacts. Generally, STM has not been particularly successful at biological imaging, also because interpretation of the image contrast that is generated is difficult to relate to molecular structure. A novel method by Guckenburger's group demonstrated that STM can be performed on insulating mica surfaces that are sufficiently hydrated [46], where the best imaging conditions are 60% RH with bias voltages of 5 V or more [47]. This method was coined hydration STM and the conduction mechanism is thought to occur via protons in the water layer and the image contrast of DNA on the surface is generated through path-length differences from the STM tip to the sample electrode. Hydration STM has also not been taken up by the wider scientific community. This development came somewhat later, when tapping-mode AFM was already becoming an established technique, and early STM studies of biomolecules often still required metal coatings, which obviated the potential advantage of SPM over scanning electron microscopy (SEM) [48].

The first comparative study between the performance of AFM and STM to image uncoated DNA under water came from Lindsay et al. [49] in 1989. Lindsay's group preferred gold as the support surface and used an electrochemical deposition method to hold the DNA firmly in place. This improved the reproducibility of the STM method and allowed evaluation of the relatively crude first AFM images. The early AFM studies up to about 1993, all used contact-mode imaging since AC techniques such as tapping mode were yet to be developed. The shear forces generated in contact mode required good immobilization of DNA and new methods were developed, such as covalent immobilization to a fatty-acid monolayer, to try and image single bases in ssDNA under water [50]. Single-stranded DNA is particularly difficult to image because of its tendency, in common with RNA, to fold up into structures with more complex 3D topology. The attempted high-resolution studies have not really stood the test of time and most studies today involve imaging to obtain resolution of simply the dsDNA backbone alone, usually using tapping mode. This approach is well suited to studying the influence of proteins on gross DNA structure (see Sect. 15.5.3.4 below). The best resolution reliably achieved to date on DNA alone used contact-mode imaging of a continuous monolayer of DNA molecules on a cationic lipid bilayer support surface under aqueous fluid, where a regular repeating helical pitch of 3.4 nm was detected [51]. Despite the advances in sample preparation and DNA immobilization for contact-mode techniques [52], once tapping mode was demonstrated to be effective in most sample preparation cases and in particular in the study of proteins bound to DNA, contact mode fell out of favor, principally because of the complexity of sample preparation.

However, before we review progress from 1993 onwards (post development of tapping mode so to speak), it is worth taking an aside to talk more generally about imaging of DNA in contact mode. Particularly in light of recent developments in fast scan systems development for AFM, one of which is based on contact-mode principles with static deflection sensing [53] (see Outlook Sect. 15.6 below). Bustamante's group discovered that DNA could be routinely imaged using contact mode on mica provided that the sample was dried [18]. They appreciated that under these ambient conditions the capillary neck of water that forms between the tip and the sample is responsible for creating a large proportion of the shearing force that creates sample distortion and damage, and ultimately prevents imaging [54]. They also realized that treating the mica with Mg(II) or making it positively charged through glow-discharge would assist DNA binding [55]. Along with other groups they found that the imaging humidity had to be kept below a critical value to avoid high forces, caused by the capillary neck sucking the AFM tip down onto the surface. On mica, the humidity had to be below 35% RH [56,57] whereas on glow-discharged mica the humidity could be as high as 45% RH [18]. Recent work in our laboratory suggests that the structure of surface-bound DNA is not strongly compromised when imaged at RH as low as 7% (Crampton and Thomson; unpublished) leading to the possibility that useful structural studies could be carried out using contact-mode imaging in dry conditions or possibly even under moderate vacuums.

Imaging DNA using contact mode under aqueous fluid is difficult because although the capillary forces are no longer present the DNA must be bound strongly enough not to diffuse across the surface. Collaboration between the Hansma and Bustamante groups discovered that imaging under alcohols, such as propanol, gave reproducible imaging, because alcohols precipitate DNA [58]. The DNA was sufficiently strongly bound to the mica that plasmid DNA could be incised by the AFM tip by increasing the contact force in a small scan area. However, the DNA was previously dried onto Mg(II)-treated mica before imaging under propanol and therefore this approach did not hold much promise for studying dynamic processes by contact-mode AFM. This appeared to be the limit in the usefulness of contact-mode AFM to image DNA structure because AM imaging soon took over.

Contact-mode AFM of DNA under water could be achieved by first drying the sample and rehydrating it [59], which was also helped by functionalizing the surface with aminosilane [60]. The major breakthrough came, however, with the introduction of tapping mode that first allowed reproducible imaging of DNA in air [59] and then ultimately under bulk aqueous liquid [61]. This advance opened up the possibility to study a wide range of structural and dynamic problems where other molecules interacted with the DNA.

15.5.3 Applications of Tapping-Mode AFM to DNA Systems

This section reviews the applications that tapping-mode AFM has made to date on various DNA systems: DNA alone; DNA after interactions with small organic molecules, such as drugs; DNA condensation in the presence of multivalent cations and on surfaces; and DNA-protein systems. It finishes by highlighting achievements of using AFM to study DNA dynamical processes in situ.

15.5.3.1 DNA Structure

As mentioned already, typical AFM resolution limits imaging to the mapping of the overall morphology of the dsDNA backbone. Contour-length measurements from the images along the DNA backbone allow one to infer something of the internal structure of the DNA. For instance, careful measurements of several hundred molecules on air-dried samples, allowed Rivetti and Codeluppi [62] to infer that 15 to 30% of the length of linear DNA fragments was in the A-form (compared to the expected hydrated B-form) when it is bound to the mica surface and imaged under typical ambient laboratory conditions (e.g. 30 to 40% relative humidity). Other work indicates that the drying is at least partly responsible for the shortening, since imaging of a 1036 bp linear dsDNA fragment under buffer and air led to apparent contractions of 2% and 6%, respectively [63]. Lindsay's group have demonstrated that a full B to A transition can occur under aqueous solutions if the DNA is bound to a mica surface cooled to 2 °C [64]. This is not a sequence specific event and requires both the presense of the surface and the lower temperature. Introduction of ethanol into bulk aqueous fluid can also induce full contraction to the A-form, when the ethanol concentration goes over 25% [65]. Again the presence of the mica surface helps to induce this transition, since in bulk liquid it occurs at > 70 to 85% ethanol. As we will see later, accurate contour-length measurements are important in understanding structural interactions between proteins and DNA.

Different DNA structures can be distinguished in the AFM on the basis of morphology alone. To date, ssDNA, dsDNA and triple-stranded DNA [66] and quadraplex DNA [67] have been unambiguously identified on air-dried samples. Triple-stranded DNA arises through Hoogsteen base-pairing in DNA strands each containing only purines or pyrimidines that are present in a ratio of 2:1 or 1:2 [8], whereas quadruplex can only occur in long repeat sequences of guanine bases, where four poly G strands hydrogen bond to form a structure known as a G-wire. These have a channel running down the middle and are found in the telomeres of chromosomes [67]. Telomeres are the structures at the end of linear chromosomes and in higher eukaryotes are made up of multiple repeats of the sequence 5′-TTAGGG-3′. These are disposable and act as a buffer zone during DNA replication as they get shortened through repeated cell division. They are therefore important structures to study in relation to the aging process.

As described before, ssDNA folds up to form compact structures with globular submorphology [68], which makes interpretation of how other molecules interact with it difficult from AFM topographical images. Triple-stranded DNA is morphologically similar to dsDNA but appears twice as high [66]. Quadraplexes appear more rod-like due to their increased stiffness but the apparent heights varied depending on the formation conditions of the buffer, ion concentrations etc. [67].

For dsDNA, lengths of linear fragments that have been imaged vary from the very short (100 or 200 bp) [66] up to extremely long λ DNA from bacteriophages (48.5 kbp) [29]. The kinetics of binding will vary with length since transport to the surface is often diffusion limited (see Sect. 15.4.3), therefore longer fragments will require longer surface incubation times. Longer fragments may also be affected more by rinsing procedures. Indeed, there is a technique known as molecular combing in which long DNA fragments, bound at one end to hydrophobic silane surfaces, are overstretched on the surface due to the passing of the water meniscus [69,70]. To gain statistics on linear DNA samples, model templates around 1 kbp are used to optimize the efficiency of imaging enough molecules. Typically, this would yield around 50 equilibrated molecules per square micrometer. As well as linear DNA templates, circular DNA templates such as plasmids can also be readily imaged using AFM [18]. Comparison of DNA in various states of supercoiling can be also be studied using air-dried samples [71] or under aqueous fluid using aminosilane treated mica [72], while the dynamics of supercoiled molecules can also be probed [73,74] (see Sect. 15.5.3.4 below). A combined AFM and EM study showed that sample preparation, i.e. adsorption onto a surface, may affect the appearance of supercoiled DNA [75], particularly in AFM studies where the DNA is adsorbed on mica and prefers to relax into the 2D plane, interchanging writhe for helical twist. The three-dimensional information in AFM images allows one to assign the chirality of supercoiling [76], which may be advantageous when studying topoisomerase proteins. More unusual structures have also been studied mostly by Lyubchenko's group and these include hairpins [77], loops and hemiknots [78], three-way junctions [34], and cruciforms (4-way junctions) [79, 80], which have structural similarities to Holliday junctions. The three-way junctions and cruciforms were seen to be non-planar structures that exhibit dynamical movement [79]. It has been shown that the structural transition between different cruciform conformations can influence communication at a distant point on plasmid DNA mediated by superhelical stress [79]. AFM has also shown that p53 protein, a transcription factor involved in tumor supression, binds exclusively at a cruciform on supercoiled DNA templates [81]. Holliday junctions are formed in the cross-strand exchange process of genetic recombination, where two dsDNA molecules with homologous sequences exchange one of their strands. AFM has the ability to investigate the dynamics of these structures (see Sect. 15.5.3.5).

The tertiary structures of DNA described above are consequences of sequence-dependent effects, where competition of internal base-pairing between the single strands affects the structural outcome. More subtle sequence-dependent effects are also possible to pick up, such as flexibility differences and intrinsic curvature of the DNA double helix. AFM has determined the average bend angle of an A-tract sequence, which contains six A–T bps in a row, to be 13.5° [82]. Linear DNA molecules with palindromic symmetry have been used to demonstrate that the apparent DNA conformation arises from intrinsic curvature and local flexibility and such a symmetric construct allows the separation of these two effects [83]. Sequence-dependent flexibility was also confirmed to be strongly influenced by A–T rich sequences [84]. Further, polymer-chain statistical models have been developed for DNA sequences containing intrinsic curvature, discrete bends or different regions of persistence length (e.g. dsDNA and ssDNA) [82]. Minicircles of dsDNA as small as 168 and 126 bps have been imaged and these have been shown to contain kinked structures in the presence of Zn(II) ions [85]. The kinks are thought to arise because of the highly strained conformations of these minicircles.

The highest resolution on DNA that has been achieved in both contact mode and tapping mode is to resolve the helical pitch of the DNA (around 3.4 nm). Tapping mode achieved this infrequently on single isolated molecules under propanol [68], while contact mode on an ordered 2D layer of DNA bound to the surface of positively charged lipid bilayers allowed detection of the helical pitch over the whole molecular array [51]. Non-contact AFM has also been claimed to resolve the helical pitch in UHV conditions [86], but the importance of this result for understanding biology may be limited, since the DNA should be dehydrated.

It should be noted here that RNA is more difficult to image than DNA using the AFM and very few such studies are published. RNA is analogous to ssDNA, it is single-stranded and base-pairs with itself to fold up into complex 3D morphologies that are difficult to adsorb to surfaces reliably and are also inaccessible to the AFM tip. Some palindromic RNA sequences that form recognizable structures, such as stems and loops, have been successfully studied [87], but in any case RNA lies outside the scope of this chapter.

15.5.3.2
Small Organic Molecule Interactions

The binding of small organic ligands, such as drugs or intercalators, to DNA can be monitored indirectly through observation of DNA contour length changes. Molecules that intercalate between the base-pairs will increase the length of linear DNA, allowing discrimination between those that intercalate and those that bind by other means, e.g. minor groove binders [88]. Accurate length determination can give information on site occupancy, binding affinity, exclusion numbers and cooperativity. Distinctions between intercalators and minor groove binders can also be made using AFM in force-spectroscopy mode, on the basis of differing mechanical responses [89]. Intercalators all contain closed-ring structures that are planar and hydrophobic and interact well with the planar hydrophobic faces of the DNA nucleotides. These molecules insert between base-pairs, some specific to sequence [90], and some non-specifically, e.g. ethidium bromide that is used to fluorescently stain DNA for detection in agarose gel electrophoresis. As well as causing length increases on linear DNA molecules, intercalators can cause topological changes in closed circular molecules, due to build up of stress. For instance, AFM has been used to show that anthracycline intercalators, chemotherapeutic antitumor agents, induce positive superhelical coiling, such that negatively supercoiled plasmid DNA becomes relaxed and relaxed plasmid DNA becomes positively supercoiled [91]. The effect on DNA structure of a number of other intercalators, used as anticancer drugs, has been studied using AFM [92–94].

The tertiary structure of plasmid DNA (how the DNA helical axis is arranged spatially), i.e. its supercoiling, has been studied in detail for various stoichiometries of binding of ethidium bromide [71]. It was found that increasing amounts of the intercalator first induced toroidal loops in the DNA before plectonemically supercoiled structures, where the DNA helix is tightly twisted around itself, were nucleated. At the highest concentrations of ethidium bromide studied, where 8 intercalator molecules were available per bp, the plasmid DNA was in a fully plectonemic structure.

Some intercalators, such as psoralen, have been used in a novel way to immobilize DNA to surfaces. Psoralen will covalently crosslink to pyrimidine bases upon radiation with soft UV light (320–400 nm). Shlyakhtenko et al. [95] used a derivatized psoralen that could then crosslink to the amine groups on an APTES mica surface. This was seen as a way of periodically attaching the DNA to a surface covalently such that dynamics of segments between crosslinks could be investigated by AFM.

Higher concentrations of intercalators, such as YOYO dyes can cause more severe tertiary and quaternary effects, such as the formation of toroids under acidic pH conditions [96]. YOYO carries a +4 charge that presumably initiates this condensation process despite YOYO containing two sets of ring structures that intercalate between DNA bases. We will move on in the next section to consider what structural changes occur when DNA interacts with larger polymeric molecules (polyelectrolytes) and more highly charged small-molecule species. Relatively low concentrations of highly positively charged ionic species will cause various degrees of DNA condensation.

15.5.3.3
DNA Condensation

Condensation is the process whereby either individual DNA molecules self-aggregate or a number of DNA molecules self-assemble to form structures with well-defined morphologies and sizes, such as globular aggregates, rods and toroids. As we have said, DNA is a negatively charged polyelectrolyte that will only interact with itself if this charge is compensated for or neutralized. The structure that the condensed DNA adopts is determined through interplay between the DNA mechanical properties and the forces involved in the condensation; stiffer polyelectrolytes will form toroids, whereas more flexible ones form disordered globules [97]. The forces of attraction between two neighboring DNA segments arise due to correlation of fluctuations of condensed counterions on the DNA surfaces [98–100]. The helical structure of the DNA plays an important role in determining the outcome of the condensation process [100], while the size limits for the toroids are determined by both kinetic and thermodynamic factors within a nucleation and growth framework [101].

To induce condensation in solution the DNA must be incubated with multivalent cations with a charge of greater than +2. Relatively high concentrations of divalent metal cations, such as Mg(II), can begin the process of condensation. Increasing the valency of the cation will cause condensation at lower co-ion concentrations. The evidence from many different AFM studies seems to imply that this process occurs initially through DNA looping, where two distant points on the DNA helix contact each other. As many of these contacts are made, the aggregates take on a flower-like appearance and at higher counterion concentrations eventually toroidal structures of well-defined size are formed [102]. Examples of multivalent cations that can induce DNA condensation include: metal cations [103] and their complexes [104]; positively charged proteins, such as protamine [105]; polyamines, such as spermine and spermidine [106–108]; positively charged polypeptides such as poly L-lysine [109, 110]; polysaccharides [111]; cationic silanes [112, 113]; or positively charged synthetic polymers [114]. For a review of recent studies on the formation of DNA toroidal condensates see [101]. AFM is particularly suited to studying DNA condensation since it involves larger hierarchical structure where the limits of the microscope resolution are not being tested so strongly.

There is a great practical interest in DNA condensation techniques for gene-therapy applications. To replace a gene within a cell, the therapeutic DNA has to get through the outer cell membrane, across the cell, through the nuclear envelope and inside the nucleus. Notwithstanding the practical issues of delivering the DNA to a target tissue within the body, the DNA must in any case be condensed to prevent it being degraded enzymatically or chemically by the body before it reaches the target. AFM has been used to characterize a variety of putative gene-therapy vectors, where the DNA has been condensed with polyethylimines [115], methacrylates [116], cationic liposomes [117], poly L-lysine [118] or a protein/lipid system that can condense bacterial artificial chromosomes about 150 kbp large enabling them to transfect cells [119]. There are a whole variety of positively charged polymeric systems that can be used, one of the most popular being lipid-based systems, because of their biocompatibility. A recent paper reviews the use of SPM and particularly AFM for studying their structure [120].

We turn now to surface-directed condensation of DNA. On surfaces where there are mobile multivalent cations it is possible for DNA to condense on the surface. This has been a prevalent feature of surfaces modified with aminosilanes used to try and bind DNA strongly. Fang and Hoh [121] discovered that DNA toroids and rods formed on silicon surfaces treated with aminosilanes and attributed this to non-polymerized silane molecules remaining on the surface. The surface induced condensation was eliminated after treatment of the silane surfaces with vacuum storage or baking at 100 °C for 2 h. The majority of studies that use aminosilanes are made on mica from vapor phase following the protocol of Lyubchenko and coworkers [29, 60, 122]. Recent work has shown that the relative humidity in the deposition chamber must be kept below 25% RH to prevent surface-induced DNA condensation by APTES [30]. In this study, baking at 150 °C for 2 h was ineffectual at eliminating the condensation on surfaces made in RH > 25% because the silane layer underwent a dramatic structural reorganization.

Interestingly, in the study of imaging DNA under various concentrations of ethanol to induce a B-to-A transition, it was also discovered that high concentrations of ethanol (> 20%) in aqueous media induce DNA condensation, including toroid formation [65]. This observation is consistent with recent theoretical work that describes DNA condensation in terms of poor solvents rather than counterion correlations [123]. As we learned earlier, other alcohols such as propanol were used to increase the adhesion of DNA to mica surfaces to obtain reproducible contact-mode imaging [58]. DNA is not soluble in alcohols and they are used routinely to precipitate it from solution.

Binding of DNA onto the surface of positively charged supported lipid bilayers has also been described as a 2D condensation process [124, 125]. In this description, however, the DNA strands are packed closely together (the interhelical distance being 4 to 6 nm), but aligned in a 2D monolayer array across the top of the bilayer. As we have seen, these close-packed arrays have allowed high-resolution imaging of DNA using contact mode, detecting the helical repeat [51], in an analogous way that high-resolution images of 2D protein crystals can be made.

As we shall see in the next section, DNA is usually condensed inside the nucleus of the cell in conjunction with a range of proteins, as a way of storing and protecting the genome of the organism.

15.5.3.4
DNA-Protein Interactions

The AFM is particularly adept at studying DNA-protein systems, since proteins that interact with DNA are globular and compact, and therefore can be distinguished from the DNA polymer on the basis of morphology alone. Figure 15.12 shows an image of a single DNA gyrase protein molecule bound to a 1070 bp linear dsDNA molecule. The images can be used to visualize how a single protein modifies local DNA structure, either through a change in bend angle [126] or through wrapping of the DNA around the protein. The ability of the AFM to accurately measure the DNA contour length can give information about wrapping of DNA around bacterial molecular motors such as *E. coli* RNA polymerase [126] and *E. coli* DNA gyrase that is a structural requisite for these enzymes' work [42]. Where two proteins bind

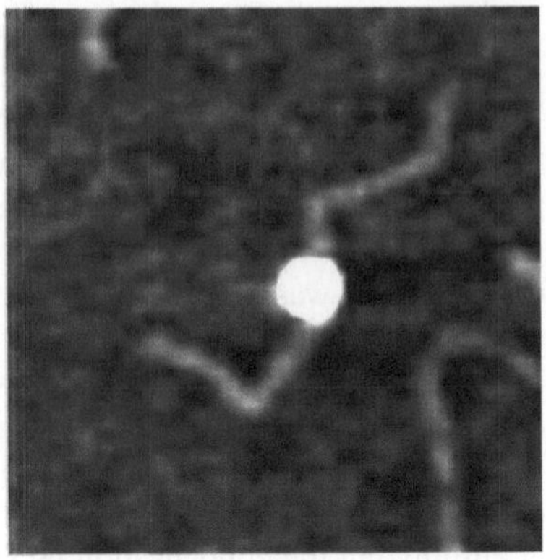

Fig. 15.12. Tapping-mode AFM image of a complex of a single *E. coli* DNA gyrase molecule bound to a linear dsDNA template 1070 bp long. The sample has been dried and imaged under ambient laboratory conditions. The DNA gyrase protein is the globular feature bound to the center of the DNA molecule. Image size = 250 × 250 nm. (Data: Thomson NH, Heddle JG, Maxwell A)

distantly on a single DNA template their interactions can often be visualized through the formation of loops in the DNA. This situation may occur, for example, during the regulation of DNA transcription through the binding of transcription factors to RNA polymerase, when the polymerase is at the promoter region [127, 128].

Complexes of RNA polymerase on linear DNA templates have demonstrated that this protein can induce a bend in the DNA and requires wrapping of the DNA around itself, to a greater or lesser extent, at various stages in the transcription cycle [129]. Bend-angle distributions tend to be very broad, and therefore are not all that informative about the local DNA deformation, mainly because the finite size of the AFM tip means that there is a "shadow" around the protein where the tip cannot contact the surface to profile the path of the DNA backbone close to the protein. However, it has been shown that the DNA bending of two different types of methyltransferases can be discriminated and is associated with their two different modes of action [130]. Wrapping of DNA around DNA gyrase, a topoisomerase that is unique in being able to introduce supercoils into dsDNA, depends on the conformational state the protein is in during its enzymatic cycle. Addition of a nonhydrolyzable ATP analog, ADPNP, was shown to remove DNA wrapping consistent with the enzyme performing one cycle of work [42]. The sizes of DNA that wraps around *E. coli* RNA polymerase and *E. coli* DNA gyrase are comparable, about 100 bp, but slightly larger wraps have been identified, such as the 160 bp wrap around histones in reconstitued chromatin [131]. In this case, the wrap is thought to involve two whole turns, whereas for RNA polymerase and DNA gyrase it is about one. Chromatin, the basic DNA-protein complex that makes up chromosomes, consists of DNA wound round a series of histone protein complexes. Histones are octameric-containing dimers of four different subclasses of histone proteins. The basic structural unit of DNA wrapped around the histone is known as a nucleosome, which can be studied in its native form by extraction from cells or can be reconstituted in vitro. Nucleosome structure has been shown to resemble "beads on a string" [132]. AFM has been used quite extensively to study chromatin to investigate how and what controls nucleosome positioning on the DNA [133–136].

There are numerous other investigations of DNA-protein systems in the literature, so many, that they cannot all be reviewed here. This chapter aims to give a flavor of what can be achieved with the AFM in this research area. We have seen, therefore, that the AFM has a role in performing single-molecule experiments to determine structural and mechanistic aspects of DNA-dependent molecular motors that cannot be accessed by ensemble techniques. To date it has been used mostly to study interactions between DNA and single proteins, but eventually it should be possible to apply AFM in a bottom-up manner to study more complex systems, such as transcription roadblocks, replication forks, convergent transcription and other important aspects of genetic regulation. It is very good at visualizing center-of-mass motions of proteins relative to linear DNA templates and as such the next generation of experiments are likely to feature systems with two or more proteins acting on a single DNA template.

15.5.3.5
Studies of DNA Dynamic Systems

As already alluded to, the study of DNA processes in situ within the AFM are the most difficult of all DNA systems to study. The main reasons for this are the difficulties of achieving suitable surface binding coupled with the relatively slow scan speeds currently available in most AFM technology. To study DNA dynamics there is a choice between adjusting buffer conditions on mica to have the DNA transiently bound or using a positively charged surfaces at neutral pH, such as aminosilanes, so that the imaging buffer can be varied to some extent without affecting the DNA surface-binding characteristics. AFM studies of this kind are usually called "real-time" or "time-lapse" imaging. Time-lapse is perhaps the most realistic description at present because the images are built up line-by-line in a raster scan pattern, and with the typical scan rates available, the time between the same point in successive images can be up to a minute or more (depending on the image size). As faster scan technologies are developed (see Sect. 15.6) then the term "real-time" may become more appropriate.

After the realization that divalent metal cations could enhance DNA adhesion to mica to the extent that samples could be dried for air imaging [24], this procedure was taken forward for imaging under bulk aqueous fluid [36]. Transition-metal cations had been shown to be most effective for increasing DNA-mica adhesion, therefore, Zn(II) was chosen in a gravity-fed flowthrough system where the buffer is continuously flowing through the AFM imaging cell and can be switched between buffers with and without Zn(II). In this manner, the DNA could be reversibly bound and released from the mica surface on the timescale of tens of seconds. In hindsight, this was probably not the best divalent cation to choose, since Zn(II) precipitates as zinc hydroxide around pH 8 and it was clear that the DNA adhesion was in part due to a precipitate salt layer forming on the surface. More recently, Ni(II) has become the preferred co-ion for DNA adsorption under liquid and it has been demonstrated that mica can be pretreated with Ni(II) solutions to enhance DNA binding [137]. This is because the Ni(II) ions are irreversibly bound into the surface sites on mica that allows experiments to be performed without the Ni(II) present in the imaging solution. This is important for biological processes, since transition-metal cations can poison the action of many enzymes. For example, RNA polymerase contains two Mg(II) ions that are required for its mechanistic action, which can become displaced by transition-metal cations.

Figure 15.13 shows a sequence of tapping-mode AFM images taken under a Mg(II)-containing buffer on mica pretreated with nickel chloride. One can see that there are portions of the molecules that are not firmly bound to the mica surface and move underneath the scanning tip. The molecules remain in the same approximate conformations for timescales of minutes but there are local motions of the DNA that occur from one frame to the next, as highlighted with the arrows. In one instance, there is a concerted motion where the movement of one DNA molecule dramatically affects the conformation of a neighboring one (see white arrows, Figs. 15.13b,h). It is apparent that the motions of the DNA molecules are not strongly influenced by the scanning tip. The center-of-mass of DNA molecules can be tracked under similar conditions where they exhibit diffusion in the 2D plane of the surface to show that the AFM tip in tapping mode does not influence the motion [138]. As we will see for the enzyme studies below, is it crucial that the DNA can equilibrate into the 2D surface plane of mica but still retain some motion. Recent theoretical and experimental work that has already been discussed has gone the furthest in understanding the amount of control an experimenter can impose of the dynamics of the DNA on mica surfaces [26, 35].

We have already seen that while positively charged surfaces can trap DNA, motions of local DNA structures such as three- and four-way junctions can be observed on the aminosilane surfaces [34, 79]. Importantly, whole-molecule center-of-mass motions can occur if the buffer conditions are suitable. Despite the previous implication that these surfaces always trap DNA molecules, recent work in the author's laboratory shows that conditions that look like 2D surface equilibration can occur. Figure 15.14 shows a sample of linear dsDNA 1070 bp long incubated with the protein DNA gyrase, which is a bacterial enzyme that controls DNA topology. Both images show the same area of the sample before and after addition of a nonhydrolyzable analog of ATP, known as ADPNP. One can see that during the time interval between the two images (a few minutes) that the gyrase molecules (globular features) have stayed stationary, whilst the DNA molecules have all moved across the surface, showing that even on surfaces where trapping might be expected, under these

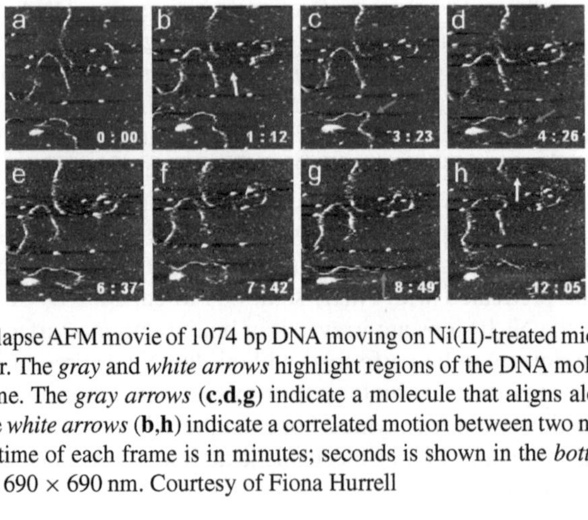

Fig. 15.13. Time-lapse AFM movie of 1074 bp DNA moving on Ni(II)-treated mica under a Mg(II) containing buffer. The *gray* and *white arrows* highlight regions of the DNA molecules that move position over time. The *gray arrows* (**c**,**d**,**g**) indicate a molecule that aligns along the fast scan direction and the *white arrows* (**b**,**h**) indicate a correlated motion between two neighboring DNA molecules. The time of each frame is in minutes; seconds is shown in the *bottom right corner*. The scan size = 690 × 690 nm. Courtesy of Fiona Hurrell

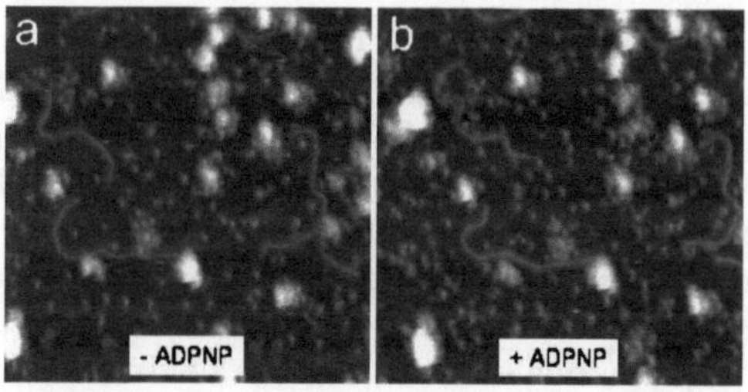

Fig. 15.14. Tapping-mode AFM images under buffer of a sample of DNA gyrase and linear dsDNA 1070 bp long. The same area of the sample was imaged before (**a**) and after (**b**) addition of ADPNP through direct exchange of buffers during imaging. The offset between the images is due to instrumental drift. While the proteins appear unchanged during the experiment the DNA molecules have moved, either as local movements or center-of-mass movement. (Data: Thomson NH, Heddle JG, Maxwell A)

buffer conditions DNA can still diffuse in the surface plane. This result illustrates the complexity of understanding DNA binding and dynamics on surfaces such as mica and aminosilanes. Aminosilane surfaces are useful for tightly binding the DNA in a kinetically trapped configuration and observing local chain dynamics of the DNA. For example, APTES-mica was used on which to observe the mobility of three-way junctions [34], while branch migration behavior of Holliday junctions was observed on APS-mica [139].

The first AFM images of motion of linear dsDNA were obtained on Ni(II)-pretreated mica and were combined with studies of the effect of nuclease DNase I, where even DNA that was well bound to the surface experienced degradation [137]. This was seen as a loss of regions of the linear DNA from the surface while the rest of the molecule remained bound. Unambiguous identification of DNase I was difficult due to its motion in solution. Recent time-lapse studies have followed the effects of DNase I on the degradation of DNA complexes condensed with polyamidoamino dendrimers [140]. These complexes are potential candidates for gene therapy and it is therefore interesting to investigate how the condensation affords protection against DNA degradation by nucleases. The image sequences revealed that the degree of protection is correlated to the amount of condensation. In some cases, the DNase I proteins can be observed diffusing across the surface encountering a DNA toroid and causing a large structural change, uncondensing the DNA.

One of the most studied proteins using these techniques is RNA polymerase (RNAP), that mediates the transcription of the genetic information from dsDNA into RNA. This is a central molecular process to gene expression and therefore generates great research efforts to understand mechanistically how this enzyme works. Early AFM studies under bulk aqueous solutions followed directly the assembly of *E. coli* RNAP onto DNA templates [141]. A few years later with great collaborative efforts from the Bustamante and Hansma laboratories, the first AFM "movies" of transcription at the single-molecule level were captured [12, 138]. In these experiments, the complexes of RNAP and DNA were formed in vitro with the RNAP in a stalled state on the DNA. These complexes were deposited in the AFM sample chamber under continuously flowing buffer that was driven by gravity to avoid vibrational interference to the AFM [36]. Switching the buffer to one containing the nucleoside triphosphates that make up the RNA transcript reinitiates transcription during continuous image acquisition. In these time-lapse "movies" the RNA transcript cannot be imaged, presumably because it is too small and too mobile, but rather one observes the relative motion between the protein and the DNA. In these cases, the RNAP is non-specifically bound to the mica surface and the DNA is translocated by the enzyme (see Fig. 15.15). Many experiments are performed to capture a few transcription events, because the interactions of the enzymes with the support surface can inactivate a large proportion of them. One can see from these experiments that the AFM is adept at measuring the center-of-mass motions of the protein relative to DNA but internal protein motions are difficult to detect with current technologies. Further experiments showed how the RNAP finds its promoter sequence, the portion of the DNA from where it starts transcribing, through a 1D diffusional search mechanism [138].

A handful of studies using similar approaches have been published on other DNA enzymes. In a dedicated home-built AFM the association, dissociation and movement of photolyase over linear dsDNA molecules was observed [142]. In another study, snap shots of the translocation and cleavage of DNA by EcoK1 endonuclease were obtained [143]. The interactions of p53 have also been studied by time-lapse AFM and indicated that the p53 could directly bind to a target sequence or bind non-specifically and search by 1D diffusional search along the DNA [144]. Most of these studies are carried out on linear fragments about 1 kbp long, where the motion of the protein relative to the DNA can be readily observed, but some have been able

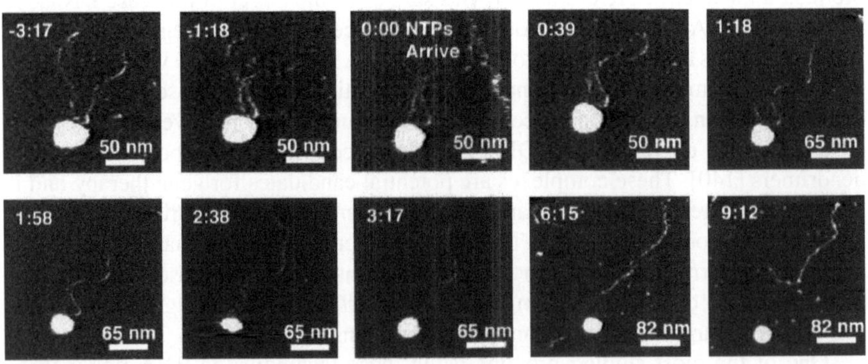

Fig. 15.15. Time-lapse AFM movie of DNA transcription by *E. coli* RNA polymerase (RNAP) at single-molecule resolution. The time of each frame is shown in minutes: seconds in the *top left corner*. The RNAP is bound to the center of a linear dsDNA template 1047 bp long. At time 0:00 the nucleoside triphosphates (NTPs) that fuel the enzyme and are building blocks for the RNA polymer that the RNAP fabricates arrive into the AFM sample chamber. As time progresses the right-hand arm of the DNA gets longer and the left arm concomitantly gets shorter, as the RNAP translocates the DNA. Eventually the RNAP releases the DNA as it transcribes off the end of the template, and the buffer is changed to a Zn(II)-containing buffer that firmly adheres the DNA to the mica. The penultimate image shows a small feature on the opposite side of the RNAP to the DNA ejection from the enzyme. This could be the RNA transcript that is formed, but better time resolution is required. Reproduced with permission from *Biochemistry* (1997) 36(3):461–468 [12]. Copyright 1997 American Chemical Society

to use supercoiled plasmids [143]. The few studies published to date highlight the difficult nature of these experiments and a lack of control of the DNA adsorption to the surface. They all require the DNA to be transiently bound to the surface, sufficiently immobilized to allow the AFM tip to track the DNA position but not fixed, at least in the surface plane in order to allow enzymes to operate on the DNA. The last section covers developments in faster scan technologies that should allow these type of experiments to become more widespread and the data gained from them to be more informative.

15.6 Outlook

It is hoped that this chapter demonstrates that the AFM has an important role to play in the research of DNA structure and interactions. In particular, its ability to study single-molecule complexes allows it to assess the heterogeneity of systems, albeit at a resolution much lower than atomic. It is expected that the bottom-up approach to increasing complexity will enable the AFM to play an even bigger role as a biophysical tool for DNA research in the future. As the macromolecular systems studied become larger and more complex, ensemble techniques such as NMR and X-ray crystallography may become increasingly challenging. AFM technology itself is still immature and there are many technical developments that can and need to be made before its full promise can be fulfilled.

Far and away the most important development necessary is to increase scan speeds of the tip to levels where image-acquisition times become comparable with enzyme turnover rates. While it has been demonstrated that AFM is a useful tool for structural studies of DNA systems, especially DNA-protein complexes, its particular advantage in terms of a high-resolution microscopy is its ability to image under aqueous solutions. Progress on increased scan speeds is already under way. There are two basic ways that this problem has been approached, although both rely on miniaturizing further one or more of the AFM components. Since the AFM is essentially a mechanical microscope, instead of a microscope that relies on the focusing of radiation (either optical or electron waves), miniaturizing components will allow them to be driven at higher speeds, since their resonant frequencies increase. Miniaturizing cantilevers has allowed images of DNA under aqueous buffer at about 8 nm pixel size to be taken every 1.7 s [145], while miniaturizing the scanners also has allowed image rates of 80 ms at a pixel size of 2.4 nm allowing the conformational switching of myosin V molecular motors to be followed in situ [146]. These approaches continue to use an amplitude-modulation imaging mode (e.g. tapping mode) as the mechanism of interacting the tip with the surface. Another approach has gone back to a type of contact mode where the deflection of the cantilever is monitored while the tip is scanned over the surface. It implements a passive mechanical feedback loop enabling it to collect images of continuous polymer films in under 20 ms [53]. This technique commercialized as videoAFM^{TM} has achieved much greater scan rates to date, but reliable operation under liquid has yet to be demonstrated. This method requires a direct force to be applied through the tip onto the surface so that the tip reliably tracks the surface.

It is still unclear how reliable this approach will be to biological systems, particularly single molecules, because the shearing interactions are not well understood at these scan rates. It has a potential advantage over the small cantilevers approach because it can image larger areas. Since AFM cantilevers need to be tilted for the tip to address the surface without the cantilever snagging, smaller cantilevers require flatter samples or are restricted to smaller areas on samples with larger topographical variations.

If a kind of real-time single-molecule enzymology is to be realized with the AFM, as well as the fast-scan approaches to the instrumentation, greater understanding and control of DNA adsorption and behavior at surfaces will be required. The recent work on counterion correlation forces between DNA and mica has allowed a greater degree of control of DNA conformations under bulk aqueous fluid. More work needs to be done on control of DNA orientation and positioning on surfaces and may benefit from techniques such as dip pen nanolithography (DPN) [147] and dielectrophoresis [148]. DPN can chemically pattern surfaces locally on nanometer scales to spatically control DNA adsorption, while microcontact printing can chemically pattern surfaces over larger areas. Molecular combing can align DNA on surfaces, however, it requires a meniscus being drawn across the surface, effectively drying it and causing irreversible binding of the DNA. This approach is unlikely to enable dynamic studies to be performed upon sample rehydration.

Gentler imaging modes in force microscopy will also assist in the goals of imaging the delicate interactions between proteins and DNA. For example, phase imaging with tapping-mode AFM under buffer can be used to follow processes at low interaction forces [149]. Using a very high setpoint where the tip is a long way from the surface, the feedback loop cannot follow DNA molecules on the mica to produce a faithful topographical image, but the phase signal is still sensitive to the motion of the molecules. Other force microscopy setups such as shear force microscopy (ShFM), also known as tranverse dynamic force microscopy (TDFM), might also be an interesting avenue to explore [150]. These instruments are interesting propects because they avoid the jump-to-contact problem of cantilevers in AFM, since they use a very stiff vertical probe and monitor the tranverse motions of the probe.

As more complex DNA-protein interactions are studied, i.e. where two or more proteins interact at the same binding site on a DNA molecule, the ability of the AFM to distinguish different proteins on a morphological basis alone will become compromised. The AFM will require molecular-recognition capability or at the least a chemical recognition capability. Chemical force microscopy (CFM), through controlled functionalization of the AFM tip, has been demonstrated on crystal surfaces [21], where chirality of surface groups can be determined, and on more complex protein–biomineral systems [151]. If the chemistry on the tip is defined and known, then knowledge of the operating environment (e.g. pH and ionic strength) allows the user to infer the chemistry on the surface. The signal for imaging can be either an adhesion event in an AC imaging technique or change in the lateral force signal (i.e. cantilever torsion behavior). The functionalization of AFM probes has been taken further, to link biological molecules such as antibodies to the tip. This approach has been called molecular recognition force microscopy (MRFM), and allows mapping of the interaction sites between the antibody on the tip and its antigen sites on the sample [22, 152]. Since antibodies can be produced against a wide variety different biological targets this approach shows a lot of promise. Generally, the resolution of CFM and particularly MRFM will be lower than AFM since the tip has molecules bound to it. However, Hinterdorfer's group have devised a method to simultaneously extract a topographic and recognition image using an AC imaging mode that can discriminate individual binding sites [152]. Development of CFM and MRFM for DNA systems will allow complexes on rougher support surfaces and with more complex topology and larger topographical changes to be investigated. Alternatively or additionally, one might also envisage combining single-molecule optical techniques with AFM to understand DNA-protein interactions in more detail. A useful surface technique is total internal reflection fluorescence microsopy (TIRFM) that allows fluorescently labeled biomolecules at surfaces to be identified and tracked.

As well as imaging, the AFM can manipulate DNA molecules, stretch them and measure inter- and intramolecular forces [153]. The limited amount of work that has been undertaken to date lies outside the scope of this chapter, but force measurements and sensor technology based on AFM cantilevers [154] has the ability to probe DNA interactions. These approaches should allow binding assays and some sequencing applications to be developed.

In conclusion, the AFM has already proved to be a versatile tool for studying the structure of DNA and its interactions with other molecules, while in situ dynamic experiments are beginning to gain more insight into the local behavior of DNA systems through direct imaging under bulk aqueous fluid. As AFM technology develops further one can see from this review of work to date that there is great potential for the range of applications to DNA research to broaden further and it is expected that the AFM will play a more important role in the areas of structural molecular biology and bionanotechnology in the future.

Acknowledgements. The author wishes to thank former researchers from his group, in particular Neal Crampton and Fiona Hurrell for their contributions to the work. I also extend thanks to current collaborators Jonathan Heddle and Tony Maxwell on the DNA gyrase project. I appreciate the critical evaluation and review of the chapter by Bill Bonass and Sarah Harris, who both stimulated insightful discussions. I also thank Sarah for providing molecular models of DNA structures in Fig. 15.2. Work from my group has been supported by the EPRSC, BBSRC and the University of Leeds.

References

1. Feynman RP (1959) There's plenty of room at the bottom. Annual Meeting of the American Physical Society, http://www.zyvex.com/nanotech/feynman.html
2. Piner RD, Zhu J, Xu F, Hong SH, Mirkin CA (1999) Science 283:661
3. Garcia R, Martinez RV, Martinez J (2006) Chem Soc Rev 35:29
4. Ulman A (1996) Chem Rev 96:1533
5. Giessibl FJ (2003) Rev Mod Phys 75:949
6. Hembacher S, Giessibl FJ, Mannhart J (2004) Science 305:380
7. Dvorak JA (2003) Methods. Academic Press, Elsevier, Science 29:86

8. Bates AD, Maxwell A (2005) DNA Topology, 2nd edn. Oxford University Press, Oxford
9. Neidle S (2002) Nucleic Acid Structure and Recognition, Oxford University Press, Oxford
10. Watson JD, Crick FHC (1953) Nature 171:737
11. Harris SA (2004) Contemp Phys 45:11
12. Kasas S, Thomson NH, Smith BL, Hansma HG, Zhu XS, Guthold M, Bustamante C, Kool ET, Kashlev M, Hansma PK (1997) Biochemistry 36:461
13. Alberts B, Bray D, Lewis J, Raff M, Roberts K, Watson JD (1994) Molecular Biology of the Cell, 3rd edn. Garland Publishing Inc., New York
14. Binnig G, Quate CF, Gerber C (1986) Phys Rev Lett 56:930
15. Gould SAC, Drake B, Prater CB, Weisenhorn AL, Manne S, Kelderman GL, Butt H-J, Hansma H, Hansma PK, Magonov S, Cantow HJ (1990) Ultramicroscopy 33:93
16. Garcia R, Perez R (2002) Surf Sci Rep 47:197
17. Zhong Q, Inniss D, Kjoller K, Elings VB (1993) Surf Sci Lett 290:L888
18. Bustamante C, Vesenka J, Tang CL, Rees W, Guthold M, Keller R (1992) Biochemistry 31:22
19. Zitzler L, Herminghaus S, Mugele F (2002) Phys Rev B 66:155436
20. Muller DJ, Fotiadis D, Scheuring S, Muller SA, Engel A (1999) Biophys J 76:1101
21. McKendry R, Theoclitou ME, Rayment T, Abell C (1998) Nature 391:566
22. Riener CK, Stroh CM, Ebner A, Klampfl C, Gall AA, Romanin C, Lyubchenko YL, Hinterdorfer P, Gruber HJ (2003) Anal Chim Acta 479:59
23. Frank-Kamenetskii MD (1997) Phys Rep 288:13
24. Hansma HG, Laney DE (1996) Biophys J 70:1933
25. Shao ZF, Mou J, Czajkowsky DM, Yang J, Yuan JY (1996) Adv Phys 45:1
26. Pastre D, Pietrement O, Fusil P, Landousy F, Jeusset J, David MO, Hamon C, Le Cam E, Zozime A (2003) Biophys J 85:2507
27. Tanigawa M, Okada T (1998) Anal Chim Acta 365:19
28. Bussiek M, Mucke N, Langowski J (2003) Nucleic Acids Res 31:e137
29. Lyubchenko Y, Shlyakhtenko L, Harrington R, Oden P, Lindsay S (1993) Proc Natl Acad Sci USA 90:2137
30. Crampton N, Bonass WA, Kirkham J, Thomson NH (2005) Langmuir 21:7884
31. Kim S, Christenson HK, Curry JE (2002) Langmuir 18:2125
32. Shlyakhtenko LS, Gall AA, Filonov A, Cerovac Z, Lushnikov A, Lyubchenko YL (2003) Ultramicroscopy 97:279
33. Rivetti C, Guthold M, Bustamante C (1996) J Mol Biol 264:919
34. Shlyakhtenko LS, Potaman VN, Sinden RR, Gall AA, Lyubchenko YL (2000) Nucleic Acids Res 28:3472
35. Pietrement O, Pastre D, Fusil S, Jeusset J, David MO, Landousy F, Hamon L, Zozime A, Le Cam E (2003) Langmuir 19:2536
36. Thomson NH, Kasas S, Smith B, Hansma HG, Hansma PK (1996) Langmuir 12:5905
37. Beaglehole D, Radlinska EZ, Ninham BW, Christenson HK (1991) Phys Rev Lett 66:2084
38. Beaglehole D, Christenson HK (1992) J Phys Chem 96:3395
39. Kohonen MM, Christenson HK (2000) Langmuir 16:7285
40. Xu L, Lio A, Hu J, Ogletree DF, Salmeron M (1998) J Phys Chem B 102:540
41. Hu J, Xiao XD, Ogletree DF, Salmeron M (1995) Surf Sci 344:221
42. Heddle JG, Mitelheiser S, Maxwell A, Thomson NH (2004) J Mol Biol 337:597
43. Binnig G, Rohrer H (1982) Helv Phys Acta 55:726
44. Driscoll RJ, Youngquist MG, Baldeschwieler JD (1990) Nature 346:294
45. Clemmer CR, Beebe TP (1991) Science 251:640
46. Guckenberger R, Heim M, Cevc G, Knapp HF, Wiegrabe W, Hillebrand A (1994) Science 266:1538
47. Heim M, Steigerwald R, Guckenberger R (1997) J Struct Biol 119:212

48. Keller RW, Dunlap DD, Bustamante C, Keller DJ, Garcia RG, Gray C, Maestre MF (1990) J Vac Sci Technol A 8:706
49. Lindsay SM, Nagahara LA, Thundat T, Knipping U, Rill RL, Drake B, Prater CB, Weisenhorn AL, Gould SA, Hansma PK (1989) J Biomol Struct Dyn 7:279
50. Hansma HG, Weisenhorn AL, Gould SAC, Sinsheimer RL, Gaub HE, Stucky GD, Zaremba CM, Hansma PK (1991) J Vac Sci Technol B 9:1282
51. Mou JX, Czajkowsky DM, Zhang YY, Shao ZF (1995) FEBS Lett 371:279
52. Allison DP, Bottomley LA, Thundat T, Brown GM, Woychik RP, Schrick JJ, Jacobson KB, Warmack RJ (1992) Proc Natl Acad Sci USA 89:10129
53. Humphris ADL, Miles MJ, Hobbs JK (2005) Appl Phys Lett 86:034106
54. Yang G, Vesenka J, Bustamante CJ (1996) Scanning 18:344
55. Vesenka J, Guthold M, Tang CL, Keller D, Delaine E, Bustamante C (1992) Ultramicroscopy 42:1243
56. Thundat T, Warmack RJ, Allison DP, Bottomley LA, Lourenco AJ, Ferrell TL (1992) J Vac Sci Technol A 10:630
57. Yang J, Mou J, Yuan JY, Shao ZF (1996) J Microsc (Oxford) 182:106
58. Hansma HG, Vesenka J, Siegerist C, Kelderman G, Morrett H, Sinsheimer RL, Elings V, Bustamante C, Hansma PK (1992) Science 256:1180
59. Hansma HG, Sinsheimer RL, Groppe J, Bruice TC, Elings V, Gurley G, Bezanilla M, Mastrangelo IA, Hough PVC, Hansma PK (1993) Scanning 15:296
60. Lyubchenko YL, Gall AA, Shlyakhtenko LS, Harrington RE, Jacobs BL, Oden PI, Lindsay SM (1992) J Biomol Struct Dyn 10:589
61. Hansma PK, Cleveland JP, Radmacher M, Walters DA, Hillner PE, Bezanilla M, Fritz M, Vie D, Hansma HG, Prater CB, Massie J, Fukunaga L, Gurley G, Elings V (1994) Appl Phys Lett 64:1738
62. Rivetti C, Codeluppi S (2001) Ultramicroscopy 87:55
63. Schulz A, Mucke N, Langowski J, Rippe K (1998) J Mol Biol 283:821
64. Feng XZ, Bash R, Balagurumoorthy P, Lohr D, Harrington RE, Lindsay SM (2000) Nucleic Acids Res 28:593
65. Fang Y, Spisz TS, Hoh JH (1999) Nucleic Acids Res 27:1943
66. Hansma HG, Revenko I, Kim K, Laney DE (1996) Nucleic Acids Res 24:713
67. Marsh TC, Vesenka J, Henderson E (1995) Nucleic Acids Res 23:696
68. Hansma HG, Laney DE, Bezanilla M, Sinsheimer RL, Hansma PK (1995) Biophys J 68:1672
69. Li J, Bai C, Wang C, Zhu C, Lin Z, Li Q, Cao E (1998) Nucleic Acids Res 26:4785
70. Bensimon A, Simon A, Chiffaudel A, Croquette V, Heslot F, Bensimon D (1994) Science 265:2096
71. Pope LH, Davies MC, Laughton CA, Roberts CJ, Tendler SJB, Williams PM (2000) J Microsc (Oxford) 199:68
72. Lyubchenko YL, Shlyakhtenko LS (1997) Proc Natl Acad Sci USA 94:496
73. Zuccheri G, Dame RT, Aquila M, Muzzalupo I, Samori B (1998) Appl Phys A 66:S585
74. Nagami F, Zuccheri G, Samori B, Kuroda R (2002) Anal Biochem 300:170
75. Cherny DI, Jovin TM (2001) J Mol Biol 313:295
76. Samori B, Siligardi G, Quagliariello C, Weisenhorn AL, Vesenka J, Bustamante CJ (1993) Proc Natl Acad Sci USA 90:3598
77. Oussatcheva EA, Shlyakhtenko LS, Glass R, Sinden RR, Lyubchenko YL, Potaman VN (1999) J Mol Biol 292:75
78. Lyubchenko YL, Shlyakhtenko LS, Binus M, Gaillard C, Strauss F (2002) Nucleic Acids Res 30:4902
79. Shlyakhtenko LS, Hsieh P, Grigoriev M, Potaman VN, Sinden RR, Lyubchenko YL (2000) J Mol Biol 296:1169

80. Shlyakhtenko LS, Potaman VN, Sinden RR, Lyubchenko YL (1998) J Mol Biol 280:61
81. Jett SD, Cherny DI, Subramaniam V, Jovin TM (2000) J Mol Biol 299:585
82. Rivetti C, Walker C, Bustamante C (1998) J Mol Biol 280:41
83. Zuccheri G, Scipioni A, Cavaliere V, Gargiulo G, De Santis P, Samori B (2001) Proc Natl Acad Sci USA 98:3074
84. Scipioni A, Anselmi C, Zuccheri G, Samori B, De Santis P (2002) Biophys J 83:2408
85. Han WH, Dlakic M, Zhu YWJ, Lindsay SM, Harrington RE (1997) Proc Natl Acad Sci USA 94:10565
86. Maeda Y, Matsumoto T, Tanaka H, Kawai T (1999) Jpn J Appl Phys 38:L1211
87. Andersen ES, Contera SA, Knudsen B, Damgaard CK, Besenbacher F, Kjems J (2004) J Biol Chem 279:22243
88. Coury JE, McFaillsom L, Williams LD, Bottomley LA (1996) Proc Natl Acad Sci USA 93:12283
89. Krautbauer R, Fischerlander S, Allen S, Gaub HE (2002) Single Molecules 3:97
90. Sun J-S, Francois J-C, Montenay-Garestier T, Saison-Behmoaras T, Roig V, Thuong NT, Helene C (1989) Proc Natl Acad Sci USA 86:9198
91. Viglasky V, Valle F, Adamcik J, Joab I, Podhradsky D, Dietler G (2003) Electrophoresis 24:1703
92. Mukhopadhyay R, Dubey P, Sarkar S (2005) J Struct Biol 150:277
93. Berge T, Jenkins NS, Hopkirk RB, Waring MJ, Edwardson JM, Henderson RM (2002) Nucleic Acids Res 30:2980
94. Saito M, Kobayashi M, Iwabuchi SI, Morita Y, Takamura Y, Tamiya E (2004) J Biochem136:813
95. Shlyakhtenko LS, Gall AA, Weimer JJ, Hawn DD, Lyubchenko YL (1999) Biophys J 77:568
96. Kuyper CL, Brewood GP, Chiu DT (2003) Nano Lett 3:1387
97. Ou ZY, Muthukumar M (2005) J Chem Phys 123
98. Borukhov I, Bruinsma RJ, Gelbart WM, Liu AJ (2001) Phys Rev Lett 86:2182
99. Grosberg AY, Nguyen TT, Shklovskii BI (2002) Rev Mod Phys 74:329
100. Cherstvy AG (2005) J Phys Condens Matter 17:1363
101. Hud NV, Vilfan ID (2005) Ann Rev Biophys Biomol Struct 34:295
102. Golan R, Pietrasanta LI, Hsieh W, Hansma HG (1999) Biochemistry 38:14069
103. Moreno-Herrero F, Herrero P, Moreno F, Colchero J, Gomez-Navarro C, Gomez-Herrero J, Baro AM (2003) Nanotechnology 14:128
104. Liu D, Wang C, Li JW, Lin Z, Tan ZK, Bai CL (2000) J Biomol Struct Dyn 18:1
105. Allen MJ, Bradbury EM, Balhorn R (1997) Nucleic Acids Res 25:2221
106. Dunlap DD, Maggi A, Soria MR, Monaco L (1997) Nucleic Acids Res 25:3095
107. Lin Z, Wang C, Feng XZ, Liu MZ, Li JW, Bai CL (1998) Nucleic Acids Res 26:3228
108. Fang Y, Hoh JH (1998) J Am Chem Soc 120:8903
109. Wolfert MA, Seymour LW (1996) Gene Therapy 3:269
110. Hansma HG, Golan R, Hsieh W, Lollo CP, Mullen-Ley P, Kwoh D (1998) Nucleic Acids Res 26:2481
111. Strand SP, Danielsen S, Christensen BE, Varum KM (2005) Biomacromolecules 6:3357
112. Fang Y, Hoh JH (1999) FEBS Lett 459:173
113. Liu Z, Li Z, Zhou H, Wei G, Song Y, Wang L (2005) J Microscop (Oxford) 218:233
114. Rackstraw BJ, Martin AL, Stolnik S, Roberts CJ, Garnett MC, Davies MC, Tendler SJB (2001) Langmuir 17:3185
115. Kleemann E, Dailey LA, Abdelhady HG, Gessler T, Schmehl T, Roberts CJ, Davies MC, Seeger W, Kissel T (2004) J Control Rel 100:437

116. Chim YTA, Lam JKW, Ma Y, Armes SP, Lewis AL, Roberts CJ, Stolnik S, Tendler SJB, Davies MC (2005) Langmuir 21:3591
117. Wangerek LA, Dahl HHM, Senden TJ, Carlin JB, Jans DA, Dunstan DE, Ioannou PA, Williamson R, Forrest SM (2001) J Gene Med 3:72
118. Chan CK, Senden T, Jans DA (2000) Gene Therapy 7:1690
119. Montigny WJ, Houchens CR, Illenye S, Gilbert J, Coonrod E, Chang YC, Heintz NH (2001) Nucleic Acids Res 29:1982
120. Mozafari MR, Reed CJ, Rostron C, Hasirci V (2005) J Liposome Res 15:93
121. Fang Y, Hoh JH (1998) Nucleic Acids Res 26:588
122. Bezanilla M, Manne S, Laney DE, Lyubchenko YL, Hansma HG (1995) Langmuir 11:655
123. Pereira GG, Williams DRM (2001) Biophys J 80:161
124. Fang Y, Yang J (1997) J Phys Chem B 101:441
125. Fang Y, Yang J (1997) J Phys Chem B 101:3453
126. Rivetti C, Guthold M, Bustamante C (1999) EMBO J 18:4464
127. Rippe K, Guthold M, von Hippel PH, Bustamante C (1997) J Mol Biol 270:125
128. Wyman C, Grotkopp E, Bustamante C, Nelson HCM (1995) EMBO J 14:117
129. Rivetti C, Codeluppi S, Dieci G, Bustamante C (2003) J Mol Biol 326:1413
130. Garcia RA, Bustamante CJ, Reich NO (1996) Proc Natl Acad Sci USA 93:7618
131. Sato MH, Ura K, Hohmura KI, Tokumasu F, Yoshimura SH, Hanaoka F, Takeyasu K (1999) FEBS Letters 452:267
132. Leuba SH, Yang GL, Robert C, Samori B, Vanholde K, Zlatanova J, Bustamante C (1994) Proc Natl Acad Sci USA 91:11621
133. Yodh JG, Lyubchenko YL, Shlyakhtenko LS, Woodbury N, Lohr D (1999) Biochemistry 38:15756
134. Yodh JG, Woodbury N, Shlyakhtenko LS, Lyubchenko YL, Lohr D (2002) Biochemistry 41:3565
135. Leuba SH, Bustamante C, van Holde K, Zlatanova J (1998) Biophys J 74:2830
136. Leuba SH, Bustamante C, Zlatanova J, van Holde K (1998) Biophys J 74:2823
137. Bezanilla M, Drake B, Nudler E, Kashlev M, Hansma PK, Hansma HG (1994) Biophys J 67:2454
138. Guthold M, Zhu X, Rivetti C, Yang G, Thomson NH, Kasas S, Hansma HG, Smith B, Hansma PK, Bustamante C (1999) Biophys J 77:2284
139. Lushnikov AY, Bogdanov A, Lyubchenko YL (2003) J Biol Chem 278:43130
140. Abdelhady HG, Allen S, Davies MC, Roberts CJ, Tendler SJB, Williams PM (2003) Nucleic Acids Res 31:4001
141. Guthold M, Bezanilla M, Erie DA, Jenkins B, Hansma HG, Bustamante C (1994) Proc Natl Acad Sci USA 91:12927
142. van Noort SJT, van der Werf KO, Eker APM, Wyman C, de Grooth BG, van Hulst NF, Greve J (1998) Biophys J 74:2840
143. Ellis DJ, Dryden DTF, Berge T, Edwardson JM, Henderson RM (1999) Nat Struct Biol 6:15
144. Jiao YK, Cherny DI, Heim G, Jovin TM, Schaffer TE (2001) J Mol Biol 314:233
145. Viani MB, Schaffer TE, Paloczi GT, Pietrasanta LI, Smith BL, Thompson JB, Richter M, Rief M, Gaub HE, Plaxco KW, Cleland AN, Hansma HG, Hansma PK (1999) Rev Sci Instrum 70:4300
146. Ando T, Kodera N, Takai E, Maruyama D, Saito K, Toda A (2001) Proc Natl Acad Sci USA 98:12468
147. Piner RD, Zhu J, Xu F, Hong SH, Mirkin CA (1999) Science 283:661
148. Germishuizen WA, Tosch P, Middelberg APJ, Walti C, Davies AG, Wirtz R, Pepper M (2005) J Appl Phys, p 97
149. Argaman M, Golan R, Thomson NH, Hansma HG (1997) Nucleic Acids Res 25:4379

150. Antognozzi M, Szczelkun MD, Round AN, Miles MJ (2002) Single Molecules 3:105
151. Smith DA, Connell SD, Robinson C, Kirkham J (2003) Anal Chim Acta 479:39
152. Stroh CM, Ebner A, Geretschlager M, Freudenthaler G, Kienberger F, Kamruzzahan ASM, Smith-Gil SJ, Gruber HJ, Hinterdorfer P (2004) Biophys J 87:1981
153. Zlatanova J, Lindsay SM, Leuba SH (2000) Prog Biophys Mol Biol 74:37
154. McKendry R, Zhang JY, Arntz Y, Strunz T, Hegner M, Lang HP, Baller MK, Certa U, Meyer E, Guntherodt HJ, Gerber C (2002) Proc Natl Acad Sci USA 99:9783

16 Direct Detection of Ligand–Protein Interaction Using AFM

Małgorzata Lekka · Piotr Laidler · Andrzej J. Kulik

Abbreviations

AFM	atomic force microscopy
ASA	arylsulfatase A
aPAP	monoclonal antiprostatic acid phosphatase antibody
BFP	biomembrane force probe
bFGF	basic fibroblast growth factor
CaY	carboxypeptidase Y
Con A	concanavalin A, lectin from *Canavalia ensiformis*
Du-145	human prostate carcinoma from the metastatic central nervous system lesion
DFS	dynamic force spectroscopy
ECM	extracellular matrix
FFM	friction force microscopy
HCV29	nonmalignant transitional epithelial cells of ureter
ICAM-1	intercellular adhesion molecule-1
IgG	immunoglobulin G
LFA-1	leukocyte function-associated antigen-1
LNCaP	human prostate carcinoma from the left supraclavicular lymph node metastasis
NeuNAc	*N*-acetylneuraminic acid or sialic acid
PAP	prostatic acid phosphatase
PC-3	human prostate carcinoma cell line from metastasis to bone
PEG	polyethylene glycol
PHA-L	leucoagglutinin from *Phaseolus vulgaris*
PMA	phorbol myristate acetate
PSMA	prostate-specific membrane antigen
PSD	position-sensitive detector
sLeX	sialyl Lewis X tetrasaccharide
SNA	lectin from *Sambucus nigra*
T24	transitional cell cancer of urine bladder
3A9	a murine T-cell hybridoma

16.1 Cell Structures and Functions

16.1.1 Membranes and their Components: Lipids and Proteins

To carry out all living functions every cell as an entity has to keep its integrity and to limit and control the contacts between its interior and surrounding. Analogously, within a cell, highly specialized, complex and very often quite opposite processes or activities have to be localized in subcellular organelles. Such arrangements are made possible by a crucial evolutionary development of a system of biological membranes. This allowed the separating and integrating of a part of a system from the rest of the environment.

Basic components of every biological membrane are complex lipids like various glycerophospholipids and sphingolipids (sphingomyelins and glycolipids). Due to their amphipathic (partly hydrophobic and partly hydrophilic) properties they form – in water or water solutions – a sheet-like structure called a lipid bilayer, which is an impermeable barrier for the majority of solutes present in biological fluids. In effect, a lipid bilayer physically separates (seals off) the interior of a cell from its surrounding. This is the biological role of the lipid component of plasma and other membranes. Simple and complex ions, hydrophilic structures (such as glucose, for example), and even some hydrophobic compounds such as fatty acids, can simply not pass through most membranes [1].

Therefore, to enable communication between a cell and its surroundings additional molecules must be present in the plasma membrane (Fig. 16.1). In fact, any function of a membrane (active role) except for its insulator property (passive role) is dependent (requires involvement of) on specialized proteins [1]. These are the numerous proteins that serve as pumps, carriers, transporters, and receptors, etc., enabling a more or less selective recognition of solutes and their translocation in and out of the cell. Interaction between a plasma membrane receptor and its ligand usually triggers a complex sequence of events inside a cell known as signal transduction that more or less affects cell behavior: proliferation, motility, apoptosis, or metabolism.

Understanding the character of interactions between proteins and their ligands and between proteins themselves, needs a deeper insight into protein structure, including all its components. Proteins are built of 20 common alpha L-amino acids linked to each other by typical covalent peptide bonds (ca. 300–400 kJ/mol). They are formed in effect, due to a reaction between the amino and carboxyl groups. The amino acids are arranged in various genetically encoded linear sequences and present in various relative amounts with respect to each other. In effect, they form polypeptides of the lowest free energy in different compositions and three-dimensional structures (referred to as a tertiary structure). This level of organization also depends on the character of the solution in which the protein naturally exists. The basic folding pattern of a polypeptide in water solution is recognized as its native structure. It is stabilized by: (1) a relatively small number of disulfide bridges (–S–S–, weak covalent bonds, ca. 200 kJ/mol), (2) hydrophobic interactions arising from the placement of large number of nonpolar amino acid side chains in water solution

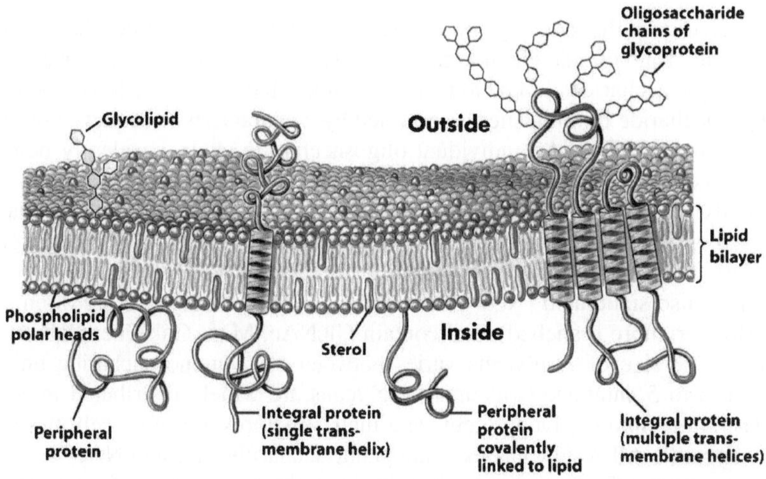

Fig. 16.1. Fluid mosaic model for membrane structure (taken from [1]). Integral proteins that allow communication between interior and exterior of a cell float in the sea of lipids, held by hydrophobic interactions with their nonpolar amino acid side chains. The carbohydrate moieties attached to some proteins and lipids of the plasma membrane are exposed on the extracellular surface of the membrane

and that are the driving force behind the folding process, and (3) a large number of relatively weak (5–25 kJ/mol) bonds/interactions of electrostatic origin, such as hydrogen bonds, van der Waals bonds, ionic bonds and salt bridges that allow fine tuning of the overall structure. Due to the folding of a nascent polypeptide, driven by the interactions between amino acid side chains and their contact with a water environment, a typical globular protein has hydrophobic side chains (mainly trp, phe, met, leu, ile, val) hidden in the interior of a globule. The hydrophilic and ionized side chains of amino acids (e.g. ser, gln, asn, glu, his, lys) are exposed on the outer face (side) of a globule [1,2].

16.1.2 Glycoproteins

Many integral membrane proteins contain an oligosaccharide component in addition to their polypeptide part. Oligosaccharide component of such complex proteins or glycolipids, alternatively called due to its origin carbohydrate moiety, is always exposed on the extracellular side of the membrane (Fig. 16.1). Due to the incorporation of an oligosaccharide component, often referred to as glycans, these proteins bear the name glycoproteins [3].

There are two major types of glycoproteins based on the kind of connection between their oligosaccharide and polypeptide components. Oligosaccharides can be covalently linked through either *O*-glycosidic or *N*-glycosidic bonds [3]. In the former case, C1 of a (first) monosaccharide at the reducing end of the oligosaccharide – *N*-acetylgalactosamine (GalNAc) – is bound through a hydroxyl group of a serine or threonine side chain. *N*-glycoproteins are formed if C1 of a (first) monosaccharide at the reducing end of oligosaccharide – usually GlcNAc – is bound by an amide group of side-chain asparagines. A protein may have only one (rarely) or more oligosaccharides linked to its polypeptide, since it usually has a few potential oligosaccharide binding sites determined by a characteristic sequence of amino acids – Asn-X-Ser/Thr. An individual oligosaccharide chain covalently bound to a protein is called a glycan.

Usually, *O*-glycans are short and not significantly branched oligosaccharides composed of a few monosaccharide residues such as *N*-acetylgalactosamine, galactose (Gal), *N*-acetylglucosamine (GlcNAc), fucose (Fuc) and *N*-acetylneuraminic acid called also sialic acid (NeuNAc). On the contrary, *N*-glycans are often much larger (longer, more branched), and contain GlcNAc, Man, Gal, Fuc, GalNAc and NeuNAc. The size of *N*-glycans varies between 6–15 monosaccharide units arranged in 2 to 5 antennae structures. *N*-glycans are widely distributed in soluble and plasma membrane glycoproteins and their structures are often cell, tissue and species specific. All *N*-glycans have the pentasaccharide $Man_3GlcNAc_2$ as a common "core structure". According to their structure and the location of the extra sugar residues attached to the core, *N*-glycans are further divided into a few different types: (a) a high mannose-type that contains only mannosyl (Man) residues attached to the core; (b) a complex-type that has "antennae" or branches attached to the core. The antennae are composed of GlcNAc, Gal, Fuc, GalNAc, NeuNAc and sulfate. The number of antennae in mammal glycoproteins ranges from two (biantennary) to four/five (tetra-/pentaantennary); (c) a hybrid-type that only has mannose residues on one arm of the core (the Man α-6 arm) and one or two antennae on the second arm (the Man α1–3 arm); (d) a poly *N*-acetyllactosamine type which contains repeating

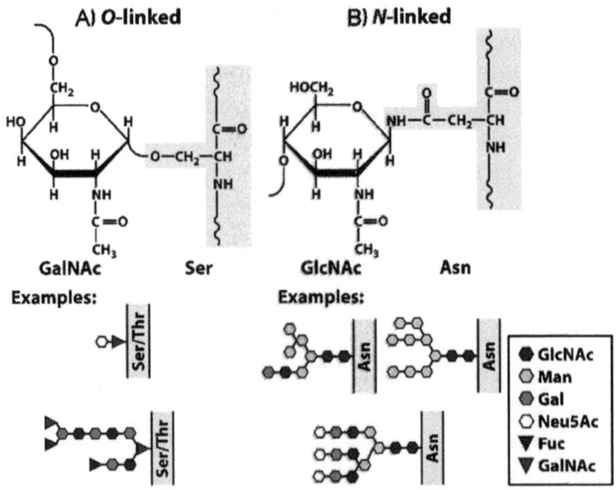

Fig. 16.2. (**A**) *O*-linked oligosaccharides have a glycosidic bond to the hydroxyl group of Ser or Thr residues, illustrated here with GalNAc as the sugar at the reducing end of the oligosaccharide. One simple chain and one complex chain are shown. (**B**) *N*-linked oligosaccharides have a *N*-glycosyl bond to the amide group of an Asn residue, illustrated here with GlcNAc as the terminal sugar. Three common types of oligosaccharides that are *N*-linked in glycoproteins are shown. A complex description of the oligosaccharide structure requires specification of the position and stereochemistry (α or β) of each glycosidic linkage. Taken from [1]

units of lactose (Galβ1-4GlcNacβ1–3), attached to the core. This repeating unit may be further branched. In addition all the N-glycans may have the bisecting GlcNAc linked β1–4 to the trimannosyl core (Fig. 16.2).

Glycans of glycoproteins fulfill many important structural and functional roles. They are responsible for increased solubility and stability of a number of proteins, e.g. as in the case of blood plasma proteins. Oligosaccharides differ from proteins and nucleic acids in a few characteristics – they are usually highly branched and their monomeric units are bound to one another by different linkages (α- or β-1,4-, 1,2-, 1,3-, 1,6-, 2,3-, etc.). Due to the branching and numerous alternative linkages these types of oligosaccharides are able to carry more information than other biological molecules. Therefore, their varying structural motives serve as a part of a recognition system, like in antigen–antibody and receptor–hormone reactions, bacterial infection, cell–cell, and cell–extracellular protein interactions. The most important "role of glycoproteins" from the point of view of their presence in plasma membranes, is that glycans form parts of: (1) antigenic determinants, which create individuality of a cell and (2) cell–cell and cell–extracellular matrix proteins recognition systems [3,4].

16.1.3 Immunoglobulins

All living organisms are constantly subjected to contact with uncountable intruders – bacteria and viruses (pathogenic microorganism) that are the threat for integrity and sometimes survival of invaded organism. To defend against such invaders vertebrates developed very elaborate and effective defense system. Pathogens that occasionally broke through the physical barriers are recognized as foreign and destroyed by various components of so-called immune system [5, 6]. One of the major weapons against pathogens is immunoglobulins. These are proteins, called due to their function, antibodies. They are synthesized by lymphocytes, highly specialized cells involved in the differentiation between self and nonself and destruction of the latter one. Various types (classes) of immunoglobulins are either exposed on plasma membrane of some types of lymphocytes or present in soluble form in body fluids (Fig. 16.3a).

Antibodies specifically recognize fragments of various structures such as simple or complex proteins, carbohydrates, nucleic acids, etc., due to their ability to trigger immune response, are called antigens. Regions of antigens that bind to antibodies are known as antigenic determinants or epitopes. Larger proteins or glycoproteins may possess a few epitopes. Each epitope is recognized by different antibody. As a result they usually bind each other, with a K_a in the range of 10^4–$10^{10}\,\mathrm{M}^{-1}$, which corresponds to binding energies of 25–65 kJ/mole. The interaction between an antigenic determinant and its specific antibody (immunoglobulin), is an excellent example of intermolecular interactions, this irrespective of whether the antigenic determinant is of a solely protein, protein/carbohydrate or other origin (Fig. 16.3b). Furthermore this binding of two or more macromolecules is determined by typical intermolecular weak forces (a few to few tens of kJ), like hydrophobic interactions, hydrogen bonds and various electrostatic interactions such as van der Waals, ionic, and salt bridges [5,6].

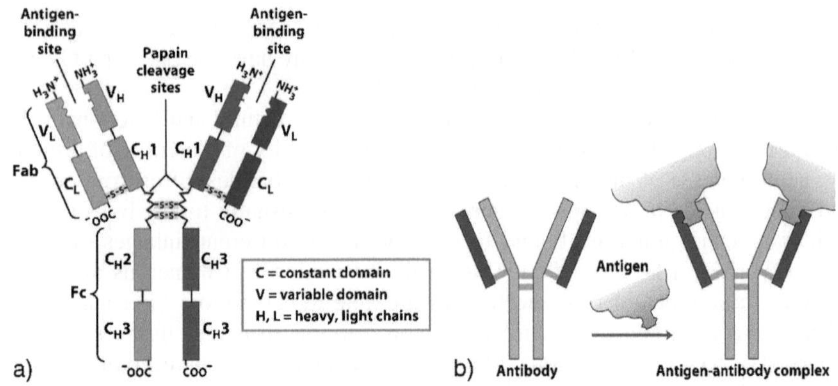

Fig. 16.3. (**a**) Structure of immunoglobulin G (IgG, from [1]). Pairs of heavy and light chains combine to form a Y-shaped molecule. Two antigen-binding sites are formed by a combination of variable domains form one light (V_L) and one heavy (V_H) chain. Cleavage with papain separates the Fab and Fc portions of the protein in the hinge region. (**b**) Scheme of binding of IgG to an antigen and induced fit in the binding of an antigen to IgG. To generate an optimal fit for the antigen the binding sites of IgG often undergo slight conformational changes

16.1.4 Adhesion Molecules

Cell–cell and cell–extracellular matrix protein interactions require adhesion molecules that are present as integral plasma membrane proteins, and exposed to the extracellular environment. All the adhesion molecules are either *N*-glycoproteins that are often heavily glycosylated or they recognize *N*-glycans (at least as a part of their specific ligands).

There are four major superfamilies of adhesion proteins: cadherins, integrins, immunoglobulin like adhesion molecules and selectins (Fig. 16.4a). They all are variously expressed by different cells of the same organism, and often by the same cells at various stages of their development, and in various metabolic circumstances [4,7].

16.1.4.1 Cadherins

The most important, in regards to maintenance of tissue integrity, are cadherins. They are involved in formation of cell–cell adherens junctions. This is the most common type of intercellular adhesion, critical for the maintaining of tissue architecture, cell polarity and limiting cell movement and proliferation. Cadherins are a superfamily of transmembrane glycoproteins located in the plasma membrane of a vast majority of cells in solid tissues. They provide strong intercellular adhesion in a Ca^{2+}-dependent manner. Cadherins were further divided into a few subgroups. The most typical members of the classic cadherin subfamily are – E-cadherin – mostly expressed in epithelial tissue of various organs and recognized as one of the suppressors of cancers, *N*-cadherin – mainly found in neural tissue and many cells at their fetal stage of development but also common to many metastatic tumors and therefore

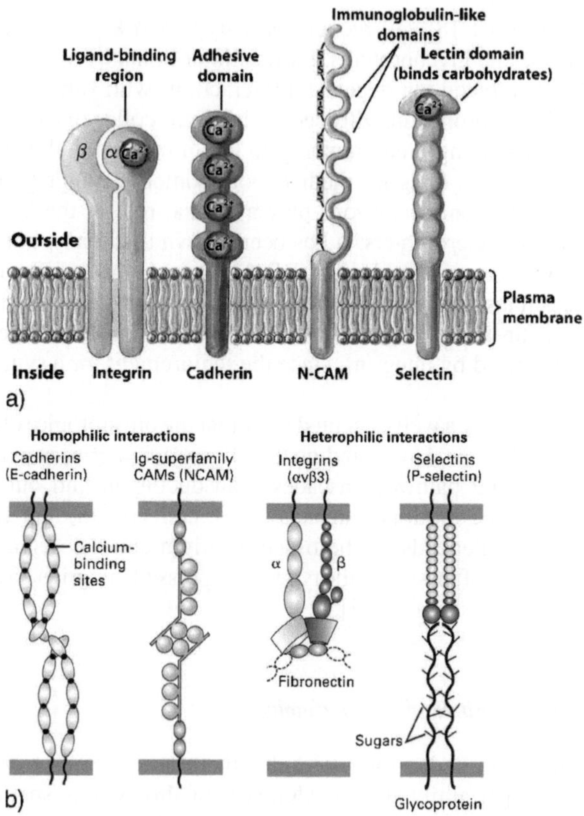

Fig. 16.4. (**a**) Four types of integral proteins participating in cell–cell interactions (integrins, cadherins, immumoglobulin–like molecules (N–CAM) and selectins, taken from [1]). (**b**) Interactions between the characteristic domains of a given type of cell-adhesion molecule [9]

known as a tumor inducer, P-cadherin – found initially in the placenta, L-cadherin from liver and V-cadherin as well as some other minor ones. Classic cadherins appear as a single polypeptide chain of different lengths (about 730 amino acids) and molecular weights, but with a high degree of homology (60%). Most of them are class I transmembrane proteins. They are composed of a large N-terminal extracellular domain, which mediates homophilic type cell adhesion, a short transmembrane domain, and a highly conserved C-terminal cytoplasmic domain (Fig. 16.4b). The latter one interacts with cytoplasmic skeleton proteins as α-, β- and γ-catenins and through them with the actin cytoskeleton. This seems to be a key event in the interaction between extracellular domains of E-cadherin and intracellular cytoskeletal proteins as well as signaling between cells and their surroundings. Cadherins play a major role in epithelial architecture, in cell differentiation upon embryogenesis, and also in the transformation and invasion of cancer cells [4, 7].

16.1.4.2
Integrins

The most diverse group of adhesion molecules are the integrins, cell-adhesion receptors, noncovalent linked heterodimers composed of α- and β-type subunits. There are at least 18 well recognized α-type and 8 β-type subunits that both are transmembrane glycoproteins. Extracellular domains of the $\alpha\beta$ complexes participate in bivalent metal ion-dependent interactions with various extracellular matrix proteins such as fibronectin, vitronectin, laminin, collagens and other numerous cell receptors belonging predominantly to the immunoglobulin superfamily as, e.g., ICAMs (Fig. 16.4b). Integrins mediate both homotypic aggregation and heterotypic cell-to-cell adhesion. Metal ions play a critical role in the ligand binding function of all integrin heterodimers. It has been shown that many, but not all integrins require an arginine-aspartate-glycine (RDG) sequence in their ligands. The presence of such an amino-acid sequence has been ascertained in, e.g., fibronectin, fibrinogen, vitronectin and collagen I. It is, however, still unclear whether this is the only signal recognized by integrins, since the requirement for a synergistic signal has also been reported.

It is now a well-accepted view that the physiologic role of integrins involves more than adhesion. Upon binding their ligands, integrins activate members of intracellular signalizing pathway cascades, transducing in and out signals that can stimulate or regulate motility and invasivnes, cell growth, and survival. The expression of integrins depends on the tissue of origin and the degree of differentiation. Cancer cells very often switch the types of expressed integrins, favoring the ones that transmit progrowth signals [4, 7].

16.1.4.3
Immunoglobulin-like Superfamily

Among the adhesion molecules there are proteins belonging to an immunoglobulin superfamily (IgSF). Members of this widely spread group of more than one hundred proteins [immunoglobulin-like domain containing cell-adhesion molecules (IgCAMs)] possess a common structural motif, an immunoglobulin fold composed of about 70–110 amino acids that form 7–9 β pleated sheets stabilized by disulfide bonds. They function as cell-adhesion and signaling receptors that transduce extracellular signals from neighboring cells or the extracellular matrix to the intracellular signaling machinery. Most of the members of this superfamily participate in the cell–cell recognition and immunological processes. These are proteins of major histocompatibility complex (MHC), T lymphocyte receptor and various cell determinants (CDs) such as CD4, CD8, N-CAM, V-CAM, ICAM-1 receptors (Fig. 16.4b). Some of them are found on almost every cell, especially on those of the immunological system and on vascular endothelium that participates in metastasis.

N-glycosylation plays a very important role in the structure and function of IgCAMs. For example, the effect of changes in the *N*-glycosylation of N-CAM (an immunoglobulin superfamily Ca^{2+}-independent adhesion protein) on the metastatic potential of some cells indicates a close relationship between the function of this heavily *N*-glycosylated adhesion molecule and the structure of its *N*-glycans. The carcinoembryonic antigen is a yet another IgCAM member, possessing 28 potential *N*-glycosylation sites, and *N*-glycans constituting even up to 50% of its molecular weight. When identified, the oligosaccharide component of this seems to correlate strongly with a stage of progression of cancer cells [4, 8].

16.1.4.4
Selectins

Most eukaryotic cells adhere firmly to cells or to the extracellular matrix in order to maintain the architectures of specific organs. One of the most important biological functions of carbohydrates with regards to an adhesion event, concerns their recognition by selectins (a family of adhesion proteins that belong to a very broad and highly diverse group of plant and animal proteins called lectins). These proteins are classified on the basis of similarities in specificity of their carbohydrate-recognition domains. Selectins are animal lectins usually present on nonadhesive blood cells (leukocytes, platelets) and on endothelial cells remaining with them in close and constant contact. Being crucial elements of the intercellular recognition system between circulating cells and alignment of blood vessels, they participate in the function of the defense system. Three members of the selectin family have been recognized. They are L-selectin (constitutively expressed on leukocytes), P-selectin (mainly expressed on platelets but also on endothelial cells) and E-selectin (inducible endothelial cell protein). All are *N*-glycoproteins with a Ca^{2+}-dependent carbohydrate recognition domain at their N-termini, followed by a single epidermal growth factor (EGF) domain, a variable number of complement-regulatory domains, a single transmembrane polypeptide, and a fairly short C-terminal cytoplasmic domain. Selectins mediate the cell–cell contacts by binding via their lectin domain, to a carbohydrate-containing counter-receptor on target cells. Like other mammalian lectins, the selectins bind selectively, but with low affinity, to particular oligosaccharides (Fig. 16.4b). All three selectins (L-, P-, E-) bind to glycans that contain lactosamine units with α2,3-linked sialic acid and α1,3- or α1,4-linked fucose. The prototype of these structures is sLex (sialo-Lewis x), a terminal component of oligosaccharides attached to glycoproteins and glycolipids on most leukocytes and some endothelial cells. The results of numerous studies strongly suggest that one of the key factors in metastasis is the presence on tumor cells of a high density of sLex structure, and ligands of E- and P-selectins. The whole complex sequence of consecutive recognition events is mediated by oligosaccharide–lectin interaction, followed by the aggregation of platelet and tumor cells. This may finally lead to the attachment of tumor cells to the endothelium, extravasation and colonization of a tissue [4, 8].

All adhesion molecules mediate numerous contacts (interactions) between cells and their surroundings (ie. other cells and proteins embedded in the extracellular matrix). The structural arrangement of protein molecules determines cells interactions (strength, lifetime, association/dissociation constants, etc.) with more or less specifically recognized ligands – substrates, effectors, proteins, antibodies, and others at the physicochemical level.

16.1.5
Plant Lectins

The interaction of lectins of plant origin and oligosaccharides is a well–known system of interaction and the most classical. In fact, due to the ability to bind fairly selectively to microbial and animal oligosaccharides many plant lectins are very toxic. Perhaps they ought to be viewed, among their other functions, as members of an evolutionary old plant-defense system.

Nonetheless, lectins are an excellent and a very useful and extensively applied tool in the characterization of animal cell glycoconjugates. The presence of proteins in plant seeds that are capable of binding to and agglutinating cells was determined during the last century. In 1919 – which means a few years before the historical crystallization of the first enzyme, urease – Sumner crystallized the first lectin – Convanavalin A from *Canavalia ensiformis*. However, the term lectin was adopted in the mid-1960s. Since that time lectins have been defined as a class of proteins of nonimmune origin that bind carbohydrates without modifying them. Until recently hundreds of different plant lectins have been identified and characterized [1,3].

The best characterized family of lectins is the Leguminosae. This family includes lectins such as ConA, soyabean agglutinin, and lentil lectin. Most leguminous lectins are metalloproteins with tightly bound Ca^{2+} and Mn^{2+}, which are essential for carbohydrate binding activity. The domain responsible for binding those ions contains conserved valine and aspartic acid. Many leguminous plant lectins are glycoproteins containing *N*-glycans that are synthesized by the typical pathway found in animal cells. The oligomeric nature of most lectins generates multivalency that increases the affinity towards their specific ligands, since single subunits have a single carbohydrate binding site and usually show lower affinity strength of binding. Therefore, oligomerization enables effective ligand binding [3]. The binding between carbohydrate residues present in recognized oligosaccharides of glycoconjugates and lectin is based mainly on the set of weak and reversible electrostatic (Fig. 16.5a), and to a minor extent, hydrophobic interactions (Fig. 16.5b) as was the case in every formerly discussed intra- or intermolecular interaction.

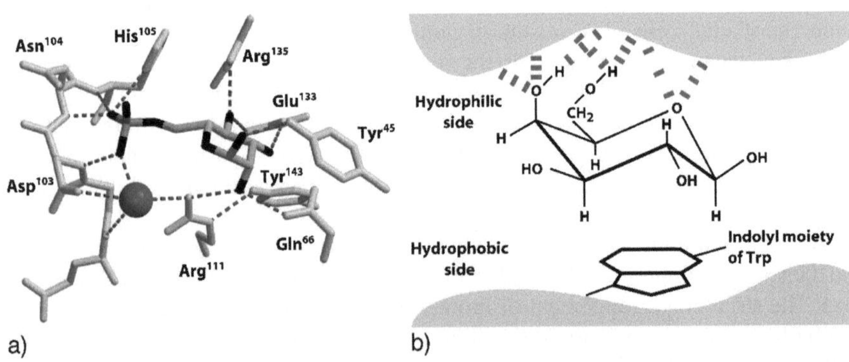

Fig. 16.5. (**a**) Structure of the bovine mannose 6-phosphate receptor complexed with mannose 6-phosphate (PDB ID 1M6P). In this complex, mannose 6-phosphate is hydrogen bonded to Arg111 and coordinated with the manganese. The His105 hydrogen-bonded to a phosphate oxygen of mannose 6-phosphate may be the residue that, when protonated at low pH, causes the receptor to release mannose 6-phosphate into the lysosome. *Black parts* indicate negative charge. (**b**) Hydrophobic interactions of sugar residues. Sugar units such as galactose have a more polar side (the top of the chair structure, with the ring oxygen and several hydroxyls), available to hydrogen-bond with lectin, and the less polar side that can have hydrophobic interactions with nonpolar side chains in the protein, such as the indole ring of tryptophan. Taken from [1]

The recognition of the differences in expression of adhesion proteins between normal (reference) cells and pathologically changed cells (e.g. cancer cells), and more precisely, their characterization (qualitative and quantitative) may have a significant impact on the development of intellectual as well as methodological approaches to the treatment of diseases, including cancer.

16.2 Forces Acting Between Molecules

Forces between biological molecules are not different from those arising between any other molecules or surfaces. They are electrical in origin and they not only determine the properties of solids, liquids or gases, but also they govern chemical reactions and the organization of biological structures. Usually, a biological interaction is very different from a simple chemical reaction or a physical change of a system. This is mainly due to the high degree of complexity of biological macromolecules forming complex structures like cell membranes or even whole cells.

All membranes in cells can be treated as assemblies of organic molecules localized in a thin, flexible layer of about 10 nm of thickness [9]. Such assemblies can be treated as macrocolloids that act with each other nonspecifically through the classical, long-range interactions similar to that observed in particle suspension. On the other hand, the adhesive contacts, governing the recognition processes, rely on the specific, short-range interactions. Therefore, the net interaction can be described by a physical potential that cumulates the action of all forces from macroscopic separations between surfaces to microscopic contact. Such potential will represent the energetic balance between the attractive and repulsive interactions that can be classified according to their range of action (see the hypothetical potential for a soluble protein and a surface, Fig. 16.6).

The long-range interactions include the attractive or repulsive electrostatic forces and the van der Waals attraction that follows the inverse power law. The van der Waals force arises mainly from the hydrocarbon core of the lipid bilayer and it dominates at large separations to bring membranes into adherent contact. The short-range force is often referred to as a hydration force only because it opposes condensation of all hydrated molecular interfaces, e.g. DNA, proteins, and lipid molecules. At small separations, repulsion between lipid bilayers becomes huge. In addition, the surface of the plasma membrane is covered by polymer-like molecules (i.e. different types of oligosachcharide's structures) that establish a significant steric barrier that is also enhanced by the electrostatic repulsion [12].

An interaction of proteins with different types of ligands, other proteins, or surfaces involves many types of forces (both short- and long-ranged), acting in liquid environment composed of water molecules and different ions that are necessary for supporting all biological functions. For many biological functions the interaction between cells involves the attachment of two complementary molecules. Thus, the attraction is limited to focal adhesions, i.e. with no apparent long-range contribution from van der Waals forces. In that manner, cells seem to be well stabilized against nonspecific interaction. Adhesive contacts between cells are often irregular, with many unbound regions and infrequent attachments between surfaces. These contacts will define the overall strength of the cell adhesion [11]. Unlike long-range colloid forces that are distributed uniformly over cell surface, the bonding between cell surface receptors and corresponding ligands is limited to microscopic sites. The resulting attractive interaction is short ranged and weak. It involves forces such as van der Waals, hydrophobic and electrostatic (i.e. present in ionic bonds) attraction, and hydrogen bonds that are a specific case of dipole–dipole interaction (Fig. 16.6a). The receptor–ligand interaction is strengthened by the complementary shape of the binding sites of both interacting molecules, leading to the formation of a strong

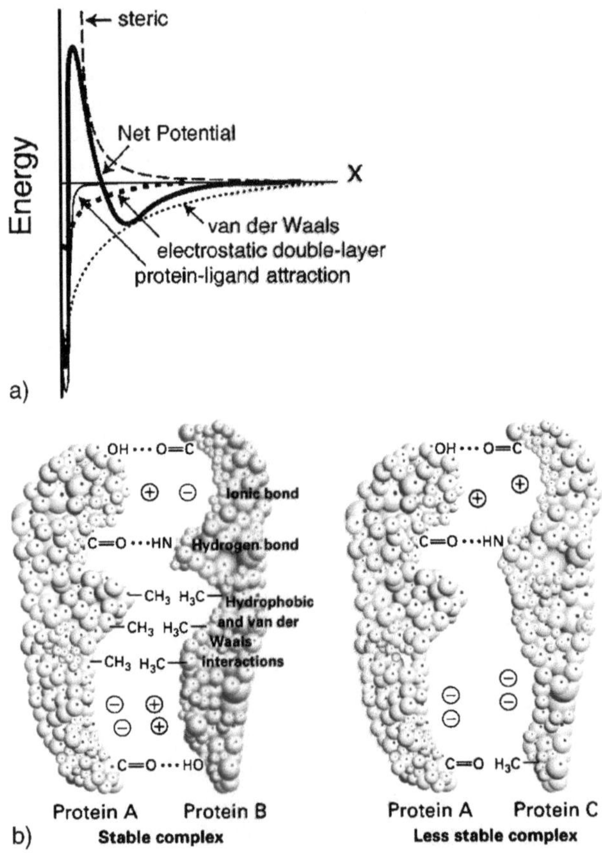

Fig. 16.6. (**a**) Hypothetical interaction potential between a soluble protein and a surface (Reprinted with permission from [13], Copyright (2000), Annual Reviews). In this case, the net potential profile (*bold solid line*) is a superposition of the van der Waals potential, attractive or repulsive double-layer potential, steric repulsion, and specific, short-range interactions. The relative ranges and magnitudes of these interactions can give rise to complicated potentials that exhibit multiple minima and energy barriers. (**b**) Binding of proteins through multiple noncovalent interactions (taken from [9]). The complementary shape of the binding sites together with these interactions results in a strong interaction, forming stable complex (proteins A and B). Any variations lead to a less stable complex and, in consequence, a weaker interaction (proteins A and C)

interaction (Fig. 16.6b, interaction between protein A and B). Any variations lead to less stable complex and, in consequence, a weaker interaction (Fig. 16.6b, proteins A and C).

16.2.1 Repulsive Forces

Repulsive forces occurring between cell membrane are long-range forces that protect living cells against sticking together. The most important repulsive interactions include an electrostatic "double-layer" force, hydration and steric forces.

16.2.1.1 Electrostatic "Double-Layer" Force

When two surfaces are immersed in aqueous electrolyte solution they are often charged due to ionization or dissociation of surface chemical groups, or due to binding of ions from the solution. Oppositely charged ions are attracted by the surface charge, while the ions with the same charge are repelled, leading to the formation of a layer composed of surface charge and ions (Fig. 16.7a). As a consequence, a repulsive force called the *electrostatic "double-layer" force* (F_{dl}) arises between surfaces [10]. It decays exponentially ($F_{dl} \sim \exp -r/\lambda_D$) with the characteristic decay length λ_D (a so-called Debye length), which depends on the concentration and the valency of ions present in the electrolyte solution (Fig. 16.7b):

$$\lambda_D = \sqrt{\frac{\varepsilon\varepsilon_0 k_B T}{2ce^2}} ,$$

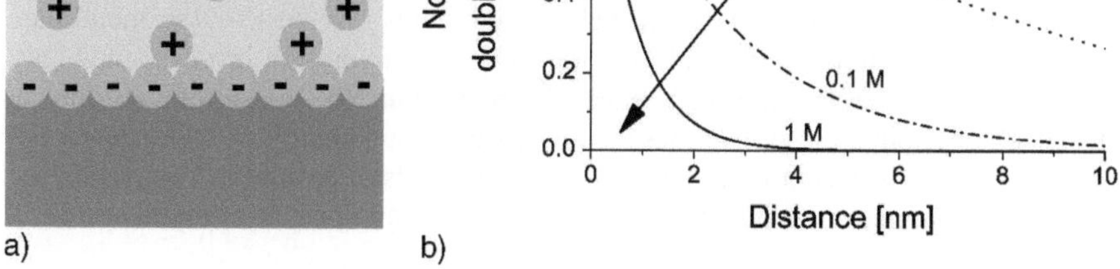

Fig. 16.7. (**a**) The origin of the repulsive electrostatic "double-layer" force acting between two surfaces charged in electrolyte solution due to ionization or dissociation of surface chemical groups or due to the binding of ions from the solution. (**b**) The dependence of the "double-layer" force on the divalent ion concentration in solution ($0.176/\sqrt{c}$, where c is an ion concentration)

where ε_0 and ε are dielectric constants of vacuum and solution (respectively), c denotes the salt concentration in mol per liter, e is a valency of ions, T is the temperature and k_B is Boltzmann's constant.

The exact form of the power law can be calculated using continuum theory where the potential distribution between two surfaces is determined from the Poisson–Boltzmann equation [10]. To solve the equation, one of the two types of boundary conditions should be assumed: either the surface charge or the surface potential must remain constant during surfaces approaching.

The electrostatic "double-layer" force between cells arises due to the negative charges present on their surface. The magnitude of the repulsion is dependent on the nonhomogeneous charge distribution on a plasma membrane. Due to the screening by ions in the electrolyte solution, the repulsion acts primarily at short distances that are comparable with the decay length; only charges that lie within a decay length from the end of the glycocalyx will contribute to the repulsion.

16.2.1.2
Hydration Force

In aqueous solution, between two hydrophilic surfaces, arises a repulsive, exponentially decaying force. This force is known as hydration repulsion and it is associated with strongly hydrated ions bound at surfaces. If two surfaces come into close proximity, water molecules must be removed from the contact area. Therefore, hydration force dominates at small distances, below 3 nm. Such an effect is more prominent for divalent ions that have more water molecules attached. The magnitude of the hydration force $F_{\mathrm{hydration}}$ can be described using the following equation:

$$F_{\mathrm{hydration}} = F_0 \cdot e^{\left(\frac{-r}{\lambda_{\mathrm{hydration}}}\right)} ,$$

where $\lambda_{\mathrm{hydration}}$ is a characteristic length of about 1 nm [10].

Hydration forces play an important role in biological systems in diverse processes, such as protein folding, ion transport, and the undesirable fouling of synthetic surfaces. In biological systems hydration forces arise from the work needed to remove water molecules bound to the polar lipid groups of a bilayer. The hydration force prevents the spontaneous attachments and fusion of membranes. Without this short-range repulsive interaction, two membranes (DNA and many proteins in solution as well) would collapse to dehydrated states that destroy the membrane structure and prohibits biological function. Since the hydration layer extends only a short distance beyond the membrane, the hydration force is significant only at distances of about 2 nm or even less. It increases exponentially as the distance between membranes decreases and its magnitude depends on several factors, including membrane roughness, radius of curvature, or its lipid and protein composition [12].

16.2.1.3
Steric Forces

When two layers having on its surface polymer-like molecules are close enough, the repulsive force arises as a result of the compression of the polymer layers.

Such a force is called a steric force. It describes the resistance of the polymer tails, counteracting the decrease of entropy when they are tangled with each other.

Steric forces interact at small distances, from 1 to 10 nm. The theoretical model of the distance dependence of steric force does not exist. However, force-value decay was experimentally determined to be inversely proportional to the twelfth power of the distance between the interacting molecules [13]. The dimensions of polymer-like molecules (i.e. the molecular weight and the grafting density) define the range of interaction. Steric forces are depended on such parameters as pH, temperature and ion concentration in solution. Usually, any increase in these parameters leads to a larger value of the steric force.

The cell surface is coated with dense layers of polymeric-like molecules, such as oligo-, polyzaccharides, long lipids, and ligand groups, exposed outside and surrounded by water molecules. These molecules will repeal during the contact between cells resulting in the steric forces. They play a similar role as the electrostatic repulsion i.e. they prevent collapse of the membrane.

16.2.2 Attractive Forces

The attraction of two specific molecules involves usually weak, short-range forces, with an interaction range limited to the contact area. The major interactions involved are van der Waals, hydrophobic, attractive electrostatic (ionic) interactions, and hydrogen bonding.

16.2.2.1 Van der Waals Force

The van der Waals force is always present between any interacting bodies. It results from the interaction of electric-dipole moments (either permanent or induced), arising due to different charge distributions in molecules. Depending on the type of dipole–dipole interaction, van der Waals forces can be divided into three types: i) the Keesom force arising between two molecules having permanent dipole moments, ii) the Debye force coming from the interaction between permanent and induced dipole moments, and iii) the London (dispersion) force resulting from the interaction between induced dipole moments [10]. The dispersion force is the most important contribution, because it appears between all molecules and atoms on distances ranging from 10 to 0.2 nm. The van der Waals interaction occurring between isolated pair of molecules is described by the following equation obtained for two interacting atoms or molecules at the distance r [13]:

$$F_{\mathrm{vdW}} = \frac{-6 \cdot C_{\mathrm{vdW}}}{r^7} \, .$$

The constant C_{vdW} denotes a quantity that depends on the optical properties and the geometry of the interacting molecules.

The interaction between macroscopic bodies ($\varepsilon_1, n_1, \varepsilon_2, n_2$) is generally described in terms of the Hamaker constant A that can be calculated for any two bodies acting across medium (ε_3, n_3) using the following formula obtained on the basis of Lifshitz theory:

$$A \approx \frac{3}{4} k_B T \left(\frac{\varepsilon_1 - \varepsilon_3}{\varepsilon_1 + \varepsilon_3}\right)\left(\frac{\varepsilon_2 - \varepsilon_3}{\varepsilon_2 + \varepsilon_3}\right) + \frac{3h\nu_e}{3\sqrt{2}} \frac{\left(n_1^2 - n_3^2\right)\left(n_2^2 - n_3^2\right)}{\sqrt{\left(n_1^2 + n_3^2\right)}\sqrt{\left(n_2^2 + n_3^2\right)}\left\{\sqrt{\left(n_1^2 + n_3^2\right)} + \sqrt{\left(n_2^2 + n_3^2\right)}\right\}} ,$$

where ε_i, n_i are corresponding dielectric constants and refractive indices, T is a temperature, ν_e is a resonant frequency, h and k_B are Planck and Boltzmann constants, respectively. The positive value of the Hamaker constant indicates attraction, while its negative value shows a repulsive character of the force. Therefore, the two following statements are valid. The first one tells us that the van der Waals between two identical bodies in medium is always attractive, whereas the force between two different bodies can be attractive or repulsive. The second statement is that the force in vacuum or in air is always attractive between any bodies. The Lifshitz theory assumes that each interacting body is a structureless continuum. Therefore, the exact equations describing of the van der Waals force are delivered for macroscopic bodies interacting at relatively large distances. The power law for the van der Waals force can be written as follows:

$$F_{vdW} = A \cdot f(G, r) ,$$

where $f(G, r)$ describes the contact geometry of the interaction. In principle, this equation should not be applied at the molecular scale.

Van der Waals interactions are relatively weak in comparison with other, usually dominant forces involved in molecules binding, such as ionic, hydrogen bonding, and hydrophobic forces.

16.2.2.2
Hydrophobic Forces

The main feature of hydrophobic surfaces is their inability to bind water molecules, which leads to the rise of a force resulting from the entropically unfavorable orientation of water. The interacting surfaces try to minimize the number of ordered water molecules required to surround hydrophobic regions. The resulting force (called the *hydrophobic force*) is attractive in character and, naturally, it increases with the increase of surface hydrophobicity.

Hydrophobic forces were first observed between two curved mica surfaces immersed in surfactant solution (cetyltrimethylammonium bromide). Instead of the expected repulsive force arising between negatively charged mica surfaces, the attractive force was observed since after adsorption of cationic surfactants both surfaces become hydrophobic. The hydrophobic force F_h decays exponentially with the characteristic length λ_h (range of 1–2 nm) and it can be described using the following empirical equation [10]:

$$F_{\mathrm{h}} = -2 \cdot \gamma \cdot e^{\left(\frac{-r}{\lambda_{\mathrm{h}}}\right)},$$

where γ is an interfacial tension of the surface with water.

The hydrophobic attraction plays an important role in biology since many molecules have both hydrophilic and hydrophobic regions on their surface. The structure of these molecules (proteins and lipids) is stabilized through such hydrophobic interaction.

16.2.2.3 Attractive Electrostatic Forces

The electrostatic force (or Coulomb interaction) arises between two oppositely charged atoms or molecules. This force is proportional to the charges Q_1 and Q_2 involved and inversely proportional to the square of the distance r [13]:

$$F_{\mathrm{el}} = \frac{Q_1 \cdot Q_2}{4 \cdot \pi \cdot \varepsilon_0 r^2},$$

where ε_0 is a vacuum permittivity ($8.854 \times 10^{-12}\,\mathrm{F/m^{-1}}$).

On the molecular scale, this type of interaction is present in an ionic bond which is an electrostatic attraction between oppositely charged groups. Such interaction does not have any specific orientation since the electric field around a single ion is uniform in all directions.

16.2.2.4 Hydrogen Bonding

Hydrogen bonding is a special case of dipole–dipole interaction that mainly determines unique properties of water. In this interaction, a hydrogen atom is shared between two other atoms called H donor and H acceptor. The acceptor atom possesses a partial negative charge that attracts the H atom and this effect is the basis of the interaction. The donor atom in a hydrogen bond is either oxygen or nitrogen atom that has a covalently attached hydrogen atom.

Initially, it was believed that the hydrogen bond involves only the sharing of a hydrogen atom between two electronegative atoms. However, now it is accepted that the nature of a hydrogen bond is predominantly electrostatic. The hydrogen atom is not shared, but remains closer to the more electronegative atom [10]. The hydrogen bond between two groups XH and Y is usually denoted by X–H ··· Y.

Hydrogen bonds are not only limited to water molecules but often are involved in interactions occurring between many other molecules such as DNA or between proteins.

16.3 Force Spectroscopy

The usual way of studying interaction between molecules assumes equilibrium binding. Kinetic measurements are carried out using such techniques as surface plasmon resonance [14] or quartz crystal microbalance [15]. In these techniques, interaction forces are indirectly inferred or calculated from the applied molecular model. However, such calculations are feasible for small molecules but they become more complicated and less precise for large proteins. Since the main goal of biophysical studies is to determine forces acting between molecules and to use this information to monitor or even manipulate the function and structure of proteins, the development of techniques that can directly and quantitatively measure molecular forces is of great importance.

There are several techniques that allow detection of molecular interactions with very high force resolution (at the order of nano- and piconewtons) that can be applied to study recognition phenomena occurring between single molecules (e.g. proteins or other types of molecules). Among them, a surface force apparatus (SFA, [13]), a biomembrane force probe (BFP, [16]) and an atomic force microscope (AFM, [17]) are the most popular ones. The latter, the atomic force microscope, was constructed in 1986 in the IBM laboratory in Zurich [18]. Initially, AFM was applied to study a surface topography in ambient or in vacuum conditions, but later this technique was found to be very useful for biologists since measurements in liquid, at almost physiological conditions, were enabled. Nowadays, AFM is applied not only to study the surface morphology of various biological samples, starting from single proteins and ending up in imaging of a living cell surface. Thanks to its possibility to determine such mechanical properties as stiffness, adhesion or friction, together with simultaneous sample topography recording, AFM is also widely applied to characterize properties of biological samples in a quantitative way.

Determination of the mechanical properties by means of AFM usually is carried out in the force-spectroscopy mode, where so-called force curves are recorded. The force curve is the dependence between a cantilever deflection (that is converted into force) and a relative sample (or scanner) position that can be transformed to a tip–sample distance. The interaction forces can be obtained from the analysis of the retraction part of the force curve recorded during the AFM tip withdrawal from the surface. The value of interaction forces between molecules measured by AFM depends on such factors as number of bonds formed within the tip–surface contact area, time of contact, etc. Therefore, the unbinding force of a single pair of molecules is usually determined by analyzing the large number of recorded force curves (see Sect. 16.3.1.3).

16.3.1 Atomic Force Microscope

16.3.1.1 Principle of Operation

The idea of AFM operation is straightforward. A sharp delicate probing tip, mounted at the end of a compliant cantilever, is placed almost parallel to the investigated surface and then moved over it, performing a raster scan. The cantilever senses forces acting between the tip and the sample surface which results in its deflection, that is usually recorded using the optical system: the laser beam is focused at the cantilever end and then, after reflection, is detected by a position-sensitive photodiode. The position of the beam spot produces information on the cantilever displacement z that

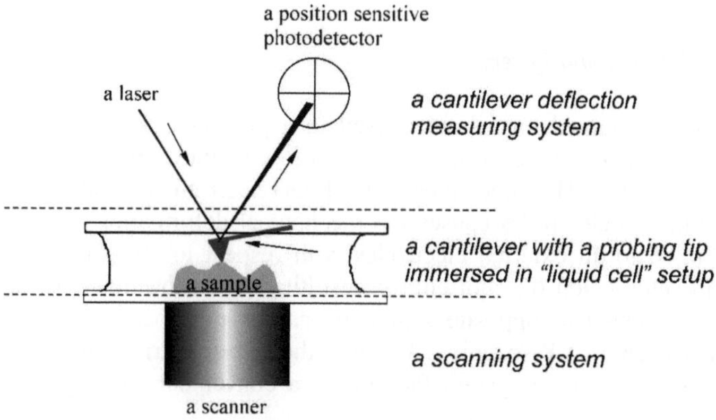

Fig. 16.8. Basic elements of an atomic force microscope (AFM): cantilever with a probing tip, cantilever deflection measuring system and a scanning system

can be converted into force F using Hooke's law ($F = k \cdot z$, where k is the cantilever spring constant).

Independent of the type of microscope, there are three elements that are common for all types of atomic force microscopes (Fig. 16.8): i) a cantilever with a tip, ii) a system that detects the cantilever deflection and iii) a scanning system.

16.3.1.1.1
Cantilevers

Usually, cantilevers are made of silicon or silicon nitride. Each cantilever is mounted on a rectangular chip that can contain one or more cantilevers with different spring constants describing their mechanical properties, typically from 0.01 N/m to 1 N/m. The probing tip is located at the free end of the cantilever. Usually, the tip has a shape of a four-sided pyramid with the height of the order of a few micrometers. The end of the tip can be characterized by the radius of curvature that varies typically from few nm to 50 nm.

16.3.1.1.2
Cantilever Deflection Measuring System

The most popular method of deflection measurement uses an optical system where the laser beam is focused at the end of cantilever. After reflection from the cantilever surface, the laser beam is detected by a position-sensitive detector (a photodiode) whose surface is divided into four quadrants. Cantilever deflection causes the movement of the reflected beam spot within the active area of the photodetector, thus changing the photocurrents of the quadrants. By proper subtraction of these signals one can obtain values proportional to cantilever deflection in a direction perpendicular and parallel to the sample surface. Summation of the signals from individual quadrants is used to correct for variations of laser intensity.

16.3.1.1.3
Scanning and Positioning System

The sample is mounted on a holder fixed to a piezoelectric scanner. The scanner has the form of a tube with outer surface segmented (parallel to tube axis) into four electrodes. The tube interior wall serves as an internal electrode. The voltage applied to electrodes causes the mechanical deformation of a tube. Equal voltage applied to all external electrodes with respect to internal one, results in scanner tube elongation (or shortening), resulting in Z movement. Voltage of opposite signs, applied to opposite segments, causes tube bending in X or Y directions that enables XY scanning. Usually, the piezoelectric scanner is mounted on the coarse positioning system facilitating a convenient and rapid sample exchange.

16.3.1.2
Tip and Surface Functionalization

The attachment of biomolecules depends on their surface properties, on the solid surface used as a substrate, and on the liquid medium. In most cases, the surface of molecules will display a higher level of complexity than the substrate or the liquid medium. Biomolecules exhibit not only an overall charge and hydrophobicity, but also a heterogeneous distribution of surface-exposed groups. The molecules of interest can be broadly grouped into nucleic acids (DNA, RNA), proteins (antibodies, enzymes, and receptors), small molecules (e.g. peptides, metabolites), and other biomolecules (e.g. carbohydrates, lipids). In addition, attached molecules should preserve their biological activity unchanged.

Mechanisms of immobilization can be divided into two major categories: adsorption and covalent binding. Adsorption relies on noncovalent interactions – mainly electrostatic, van der Waals, and dehydration of hydrophobic interfaces. It has a purely physical nature and therefore displays varying levels of reversibility. Such way of molecular immobilization results in randomly oriented molecules and a relatively weak attachment, which may influence the time of measurement. The covalent binding of specific molecule functional groups to functionalized surfaces, by definition, involves the formation of essentially irreversible chemical bonds between the molecule and the substrate surface. Therefore, it allows a very strong attachment and, in certain instances, enables the oriented molecule deposition. A variety of side groups can be easily used for covalent binding – the most common ones are amino, carboxylic, hydroxyl, and thiols groups. However, in many cases, covalent binding is enabled after a functionalization of the surface and/or biomolecule. Very often, it is realized using crosslinkers such as glutaraldehyde. For biological applications, where interaction forces are to be measured by AFM, the most common probes are made of silicon or silicon nitride. Biomolecules are usually immobilized on glass, mica, and gold surfaces. Therefore, immobilization requires the development of appropriate protocol of the attachment. Below, there are a few examples illustrating different ways of molecular deposition.

16.3.1.2.1 Protein Immobilization on Substrate or on AFM Tip

The diversity and complexity of proteins can make their deposition very difficult. Proteins have many different structures containing heterogeneous hydrophobic and charged domains. They are extremely fragile with respect to their biological activity and they can have multiple interaction sites. Additional complications arise when a correct orientation of the bound protein is required, for example to increase the number of functional groups exposed to the buffer. The immobilization protocols that have been used so far, apply both mechanisms of immobilization, i.e. adsorption and covalent binding. The appropriate protocol of the protein deposition should therefore be tailored to the specific protein used.

The adsorption of proteins depends on two main features: their surface charge and their hydrophobic domains. Both these properties enable a certain degree of control of protein deposition, however, they can result in randomly oriented molecules.

The electrostatic adsorption seems to be sufficient to assure a relatively strong attachment, but it does not have a permanent nature and it can be strongly affected by changes of solution pH and ionic strength. Therefore, there are a limited number of proteins that can be immobilized in this manner. When the hydrophobic attraction is chosen as a main source of adsorption, stronger and less reversible interaction is expected. However, on the other hand, it may result in loss of functional activity due to partial denaturation, as the protein unfolds to expose a hydrophobic interior portion to the hydrophobic surface.

Covalent binding is quite commonly used since proteins present a variety of functional groups, including amino, carboxyl, hydroxyl, and thiol that can be readily used for such binding to the different surfaces possessing complementary chemical groups. However, more care should be taken in order to avoid chemically induced denaturation of protein during the attachments process. There are many strategies for crosslinking of available functional groups. Most of them use a specialized designed crosslinkers for both attachment and physical separation of the protein from the surface, thereby allowing more of the protein functional domains to be exposed to the buffer.

Covalent binding generally produces a higher concentration of proteins than the adsorption. Proteins can also be more oriented when some techniques are applied such as i) the use of antibodies that bind proteins to their *Fc* portion, leaving binding sites free, ii) the use of biotinylation that enriches proteins with the binding site specific to streptavidin-coated surface, iii) cysteine thiol production in the protein fragment far from the binding site, allowing its deposition on a gold-coated surface, or iv) using the sugar binding molecules that binds to the oligosaccharide's moieties of proteins.

There are two most popular protocols enabling direct or indirect protein immobilization for the AFM measurements. In the first protocol, the glass or mica surface is silanized using e.g. 3-aminopropyltriethoxylsilane which enriches the surface with amino groups. Next, the silanized surface is activated using crosslinking agent such as glutaraldehyde [19]. The choice of agents depends on the proteins to be deposited. Both homofunctional and bifunctional agents can be employed in the protocol. Afterwards, proteins are delivered. Such a protein immobilization enables direct deposition of proteins on the surface. However, for some purposes, such procedure may not be desired since it blocks protein reorientation. To avoid this, proteins may be attached to the surface using a polymeric spacer. A polyethylene glycol (PEG) is a common spacer. Its thiol group is used to bind PEG onto a gold-coated surface of the silicon nitride tip. An amino group, located at the other end of the PEG molecule, attaches proteins via a covalent bond [20].

16.3.1.3
Force Spectroscopy

AFM measurements of interaction forces between molecules are carried out in a force-spectroscopy mode, where scanning is disabled and so-called *force curves* are recorded. A force curve is measured as a dependence of the cantilever deflection and a relative sample position and is usually converted into a function of force versus tip–sample distance.

A force curve consists of two parts: the one recorded during the approach of the tip to the sample surface (*an approach curve*) and the other one, collected during the opposite motion (*a retract curve*). The character (shape) of the force curve depends on the physical and chemical properties of the two interacting surfaces. Regardless of the sample type, there are several features that are present in all curves. Figure 16.9 shows the ideal curve.

16.3.1.3.1
Base Line (A)

When the cantilever is away from the surface, the cantilever deflection should be zero since there is no detectable interaction force. In fact, due to thermal vibrations, the cantilever oscillates around its free position, reflecting the noise present in a specific AFM system. When both the tip and the surface are charged of the same sign, at close distances prior to the contact the cantilever can be repulsed from the surface. It is represented by the slight rise of the baseline, as it is visible in Fig. 16.9 (region R).

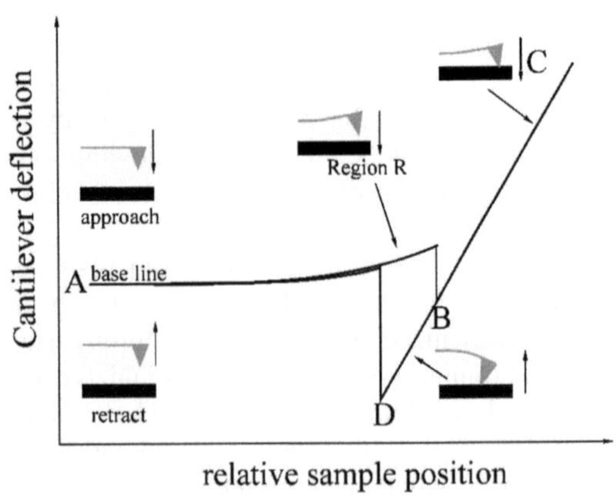

Fig. 16.9. An ideal force curve recorded by AFM. *Arrows* are indicating direction of movement of the tip

16.3.1.3.2
Jump-in (B)

The presence of attractive forces on a surface is reflected by a jump-in, i.e. the moment when the cantilever is suddenly attracted. At this moment, the gradient of the attractive force is larger than the cantilever spring constant.

16.3.1.3.3
Contact (C)

When the tip is in contact, the electron clouds of atoms of both tip and sample are overlapping and repulsing. The further approach results in cantilever bending that depends on the material properties of the investigated sample (can be linear or nonlinear, region BC). During separation of two surfaces, the interacting repulsive force decreases (region CD).

16.3.1.3.4
Pull-off (D)

During retraction the tip is not separated from the surface exactly at the same point where it started to touch the surface. Forces that are responsible for such behavior arise from adhesive properties of investigated surfaces. When the elastic force exceeds the gradient of the adhesive force, the tip is rapidly separated from the surface. The point D reflects the maximum value of the force (so-called *pull-off force*). Further separation results in cantilever fluctuations around its free position (*base line*).

16.3.2
Force Curves Calibration

In force spectroscopy, the raw dataset represent usually a photodetector response versus a voltage applied to a piezoelectric scanner. Both these quantities have to be converted into the dependence between a force (in nN) and a relative surface (or scanner) position (in nm). This requires the knowledge of the cantilever spring constant, the piezoelectric scanner nonlinearity, and the photodetector sensitivity. The detailed calibration procedure is described in [21].

16.3.2.1
Normal Spring Constant Determination

The force calibration requires the knowledge of the normal cantilever spring constant. The most widely applied method for spring-constant calibration uses the measurement of resonant frequency of thermally excited cantilever. The detailed procedure of the spring-constant determination carried out in this way was described in [22].

16.3.2.2
Scanner Linearization

The correction of the piezoelectric scanner nonlinearity and hysteresis is crucial for exact knowledge of a real displacement. In commercially available AFMs, scanner linearization is frequently an integral part of their hardware and thus scanner calibration is not needed. Real displacements of the sample in all three axes are measured using optical, capacitive or another method. A simple way to verify the scanner performance is a raw measurement of a force curve on a stiff, nondeformable surface and checking whether the approach curve is identical to the retraction one and if it is represented by a straight line. If this is true, the piezoelectric scanner is hardware-linearized. If not, the piezoelectric scanner should be linearized offline, using for example the method proposed by Jaschke and Butt [23].

16.3.2.3
Photodetector Signal Calibration

The AFM measures the cantilever deflection by monitoring the position of the laser beam reflected from the cantilever. The active area of the measuring position-sensitive detector (PSD) is usually divided into four quadrants. The cantilever deflection that is perpendicular to the investigated surface is related to the difference between signals coming from the two upper and two lower quadrants. In order to calibrate this cantilever deflection, the force curve should be recorded on a stiff, nondeformable surface (e.g. glass or mica), where the deflection directly reflects the position of the sample.

This dependence is a straight, sloped line and it is usually further employed as a reference. The slope determines the calibration coefficient between the PSD signal (in volts) and the real displacement (in nanometers).

16.3.3
Determination of the Unbinding Force

The value of the unbinding force needed to separate the two interacting molecules is delivered from the analysis of the force curve. The pull-off force $F_{\text{pull-off}}$ is determined as a difference between the force F_{base}, corresponding to the free cantilever position (when the interacting force is negligible), and the maximum value of the force F_{min}. Figure 16.10 presents the exemplary force curve (only retracting part is shown) with the marked difference between F_{base} and F_{min}.

In such studies, proper control experiments proving that the observed interaction comes indeed from investigated molecular pairs must be carried out. The common way of performing such controls is to block the studied interaction by adding to the solution the same type of molecules as those attached to the tip surface. These free molecules bind to the binding sites causing them to become inactive during probing with the functionalized AFM tip.

Figure 16.11 shows the inhibition of the specific interactions studied between concanavalin A (Con A, lectin from *Canavalia ensiformis*) and two glycoproteins

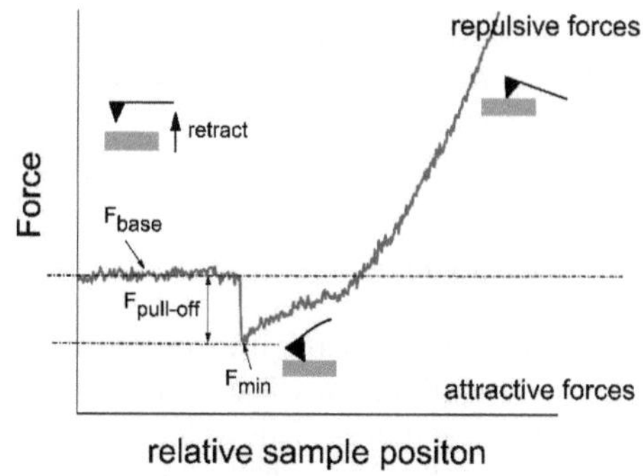

Fig. 16.10. Typical force curves recorded for HCV29 cell probed using PHA-L (lectin from *Phaseolus vulgaris*) functionalized silicon nitride tip, showing single unbinding event. The pull-off force $F_{\text{pull-off}}$ is determined as a difference between the free cantilever position F_{base} when the interacting force is negligible and the maximum value of the force F_{min}

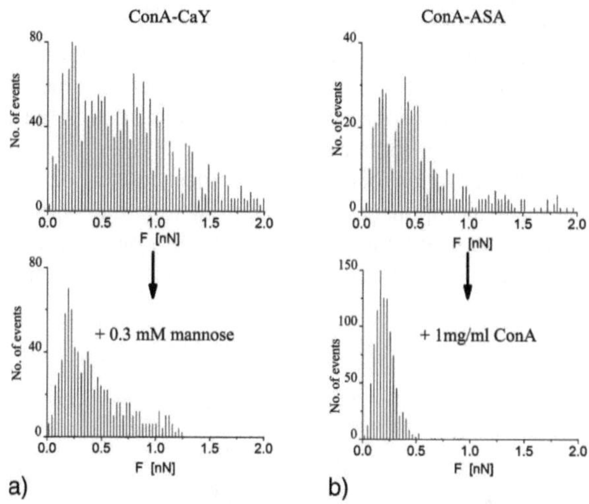

Fig. 16.11. The inhibition of the specific interaction that occurs between ConA and either CaY (**a**) or ASA (**b**) representing the lectin–carbohydrate interactions. The AFM tip was functionalized with Con A and both proteins were deposited on a glass surface. After adding to the solution either mannose or Con A, most binding sites of ASA and CaY were occupied by these molecules. As a result, a decreased number of force curves showing the adhesion events was observed (from [24], with kind permission of Springer Science and Business Media)

bearing mannose-type oligasachcarides: carboxypeptidase Y (CaY) and arylsulfatase A (ASA). After adding to the solution either mannose or concanavalin A, a number of force curves recorded for the interaction between Con A–ASA and Con A–CaY decreased significantly, thereby proving the specificity of the studied interaction [24].

16.3.4
Data Analysis

Extensive data analysis is crucial for obtaining the information about the chemical or biochemical interaction. The unbinding force that is characteristic for a specific interaction between a single pair of molecules can be obtained only after a careful analysis of a large number of recorded force curves showing adhesion events. The most popular approach is based on analysis of force histograms. Other methods use either the assumption that the number of bonds is governed by the Poisson distribution or rely on determination of the so-called separation work. All three techniques are described below.

16.3.4.1
Force Histograms

Force histograms of the investigated interaction are created using the bin size reflecting the minimum value of the detected force that is limited by thermal vibrations of a cantilever. This value can be estimated using the following equation: $F_{det} = (k_B \cdot T \cdot k)^{0.5}$, where k is a cantilever spring constant, k_B is the Boltzmann constant, and T is the temperature. For a cantilever with the spring constant of 0.03 N/m at the room temperature, the minimum of the detection force is 11.5 pN. This value is the theoretical one and it might not reflect the true force resolution. Therefore, the more practical approach is to determine this value from the noise fluctuations of the base line of force curves.

Interaction forces occurring between molecules consist of two components: i) discrete, short-range component, dominating within the binding sites, that is related to the bond of the single molecular pair and ii) component originating from long-range, distance-dependent forces dominating outside of the binding site.

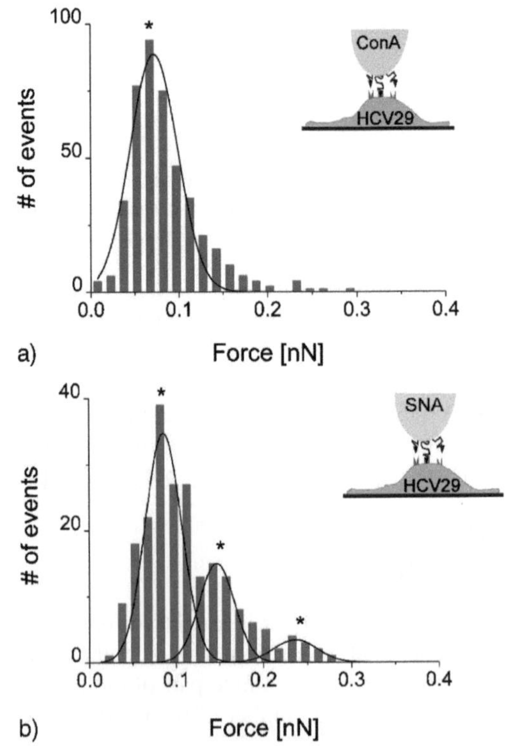

Fig. 16.12. Force histograms showing only one peak and multiple peaks that are attributed to the specific interaction occurring between glycans present on the surface of HCV29 cells and two probing lectins (adapted from [25]): Con A (**a**) and SNA (**b**). The *line* is a Gaussian fit used for determination of the unbinding force

The shape of histograms depends on the case studied and on the number of molecules present within the contact area. Histograms can show either only one peak that can be directly attributed to a single bond rupture or multiple peaks that correspond to the case of when simultaneous rupture of one or more bonds takes place (Fig. 16.12, [25]).

The common way of unbinding force determination attributes the position of the first peak in the obtained histogram to the unbinding event of a single molecular complex. The bond force is calculated as a center of the fitted Gaussian distribution and the corresponding error is a standard deviation determined from the half-width of the peak at half-height. However, this simple approach is reliable only under the assumption that the unbinding probability is lower than 30% [26]. In that case only a few molecular bonds are expected to be formed.

When multiple peaks are observed in a force histogram, they are usually attributed to the possibility of formation (and rupture) of more than one bond within the contact area of the AFM probe and the surface. Thus, the first peak corresponds to the unbinding event involving the rupture of one bond, the second is related to the simultaneous unbinding of two bonds (thus, the force values at the second maximum is doubled), the third – the unbinding of three bonds, etc. Such a force histogram can be translated into the relationship of the unbinding force determined for each consecutive peak and the peak number (i.e. bond number) and the linear dependence is expected if only one type of interaction is present (see Fig. 16.13). The superposition of a few types of interactions deviate the experimental data from the fitted line. The obtained unbinding force of a single complex should overlap (within the experimental error) with the values determined on the basis of the unbinding force histogram except in the situation where the presence of the nonspecific forces (that compete with the specific ones) dominates. These nonspecific forces can arise e.g. due to the procedure of the cantilever modification. The advantage of using linear regression is that errors are smaller and the obtained value is only slightly influenced by other nonspecific forces.

Figure 16.13 presents such a linear relation for the interaction studied between lectin SNA (from *Sambucis nigra*) and its sialic acids composed of glycans present on the surface of living HCV29 cells [25]. The exact value of the unbinding force for

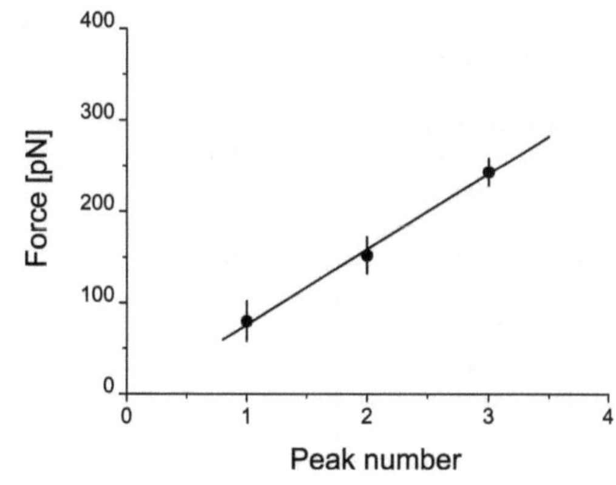

Fig. 16.13. Linear regression fitted to the unbinding force as a function of the number of succeeding peaks observed in a given histogram measured for HCV29 (*black dots*) cells probed with SNA (lectin from *Sambucus nigra*). Data points correspond to centers of Gaussians fitted to each single peak present in the force histogram from Fig. 16.12b, while *error bars* represent their standard deviations. Taken from [25]

a given individual lectin–oligosaccharide complex was determined from the slope of the fitted line. This value overlapped with that obtained on the basis of the unbinding force histogram (69 ± 11 pN versus 73 ± 28 pN).

16.3.4.2
A Statistical Approach Using the Poisson Distribution

The unbinding force, characteristic for a single molecular pair, can also be analyzed on the basis of a statistical approach using the Poisson distribution [27, 28]. The method is based on two assumptions: i) the pull-off force measured by AFM is a sum of a finite number of discrete, independent, randomly formed chemical bonds of similar strength values and ii) the distribution of the number of bonds is governed by Poisson statistics.

The mean adhesion force μ is a product of the value of the single-bond force F and the mean number of bonds n active during the contact of the tip with the surface ($\mu = n \cdot F + F_0$, where F_0 is a value of a nonspecific force). For the Poisson distribution, the variance of bond number is equal to its mean value. Thus, the dependence between the variance and the mean adhesion force is linear ($\sigma_\mu^2 = F \cdot \mu - F \cdot F_0$) and its slope determines the single-bond force F, while the intercept with the Y axis describes an additional nonspecific force F_0. If within the contact area between the AFM tip and surface only one type of bond is formed (F is constant), the dependence is linear. The more complicated interactions, like e.g., overlapping of two interaction types or when force F has some spread, will manifest in a nonlinear character.

As an example, Fig. 16.14 presents the linear relationship between the mean value of the adhesion force and its variance for prostatic acid phosphatase (PAP) interacting with its monoclonal antibody (aPAP, antiprostatic acid phosphatase from mouse ascites fluid, clone PAP–12). The slope of the line gives the value of the force required to break a single ligand–receptor pair. The result shows rather large force of about 530 ± 25 pN [24].

If a nonspecific force is present, it manifests in the linear dependence as a nonzero value of a Y-axis intercept. The nonspecific component for the studied PAP–aPAP complex, calculated from the last equation, was 406±60 pN, which is comparable to

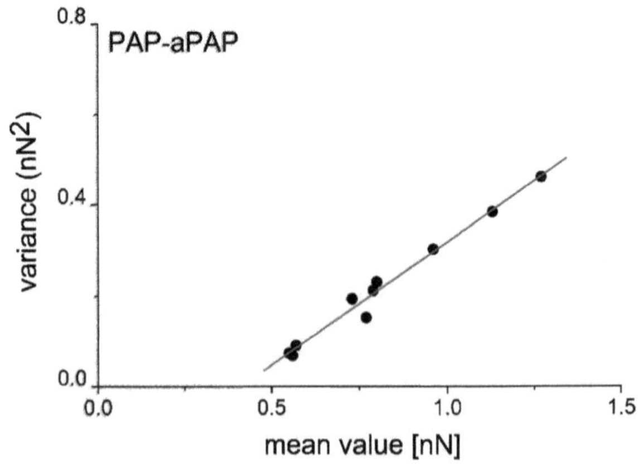

Fig. 16.14. Variance as a function of the mean value determined for prostatic acid phosphatase (PAP) interacting with its monoclonal antibody (aPAP, antiprostatic acid phosphatase from mouse ascites fluid, clone PAP–12). The slope of the line gives the values of the force required to break a single ligand–receptor pair (from [24], with kind permission of Springer Science and Business Media)

the bond force value. This may indicate a strong significance of nonspecific interactions occurring in the presence of other molecules, due to a charge on a cell surface, etc. Both (specific and nonspecific) forces are large and, when added together, produce a high value of the adhesion force (even of about 1000 pN upon the assumption that the adhesion force is a sum of specific and nonspecific forces). Therefore, the simple assumption of a constant nonspecific force value seems not to be realistic and a model refinement should be developed by introducing some distribution of a nonspecific force, represented by σ_{F0}^2. Then, the dependence between the variance and the mean value can be rewritten as follows:

$$\sigma_{\mu}^2 = F \cdot \mu - F \cdot F_0 + \sigma_{F0}^2 \ .$$

The dependence between the variance and the mean adhesion force remains linear. Its slope still determines the single-bond force but its intercept with the Y-axis is a two-term expression containing the nonspecific force and its variance. Using this approach, for PAP–aPAP pair, the average value of the nonspecific force was about 180 ± 130 pN.

16.3.4.3
Separation Work Determination

The third method used for quantification of adhesive interactions employs the concept of the so-called "separation work" [29]. The method was developed to facilitate detection of the specific interaction on a surface of living cells, where the density of membrane receptors is low. Therefore, it is used for the evaluation of the total adhesion strength per unit contact area and does not enable the determination of the unbinding force of a single pair of molecules [30]. In this method, raw force curves are converted into force–extension curves. Then, the area under such a curve is integrated in order to get a quantity with dimension of energy, i.e. separation energy. Values of the separation energy obtained in that manner form a histogram, delivering the overall value of the adhesion force. Despite the fact that this method averages all interactions over the contact area, it seems to be a good measure of the specific interactions, thus complementing the other two analysis techniques.

16.4
Detection of the Specific Interactions on Cell Surface

Many different techniques have been applied so far to investigate the unbinding processes between two molecules. These methods encompass, for example, a biomembrane force probe with pipette suction [31], a hydrodynamic flow-based method [32], and also atomic force microscopy (AFM, [33]). AFM offers a convenient way to measure forces involved in specific interactions by detection of the unbinding events between the AFM tip functionalized with one type of molecule (usually a ligand) probing its complementary partner (a receptor) that can be either immobilized on the substrate or be present on a surface of the cell plasma membrane. The quantitative determination of the strength of interaction between different pairs of molecules is usually obtained from the force-spectroscopy measurements where the molecular interaction forces can be determined with a force resolution down to tens of piconewtons. It should be pointed out that the unbinding force can be measured in two ways: i) as a measurement of the unbinding force only, giving the strength of interaction at a given experimental condition and, ii) as a function of the logarithm of loading rate (dynamic force spectroscopy) enabling the study of the unbinding pathways.

This chapter is divided into two sections. The first one is devoted to measurements performed for simple systems composed of isolated molecules immobilized on the AFM tip and substrate surfaces. The second section describes the detection of the cell-adhesion molecules present on a surface of living mammalian cells and their characterization using the unbinding force value. However, it should be mentioned here that only a small sample of possible applications of the AFM techniques to study the interaction forces occurring between single molecules is presented below.

16.4.1 Isolated Proteins

At first sight, measurements of the unbinding force between isolated proteins may have limited applicability, since in nature proteins are embedded in the plasma membrane. However, from such AFM experiments, it is possible to characterize the interaction between a single pair of molecules, which can be useful for the investigations of a wide range of biochemical or immunological questions at the molecular level.

Initially, much effort was put on the biotin–(strept)avidin interaction since it can serve as the primary model [34]. Besides the biotin–(strept)avidin complex, several other biological pairs have been investigated with the use of AFM. This group encompasses the interaction with cell-adhesion molecules and with various types of immunoglobulins. The unbinding force was determined for several model systems, where the studied proteins were isolated. In all these studies, the distributions of the pull-off forces were measured to quantify the unbinding forces occurring between single molecules. However, the direct comparison of the unbinding force obtained for different molecular pairs is difficult since the unbinding force depends strongly on the loading rate. For example, the unbinding force of the complex antiHSA–HSA (polyclonal antibody to human serum albumin and human serum albumin [20], respectively) was of about 240 pN measured for the loading rate of 54 nN/s. At the other end, one can place the interaction force measured for the insulin–insulin complex where the large unbinding force of 1345 pN was obtained for low loading rate of 7.4 nN/s [35]. Therefore, the unbinding force can be used as a parameter characterizing a given molecular complex only under the assumption that all experimental conditions are similar.

The dependence of the unbinding force on the loading rate can be explained in terms of Kramer's rate theory [36], where a binding affinity is characterized by rates of a bond association and dissociation. The rate theory was delivered for bond formation and dissociation between molecules that are freely dissolved in solution. In reality, the majority of molecules involved in cellular interactions are embedded in a cell plasma membrane and frequently dissociation depends on the external forces causing the bond to rupture and thus modifying their dissociation rate. The first theoretical description of the unbinding processing due to the action of the external force was presented in 1978 by Bell [37], who stated that such a force determines the lifetime of the bond. The first measurements performed using AFM reported by Lee and coworkers in 1994 [38] produced data that were consistent with the Bell's theory. Later, Evans and Ritchie [16] showed that the unbinding force observed in each particular experiment is a function of the applied loading rate describing how fast the external force applied to a bond changes in time. The applied external force will lower the energy barrier(s) of the energy landscape of the two interacting molecules. The loading rate is usually determined as a product of retraction velocity and the system spring constant taking into account the cantilever and bonds spring constants. Each line on the plot of the force versus logarithm of the loading rate corresponds to a single energy barrier.

The interaction potential of biotin and (strept)avidin contains several energy barriers as was predicted from molecular dynamic simulations [39]. Thus, the dependence of the unbinding force versus the logarithm of the loading rate should consist of multiple linear sections where one energy barrier should be represented by each linear section. These predictions have been confirmed by Merkel and coworkers [40] in 1999. They used a biomembrane force probe to measure the unbinding force as a function of the loading rate applied over six orders of magnitude for biotin–avidin and biotin–streptavidin pairs. The two linear sections were obtained, indicating the presence of two energy barriers in the molecular potential showing thereby that the unbinding process proceeds through the intermediate state. From these measurements, the position of the energy barrier(s) and the dissociation rate can be determined.

Dynamic force spectroscopy (DSF) was performed for many distinct pairs of molecules, bringing deeper insight into molecular mechanism of the bond breaking. Apart from the biotin–(strept)avidin complex, the DSF measurements were applied to study the selectin interactions since during rolling these types of bonds are exposed to the action of the external forces [41]. There are three types of selectins, P, L,

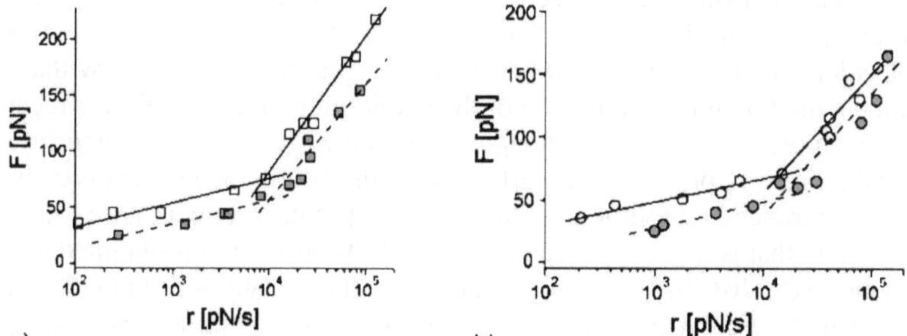

a) b)

Fig. 16.15. The unbinding force as a function of the logarithm of the loading rate obtained for selectin E (**a**) and P (**b**) interacting with sialyl Lewis X antigen [40]. Measurements were preformed in a buffer containing 2 mM $CaCl_2$. *Gray points*, in both cases, correspond to the measurements carried out after neuraminidase treatment. To each linear regions, the line was fitted and the Bell model parameters (the position of the energy barriers and the dissociation rates) were determined (adapted from [41]) and E. They are present on the surface of leucocytes that upon activation interact with ligands present on the endothelium. AFM measurements were performed in search of ways of identification of the molecular determinants within sialyl Lewis X (sLeX). It is an oligosaccharide moiety present in the structure of the endothelial protein. Its structure contains two sugars, sialic acid and fucose, participating at distinct degree in interactions with different types of selectins. The dependence between the unbinding force and the logarithm of loading rate showed two linear sections (Fig. 16.15).

The two linear sections obtained for each selectin type (P, E, and L) and their sLeX ligand indicate that the unbinding process involves overcoming of two energy barriers located at different distances from the energy minimum (bound state). In addition, results have showed that the inner barrier (detected at high loading rates) is governed by the interaction between fucose and Ca^{++} ion, whereas the lower loading regime corresponds to the outer barrier in interactions dominated by sialic acids.

16.4.2
Receptors in Plasma Membrane of Living Cells

Investigations of receptor molecules that are present on the surface of living cells produce more difficult results to be interpreted. However, after solving these problems, AFM application will in future result in the powerful technique that can be applied not only for affinity imaging of biologically active surfaces but also for obtaining the quantitative description of the interaction forces between molecules.

16.4.2.1
Affinity Imaging

Parallel to determination of the interaction strength, AFM tip functionalization has opened up the possibility of the imaging where the modified tip is scanned over a surface. The pull-off force, related to the overall adhesive interaction within the contact area, will contribute to the contrast in a recorded image, which arises from the biomolecular specific interactions. Few examples presented below show that AFM is a suitable method for visualization of the distribution of the membrane receptors.

The example of such affinity imaging was shown for the $\alpha_V\beta_5$ integrin where the distribution was probed on a surface of living murine osteoblastic cells with the use of vitronectin–modified AFM probe [42]. Vitronectin is an extracellular matrix protein that is involved in cell attachment. Adhesion maps obtained by AFM correspond to the distribution of $\alpha_V\beta_5$ integrin on the cell surface obtained from the immunohistochemical measurements. Both results showed similarities, however the obtained images were not completely the same.

The affinity imaging has been also nicely demonstrated on the population of two groups of erythrocytes [43]. The erythrocyte surface was probed using an AFM tip functionalized with the lectin from *Helix pomatia* that recognizes *N*-acetylgalactosamine-terminated glycolipids in erythrocytes of the A group. After functionalization, tip was used for imaging of a red blood sample, composed of erythrocytes coming from two groups A and 0 (Fig. 16.16).

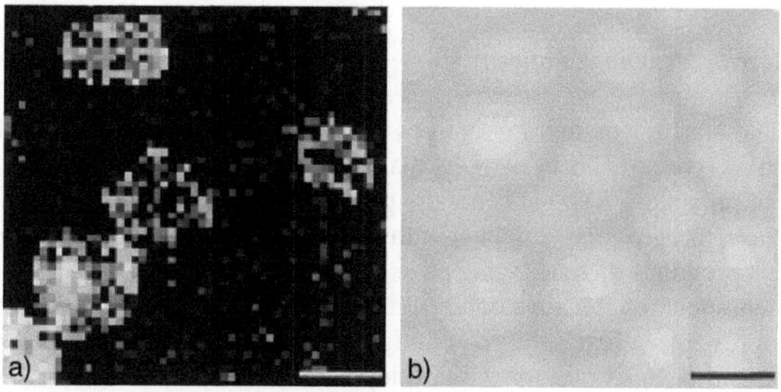

Fig. 16.16. (**a**) Adhesion image recorded on a layer of mixed group A and 0 of red blood cells recorded using the AFM tip functionalized with lectin from Helix pomatia having the specificity to the *N*-acetylgalactosamine-termined glycolipids present in the membrane of group A of red blood cells. (**b**) The corresponding topographic image of the erythrocyte layer. *Bars* = 5 μm. (Reproduced with permission from M. Granbois et al.: *Affinity Imaging of Red Blood Cells Using an Atomic Force Microscopy*, Journal of Histochemistry 48:722, 2000)

Using the modified AFM tip it was possible to distinguish between the two erythrocytes' populations coming from group A and 0 mixed within a sample. The obtained image (Fig. 16.16) showed the contrast based on the specificity of lectin from *Helix pomatia*. Only a few bonds were needed to produce the contrast necessary for affinity imaging. The unbinding force determined for this lectin was 65 pN (for group A of erythrocytes recorded at the loading rate of 1000 pN/s), which was almost twice the force recorded for the agarose beads (35 pN).

16.4.2.2
Quantitative Interaction Strength Determination

When static force spectroscopy is performed under certain experimental conditions, it provides useful information about the strength of the interaction of different types of molecules present on the surface of living cells. The expression of cell-surface molecules can be quantified by two factors. The first parameter is the unbinding force of a single pair of molecules. The unbinding force is directly measured by AFM due to its very high force resolution and its value is characteristic for a studied complex and current experimental conditions. A stronger unbinding force means the more difficult rupture of the molecular complex and thereby indicates the more stable complex formed. When a certain ligand type, recognizing the specific structural fragment of its receptor, is immobilized on the AFM probe, any variations of the force value indicate the alteration in the binding-site structure of the receptor. The second parameter, the unbinding probability, corresponds to the number of molecules that are present on the surface of living cells and can be related to the density of surface receptors. Examples presented below suggest the use of AFM as a technique alternative to other biochemical methods in characterization of the expression of cell-surface molecules. These results also highlight the applicability of AFM to investigate the interaction forces between single molecules with reference to cancer-drug discovery.

16.4.2.2.1
Prostate-Specific Membrane Antigen

The prostate-specific membrane antigen (PSMA) is nowadays widely studied in search of new types of cancer markers that allow better detection of prostate cancer. In a normal prostate, PSMA is generally present in soluble form in cytoplasm [44]. During cancer progression, cells begin to express more PSMA and the dominating form of this protein is the one that appears in plasma membranes – PSMA [45]. This unique phenomenon – the appearance of insoluble, membrane form of PSMA upon cancer progression – creates the "new antigenic" target present on cancer cell plasma membrane and brings hope for the development of successful immunotherapeutic strategies in the treatment of prostate cancer.

Investigations of the effect of certain growth factors on PSMA expression showed that basic fibroblast growth factor (bFGF) and estradiol were able to up regulate the expression of this protein [45]. Parallel to other techniques applied, AFM was chosen as a method giving additional evidence that bFGF and estradiol can restore the expression of the membrane-anchored form of PSMA in a PC-3 prostate cancer cell line that lost the ability to express this protein, probably due to cancer progression while transforming to metastatic phenotypes [46]. The presence of PSMA antigen was investigated on the surface of three types of prostate cell lines: LNCaP (androgen-dependent), PC-3 and Du 145 (androgen-independent). For the latter two cell lines, the AFM analysis showed that there were rather few PSMA molecules present in their plasma membranes.

The unbinding force of PSMA–anti-PSMA (monoclonal antibody against prostate-specific membrane antigen) interaction was determined on the basis of force histograms, where only one peak was observed, independent of the studied case. The calculated values for a single molecular pair composed of PSMA–anti-PSMA on either LNCaP or PC-3 or Du 145 cell membrane are presented in Table 16.1.

The calculated strength of a single bond of PSMA–anti-PSMA present on LNCaP, PC-3 or Du 145, regardless of bFGF or estradiol treatment (or lack of treatment), was

Table 16.1. The binding force obtained for the interaction occurring between the AFM tip functionalized with anti-PSMA (monoclonal antibody against prostate-specific membrane antigen), and PSMA present in the plasma membrane of prostate cells (LCNaP, PC-3 and Du145 cell lines), measured in different culture conditions. Cells were grown in the medium supplemented in FCS, after bFGF (basic fibroblast growth factor) and estradiol treatment [46]

Cell line	Condition	Unbinding force [pN]
LNCaP	FCS	63 ± 27
	bFGF	44 ± 17
	estradiol	58 ± 25
PC-3	FCS	61 ± 23
	bFGF	45 ± 17
	estradiol	65 ± 16
Du 145	FCS	65 ± 27
	bFGF	50 ± 22
	estradiol	60 ± 14

in the range of 45–64 pN. Taking into account the experimental error, these results suggest the antigenic identity or a very close similarity of the membrane form of expressed PSMA in all studied cell lines.

16.4.2.2.2
Surface Glycans

The feasibility of AFM to investigate the oligosaccharide's structure on the surface of living bladder cells was demonstrated with the use of lectins as probes [25]. In this work, plasma membrane oligosaccharides of two human bladder cell lines were probed directly on the surface of a living cell using AFM with the probing tip functionalized with three distinct lectins: concanavalin A (Con A, lectin from *Canavalia ensoformis*), lectins from *Sambucus Nigra* (SNA) and from *Phaseolus vulgaris* (PHA-L). Measurements were performed in order to quantify the expression of oligosaccharides on the plasma membrane of cancer bladder cells. Using AFM, their expression was quantified by the unbinding force giving the strength of interaction occurring between a single pair of molecules and by the unbinding probability indicating the number of oligosaccharide's ligands present on the surface of living bladder cells (Fig. 16.17). The obtained results showed the differences of the oligosaccharide's expression in cancer cells (T24) compared with the reference cells (HCV29).

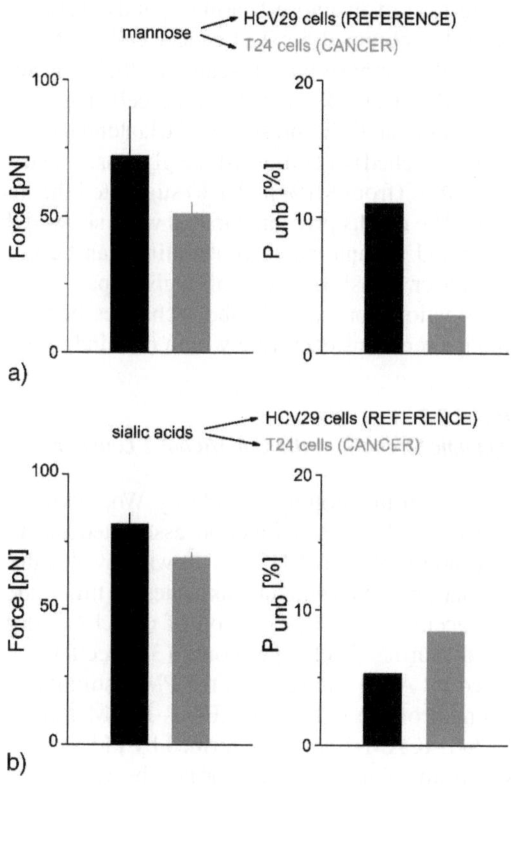

Fig. 16.17. The unbinding force and the unbinding probability determined in search of the quantification of the expression of surface glycans (composed of mannose (**a**) and sialic acids (**b**)) on the surface of living bladder cells: cancer (T24) and reference ones (HCV29). Adapted from [25]

The unbinding force corresponds to the strength of interaction occurring between a single pair of molecules. When a given adhesion molecule is probed with the same ligand, the unbinding force is expected to be the same if structures of both adhesion molecule and ligand are unchanged. Force variations can indicate changes in the structure of the binding site, either in the adhesion molecule or in the ligand. The obtained results showed small dissimilarity in the unbinding force values for ConA and SNA lectins, indicating in both cases a somewhat altered structure of the binding site of cell-surface ligands. This fact can be attributed to significant structural changes of oligosaccharide ligands caused by cancer transformation influencing the binding stability. On the other hand, the unbinding force value characterizes the interaction as weak or strong that corresponds to formation of the more or less stable complex. Higher values pointed to the formation of more stable complexes with oligosaccharides (composed of mannose and sialic acids, respectively) present in non-malignant cells (HCV29) than with those present on cancer surface (T24 cells).

The unbinding probability can be attributed to the number of single bonds that are formed between the lectin and its ligand present on a cell surface, which describes in a quantitative way the expression of cell-surface molecules. The values determined by AFM in this work were in agreement with the indirect (and rather qualitative) results obtained using standard biochemical methods. The higher content of sialic acid residues (5.3% versus 8.4% for HCV29 and T24 cells, respectively) has been reported as a consequence of an increased sialylation of cancer cells [25]. Larger expression of mannose-bearing ligands (11%) in HCV29 cells in comparison with the value obtained for T24 cancer cells (2.8%) showed a much higher amount of the high mannose-type glycans on their surface that was previously reported by Przybyło et al. [48] for the same cell line. The specific interaction with PHA-L lectin indicated the presence of triantennary or tetraantennary structures (GlcNac β1,6-branched) of complex-type glycans. The decrease of the unbinding probability of PHA-L (from 8.1% to 2.1%) suggested the loss of such structures in T24 cancer cells. The results presented in this work showed that both parameters (the unbinding force and the unbinding probability) can be used to trace changes (induced, e.g., by cancer transformation) of a given pair of molecules, bringing also additional information, for example about changes in a number of cell-surface molecules or about alterations occurring within the binding site.

16.4.2.2.3
Dynamic Strength of the LFA-1/ICAM-1 Complex

It is worth mentioning a study by Wojcikiewicz et al. [49], where the interaction between a leukocyte function associated antigen-1 (LFA-1) and an intercellular adhesion molecule-1 (ICAM-1) was investigated. Such interaction is important in regulation of adhesion in leukocytes. In this study, the mechanism of phorbol myristate acetate (PMA) promotion of the LFA-1-dependent adhesion was investigated for a murine T-cell hybridoma (3A9 cell line). The single-molecule force measured by AFM showed that the PMA stimulation does not modify the mechanical strength of the individual LFA-1–ICAM-1 interaction. The force spectrum of the LFA-1–ICAM-1 complex formed by PMA-stimulated cells is identical to the force spectrum of the complex formed by resting cells. The force required to unbind

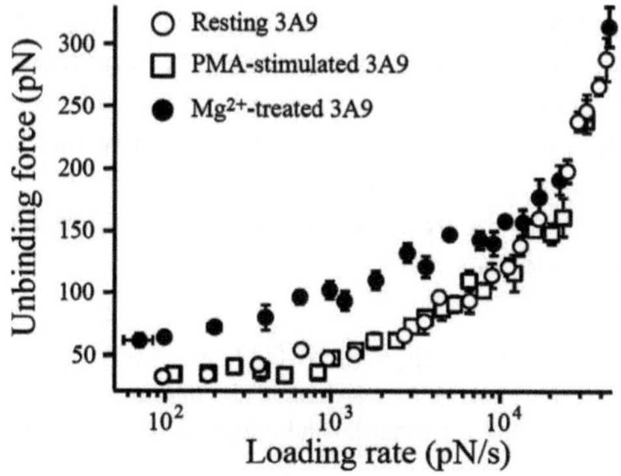

Fig. 16.18. The unbinding force of a single LFA-1–ICAM-1 complex as a function of the logarithm of the loading rate [49]. Measurements were performed for i) resting 3A9 cells (*open circles*), ii) PMA–stimulated cells (*open squares*), and iii) Mg^{2+}/EGTA activated cells (*black circles*), at loading rates between 20 and 50,000 pN/s. *Error bars* represent standard error of the mean that was calculated for five cells with an average of 100 measurements per single cell. (Reprinted with permission from J Cell Sci 2003 116:2531–2539, Company of Biologists Ltd.)

a single LFA-1/ICAM-1 bond was measured at different loading rates ranging from 20 pN/s to 50,000 pN/s. The average unbinding force of the LFA-1 and ICAM-1 complex increases by more than three orders of magnitude, Fig. 16.18.

The force versus loading rate dependence shows evidently two regimes for the resting 3A9 cells and PMA-stimulated cells. The gradual increase of the unbinding force was observed for loading rates up to 10,000 pN/s and the second regime was visible for higher rates. Cells activation with Mg^{2+}/EGTA resulted in higher unbinding forces that were pronounced at the slow loading rate regime. In addition, there were no significant differences in the dynamic response of the low and high affinity complexes for higher loading rates. This work demonstrates that the unbinding force together with the dynamic response of cells during bond breaking can be used for the complete characterization of a given molecular interaction.

16.5 Summary

A natural question arises why the receptor–ligand binding is so important in biological systems and why it is so deeply investigated. The answer is the following. First, the binding between molecules is one of the methods that cells utilize to interact with a huge variety of different types of molecules that are delivered not only from the outside environment but also with those that are generated within the cell itself. Secondly, the binding is the basis of a great number of normal biological functions such as immunological response, embryogenesis, cell growth, differentiation, gene regulation, enzyme catalysis, etc., that are crucial for our life. The loss or alterations of these interactions can affect cell functioning and may lead to many pathological states. Numerous diseases are related to malfunctioning of these molecular recognition processes. Investigations of the interaction mechanism are crucial for understanding a wide spectrum of biological processes present during inflammation or cancer metastasis, which should lead, among other applications, to the development of highly specific drugs. The biological cell adhesion is a complex process and, in most cases, it is not possible to write down a set of universal laws that explain the cell-adhesion phenomena. Such techniques as atomic force microscopy deliver only the possibility of obtaining the detailed and quantitative characterization of various types of involved interactions.

Acknowledgements. The authors are thankful to Dr Janusz Lekki, Dr Zbigniew Stachura and Dr Wojciech Zając from the Henryk Niewodniczański Institute of Nuclear Physics (Cracow, Poland) and Mr Stephan Należny for remarks and corrections of the manuscript.

References

1. Nelson DL, Cox MM (2005) Lehninger Principles of Biochemistry, 4th edn, WH Freeman and Company, New York
2. Branden C, Tooze J (1998) Introduction to Protein Structure, 2nd edn, Garland Scientific Books
3. Varki A, Cummings R, Esko J, Freeze H, Hart G, Marth J (1999) Essentials of Glycobiology, Cold Spring Harbor University Press
4. Laidler P, Lityńska A (1997) Acta Biochim Pol 44:343
5. Wilson IA, Stanford RL (1994) Curr Opin Struct Biol 4:857
6. Voet D, Voet JG (1990) Biochemistry, John Wiley & Sons, New York
7. Kreis T, Vale R (1999) Extracellular Matrix, Anchor and Adhesion Proteins, A. Sambrook & Tooze Publications, Oxford University Press, Oxford
8. Beckerle MC (2001) Cell Adhesion. In: Hames BD, Glover DM (eds) Frontiers in Molecular Biology. Oxford University Press, Oxford
9. Lodish H, Berk A, Matsudaira P, Kaiser CA, Krieger M, Scott MP, Zipursky SL, Darnell J (2004) Molecular Cell Biology, 5th edn, WH Freemann and Company, New York
10. Israelachvili JN (1992) Intermolecular and surface forces, Academic Press, London, San Diego, New York, Boston, Sydney, Tokyo, Toronto
11. Bongrand P (1999) Rep Prog Phys 62:921
12. Lipowsky R, Sackmann E (1995) Structure and Dynamics of Membranes, Elsevier, New York
13. Leckband D (2000) Annu Rev Biophys Biomol Struct 29:1
14. Gallego RG, Haseley SR, Van Miegem VFL, Vliegenthart JFG, Kamerling JP (2004) Glycobiology 14:373
15. Hu J, Yang D, Kang Q, Shen D (2003) Senss Actuators B 96:390
16. Evans E, Ritchie K (1997) Biophys J 72:1541
17. Horton M, Charras G, Lehenkari P (2002) J Recept Signal Transduct Res 22:169
18. Binning G, Quate CF, Gerber Ch (1985) Phys Rev Lett 56:930
19. Luckham PF, Smith K (1998) Faraday Discuss 111:307
20. Hinterdorfer P, Baumgartner W, Gruber HJ, Schilcher K, Schindler H (1996) Proc Natl Acad Sci USA 93:3477
21. Lekka M, Kulik AJ (2006) Quantitative nanomechanical measurements in biology. In: Bhushan B, Fusch H (eds) Applied Scanning Probe Methods II. Springer, Berlin Heidelberg New York, p 203
22. Sader JE, Larson I, Mulvaney P, White LR (1995) Rev Sci Instrum 66:3789
23. Jaschke M, Butt HJ (1995) Rev Sci Instrum 68:1258

24. Lekka M, Laidler P, Dulińska J, Łabędź M, Pyka G (2004) Eur Biophys J Biophy 33:664
25. Lekka M, Laidler P, Łabędź M, Kulik AJ, Lekki J, Stachura Z (2006) Chem Biol 13:505
26. Tees FJ, Waugh WE, Hammer DA (2001) Biophys J 80:668
27. Williams JM, Han T, Beebe TP (1996) Langmuir 12:1291
28. Lekka M, Gryboś J, Lekki J, Stachura Z, Styczeń J (2002) Acta Phys Pol A 102:40
29. Hyonchol K, Arakawa H, Osada T, Ikai A (2002) Colloids Surf B 25:33
30. Sneddon IN (1965) Int J Eng Sci 3:47
31. Evans E (2001) Annu Rev Biophys Biomol Struct 30:105
32. Pierres A, Benoliel AM, Bondrang P (1998) Faraday Discuss 111:321
33. Kienberger F, Ebner A, Gruber HJ, Hinterdorfer P (2006) Acc Chem Res 39:29
34. Chilkoti A, Ratner BD, Stayton PS (1995) Biophys J 69:2125
35. Yip CM, Yip CC, Ward MD (1998) Biochemistry 37:5439
36. Hänggi P, Talkner P, Borkovec M (1990) Rev Mod Phys 62:251
37. Bell GI (1978) Science 200:618
38. Lee G, Kidwell DA, Colton RJ (1994) Langmuir 10:354
39. Grubmuller H, Heymann B, Tavan P (1996) Science 271:997
40. Merkel R, Nassoy P, Leung A, Ritchie K, Evans E (1999) Nature 397:50
41. Zhang X, Bogorin DF, Moy VT (2004) Chem Phys Chem 5:175
42. Kim H, Arakawa H, Osada T, Ikai A (2003) Ultramicroscopy 97:359
43. Grandbois M, Dettmann W, Benoit M, Gaub HE (2000) J Histochem Cytochem 48:719
44. Izraeli RS, Powell CT, Copr. JG, Fair WR, Feston WD (1994) Cancer Res 54:1807
45. Grawer LS, Lawler KD, Marignac JL, Kumar A, Goel AS, Wolfert RL (1998) Cancer Res 58:4787
46. Diamond SM, Fair WR, Feston WD (1995) Proc Am Assoc Cancer Res 36:643
47. Laidler P, Duliñska J, Lekka M, Lekki J (2005) Arch Biochem Biophys 435:1
48. Przybyło M, Lityńska A, Pocheć E (2005) Biochemie 87:133
49. Wojcikiewicz EP, Zhang X, Chen A, Moy VT (2003) J Cell Sci 116:2531

17 Dynamic Force Microscopy for Molecular-Scale Investigations of Organic Materials in Various Environments

Hirofumi Yamada · Kei Kobayashi

17.1 Brief Overview

Atomic force microscopy (AFM) has been a most powerful, versatile technique for investigating nanometer-scale surface structures of various materials, including metals, semiconductors, insulators, and organic materials. AFM imaging can be performed in vacuum, air, and even in a liquid environment without any special treatment of samples such as staining or metal coating, which are common techniques in electron microscopy imaging. In addition, AFM allows us to study local surface properties on an atomic or molecular scale because various interaction forces between an atomic force microscope tip and a sample such as mechanical, electrical, and short-range chemical forces can be simultaneously detected.

The simple way to detect the interaction forces in AFM is to measure static deflection of the cantilever when the tip is fully in contact with the surface (contact mode AFM) [1]. The deflection is proportional to the acting force (Hooke's law). However, one can often experience that the scanning process in contact mode modifies or damages soft materials, especially biological samples, most of which are adsorbed onto a substrate by a weak van der Waals interaction. In order to reduce perturbing interactions by the contact tip, dynamic mode AFM (intermittent contact mode AFM) was developed, where the cantilever was dynamically vibrated with an instantaneous touch on the surface at a fixed frequency near the resonance [2]. Historically, the dynamic mode was proposed in the first AFM paper by Binnig et al. [1] in 1986 and was experimentally realized by Wickramasinghe et al. [3] in 1987, where the resonance enhancement of the cantilever was utilized to detect a weak van der Waals force. However, the closest position of the tip in their experimental setup was slightly above the surface, and therefore the resolution was limited.

There are two working modes in dynamic force microscopy (DFM). One is amplitude modulation (AM) AFM (often referred to as *tapping mode* AFM) using AM detection, which usually works in the intermittent contact regime, and the other is frequency modulation (FM) AFM using FM detection [4], which is mainly used in the noncontact regime. Average interaction/contact forces in typical imaging conditions of each mode are heavily reduced compared with the case of contact mode AFM and hence sample damage can be avoided.

In AM-AFM a cantilever is vibrated at a fixed frequency near the resonance frequency. The amplitude of the cantilever vibration is varied depending on instantaneous contact interactions in each oscillation cycle of the cantilever. Surface topographic data are obtained by measuring the change in the amplitude, from which the term AM-AFM is derived [5].

While the cantilever is vibrated at a fixed frequency in AM-AFM, the cantilever oscillation always tracks the resonance frequency in FM-AFM, where the cantilever works as a mechanical resonator. Interaction forces between the tip and the sample cause a shift of the resonance frequency and hence the same shift of the cantilever oscillation frequency. This frequency shift is used as a feedback signal for the z tip–sample motion, which corresponds to the topographic information. FM-AFM was originally invented by Albrecht et al. [4] for improving the slow time response of the change in the cantilever oscillation in a vacuum environment, where the high quality factor of the cantilever (approximately 10,000) makes the response extremely slow. Recently there has been great progress in FM-AFM [6]. High-resolution AFM imaging of atomically flat samples in ultrahigh vacuum (UHV) environments is becoming routine work. FM-AFM is now widely used as a high-resolution analysis method for various materials. In this chapter we describe fundamental basics and instrumentation of FM-AFM followed by recent progress in FM-AFM investigations of organic molecules.

A comparison between the two modes is made as shown in Table 17.1. These two modes are complementarily used according to the experimental conditions. AM-AFM is usually used in the intermittent contact regime in air or a liquid environment, while FM-AFM is mainly operated in the noncontact regime in a vacuum. Note that AM-AFM (FM-AFM) can be used in the noncontact regime (intermittent contact regime).

Table 17.1. Comparison between noncontact mode and intermittent contact mode atomic force microscopy (AFM)

	Noncontact mode	Intermittent contact mode
Working regime	Attractive regime	Repulsive regime
Feedback control method	FM-AFM	AM-AFM
Working environment (Q-factor)	In vacuum ($Q > 10{,}000$)	In air/in liquid ($Q = 1$–1000)
Detection bandwidth	Bandwidth of FM detector (up to $\omega_0/2\pi$)	ω_0/Q (large response time)

AM amplitude modulation, *FM* frequency modulation

17.2 Principles and Instrumentation of Frequency Modulation Detection Mode Dynamic Force Microscopy

17.2.1 Transfer Function of the Cantilever as a Force Sensor

As mentioned in the previous section, there are two working modes for the tip–sample distance regulation in DFM, which are the AM detection mode and the FM detection mode. Both modes utilize the shift of the resonance frequency of the cantilever when the tip is brought in close proximity to the sample surface.

In this section, we describe the equation of motion of the cantilever and derive the frequency response of the amplitude and the phase of the cantilever, which is oscillated around its resonance frequency. We model a cantilever end with a tip as a mass, whose weight is equal to the effective mass of the cantilever (m^*), as shown in Fig. 17.1. The mass is connected to an end of a spring whose spring constant is equal to that of the cantilever (k). The other end of the cantilever is vibrated by an actuator. The equation of motion is written as

$$m^*\ddot{s} + m^*\frac{\omega_0}{Q}\dot{s} + ks = F_0 \cos \omega_\mathrm{d} t + F_\mathrm{ts}\left(s + s_0\right) , \tag{17.1}$$

where Q is the mechanical Q-factor of the cantilever resonance. F_0 is the amplitude of the external cyclic force given by the actuator, and $F_\mathrm{ts}(z)$ is the tip–sample interaction force depending on the distance from the sample surface, z. s_0 is the equilibrium distance of the tip end from the sample surface. ω_d is the angular frequency of the actuator drive, and ω_0 is the angular resonance frequency of the cantilever, which is equal to $\sqrt{k/m^*}$. The resonance frequency of the cantilever (f_0) is

$$f_0 = \frac{\omega_0}{2\pi} = \frac{1}{2\pi}\sqrt{\frac{k}{m^*}} . \tag{17.2}$$

The solution of this equation gives the displacement of the mass in the stationary state. If we assume that F_ts is negligibly small and hence the motion of the mass is purely harmonic, the displacement of the mass is described as

$$s(t) = A_0 \cos\left(\omega_\mathrm{d} t + \varphi\right) , \tag{17.3}$$

where A_0 is the vibration amplitude of the mass, which corresponds to the vibration amplitude of the cantilever end with the tip. Inserting (17.3) into (17.1), we obtain

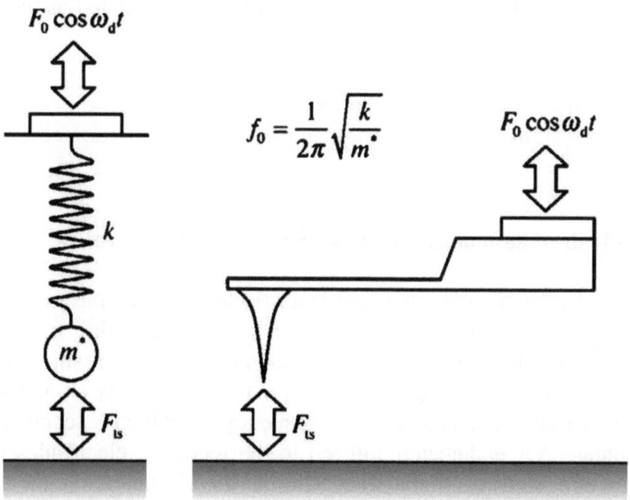

Fig. 17.1. Model of a cantilever as a mass connected to a spring

the following relationships.

$$A_0 = \frac{Q}{\sqrt{Q^2\left(\omega_0^2-\omega_d^2\right)^2+\omega_0^2\omega_d^2}}\frac{F_0}{m^*} \tag{17.4}$$

and

$$\varphi = \arctan\left(\frac{-\omega_0\omega_d}{Q\left(\omega_0^2-\omega_d^2\right)}\right). \tag{17.5}$$

Therefore, the complex transfer function of the cantilever as the force-to-deflection transducer can be written as

$$G_{\text{cantilever}}(\omega) = |G_{\text{cantilever}}(\omega)| \exp\left[i\theta_{\text{cantilever}}(\omega)\right], \tag{17.6}$$

where ω is the angular frequency of the cantilever vibration. Since $|G_{\text{cantilever}}(\omega)|$ is equal to A_0/F_0, and $\theta_{\text{cantilever}}(\omega)$ become

$$|G_{\text{cantilever}}(\omega)| = \frac{Q}{\sqrt{Q^2\left(\omega_0^2-\omega^2\right)^2+\omega_0^2\omega^2}}\frac{1}{m^*} \tag{17.7}$$

and

$$\theta_{\text{cantilever}}(\omega) = \arctan\left(\frac{-\omega_0\omega}{Q\left(\omega_0^2-\omega^2\right)}\right), \tag{17.8}$$

respectively.

For the case when Q is sufficiently larger than unity ($Q \gg 1$), $|G_{\text{cantilever}}(\omega)|$ around the angular resonance frequency ω_0 can be approximated as the Lorentzian function as

$$|G_{\text{cantilever}}(\omega)| \cong \frac{Q}{\omega_0\sqrt{4Q^2\left(\omega_0-\omega\right)^2+\omega_0^2}}\frac{1}{m^*}. \tag{17.9}$$

17.2.2 Detection Methods of Resonance Frequency Shift of the Cantilever

The interaction forces between the cantilever tip and the sample surface shifts the resonance frequency of the cantilever. A schematic illustrating the effect of the resonance frequency shift on the transfer function $G_{\text{cantilever}}(\omega)$ is shown in Fig. 17.2.

In the AM detection method, the cantilever is oscillated by an actuator such as a piezoelectric actuator at a fixed angular frequency ω_d, which is set close to ω_0. The amplitude change ΔA is detected and fed to the feedback electronics for the tip–sample distance regulation. On the other hand, in the FM detection, the cantilever is self-oscillated at the resonance frequency and the change in the oscillation frequency Δf is directly detected and fed to the feedback electronics.

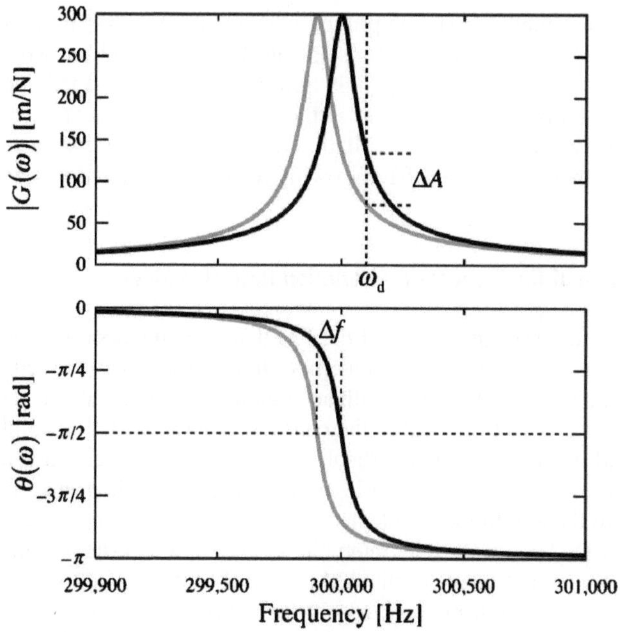

Fig. 17.2. The effect of the resonance frequency shift on the cantilever transfer function as the force-to-deflection transducer for a cantilever with $k = 10\,\mathrm{N/m}$, $f_0 = 300{,}000\,\mathrm{Hz}$, and $Q = 3000$

If the transfer function of the self-oscillation electronics including the deflection-sensing electronics and the piezoelectric actuator is given by

$$G_{\mathrm{electronics}}(\omega) = |G_{\mathrm{electronics}}(\omega)| \exp\left[i\theta_{\mathrm{electronics}}(\omega)\right] , \tag{17.10}$$

the Barkhausen criteria for self-oscillation at the angular oscillation frequency ω_{osc} are

$$\theta_{\mathrm{electronics}}(\omega_{\mathrm{osc}}) + \theta_{\mathrm{cantilever}}(\omega_{\mathrm{osc}}) = 2n\pi \tag{17.11}$$

and

$$|G_{\mathrm{electronics}}(\omega_{\mathrm{osc}})|\,|G_{\mathrm{cantilever}}\,(\omega_{\mathrm{osc}})| = 1 . \tag{17.12}$$

Equation (17.11), the phase condition, determines the angular oscillation frequency ω_{osc}. If ω_{osc} is tuned so that it becomes ω_0, where the optimum force sensitivity is achieved, and if the resonance frequency is shifted by $\Delta f = \Delta\omega/2\pi$ due to the interaction forces and the change in $\theta_{\mathrm{electronics}}(\omega)$ is negligibly small in the range $\omega_{\mathrm{osc}} \pm \Delta\omega$, the change in the oscillation frequency ($\Delta\omega_{\mathrm{osc}}$) becomes equal to $\Delta\omega$. Equation (17.12), the amplitude condition, describes that the oscillation amplitude is constant during a cycle.

The two operating modes bring the same force sensitivity, which is described later in this chapter. However, there is a major advantage of the response time in the FM detection method. The amplitude change (ΔA) responds to the change in the frequency shift with the time constant on the order of Q/ω_0 which limits the available operating bandwidth [4]. This limitation makes the operation of the AM detection method practically difficult in the high-Q environment such as an UHV condition. On the other hand, the oscillation frequency changes instantaneously with the time constant on the order of $1/\omega_0$ owing to the self-oscillation feedback loop.

17.2.3
Instrumentation of the Frequency Modulation Detection Mode

Before describing instrumentation of the FM detection method, we briefly describe instrumentation of the AM detection method. In the AM detection method, an oscillator producing a sinusoidal signal with an angular fixed frequency (ω_d) is connected to a piezoelectric actuator to dither the cantilever. The oscillation amplitude of the cantilever end is monitored by the displacement-sensing system such as an optical beam deflection sensor including a laser light source, a mirror and a photodiode. The change in the amplitude is detected by using a root-mean-square to direct current (RMS-to-DC) converter integrated circuit (IC) or a lock-in amplifier. Figure 3a shows an experimental setup for the AM detection method.

For the FM detection method, a self-oscillation circuit to keep the cantilever oscillating at its resonance frequency and an FM detector (demodulator) are required. Figure 17.3b shows an experimental setup for the FM detection method. The self-oscillation circuit is composed of two key components, which are a phase shifter for tuning $\theta_{\text{electronics}}(\omega)$ so that (17.11) is fulfilled, and a variable gain amplifier (VGA) for regulating the oscillation amplitude of the cantilever. The phase shifter is often implemented by a tunable all-pass filter whose center frequency is tuned by a variable resistor. For fulfilling (17.12), a limiter or an automatic-gain-control (AGC) amplifier is placed in the input or the output of the phase shifter for normalizing the amplitude of the input signal. The VGA is often implemented by an analog multiplier IC. The output signal of the phase shifter is fed to one input of the multiplier while a DC voltage, whose magnitude controls the gain of the VGA, is fed to the other input. There are two operating modes in DFM based on the FM detection method, which are the constant-excitation (CE) mode and the constant-amplitude (CA) mode. In the CE mode, the DC voltage fed to the multiplier is constant; hence, the oscillation amplitude may be changed by the dissipative interaction forces. On the other hand, in the CA mode, the DC voltage fed to the multiplier is dynamically controlled to keep the oscillation amplitude constant. Therefore, an RMS-to-DC converter for detecting the amplitude and a proportional-integral-derivative (PID) feedback circuit are required for the CA mode. In the CA mode, the magnitude of the DC voltage fed to the multiplier is a measure of the amount of the energy loss due to the dissipative interaction forces.

In the FM detection method, the cantilever is self-oscillated at the resonance frequency f_0; hence, the displacement of the tip can be described as $s = A_0 \cos \omega_0 t$ if F_{ts} is negligibly small. The equation of motion of the cantilever tip end becomes

$$m^* \ddot{s} + m^* \frac{\omega_0}{Q} \dot{s} + ks = F_0 \cos(\omega_0 t + \pi/2) + F_{\text{ts}}(s + s_0) , \qquad (17.13)$$

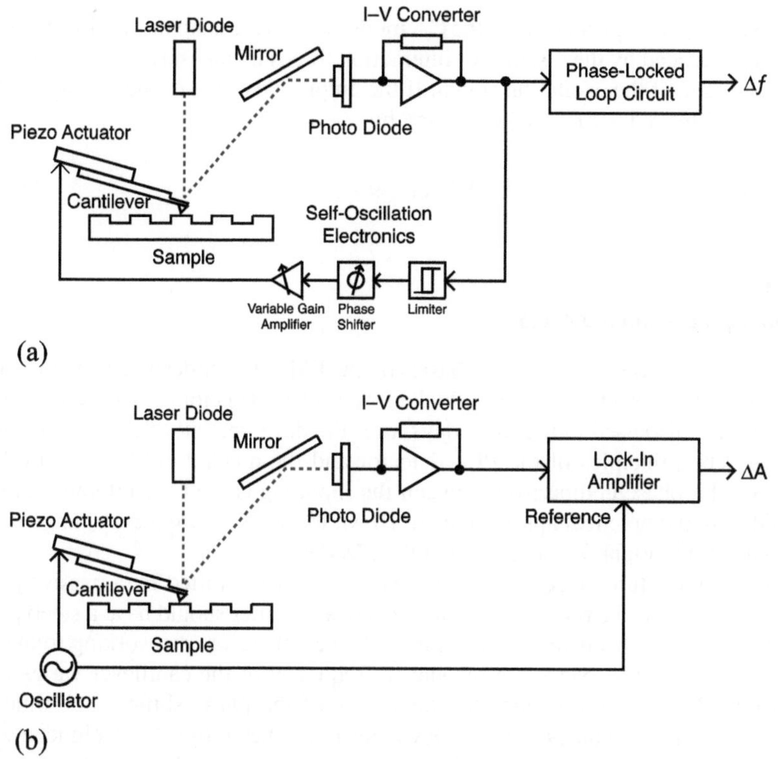

Fig. 17.3. Typical implementation of the amplitude modulation (AM) detection method and the frequency modulation (FM) detection method. (**a**) In the AM detection method, a lock-in amplifier is commonly used for detecting the change in the amplitude. (**b**) In the FM detection method, a phase-locked-loop (PLL) circuit is often used for detecting the frequency shift

or

$$\ddot{s} + \frac{\omega_0}{Q}\dot{s} + \omega_0^2 s = \frac{\omega_0^2}{k} F_0 \cos\left(\omega_0 t + \pi/2\right) + \frac{\omega_0^2}{k} F_{\mathrm{ts}}\left(s + s_0\right) \ . \tag{17.14}$$

The energy loss of the cantilever per oscillation cycle is calculated by integrating the left-hand side of (17.13) as

$$E_{\text{loss per cycle}} = \oint \left(m^*\ddot{s} + m^*\frac{\omega_0}{Q}\dot{s} + ks\right) \mathrm{d}s = \frac{2\pi}{Q}\left(\frac{1}{2}kA_0^2\right) = \frac{2\pi}{Q}E_{\text{cantilever}} \ , \tag{17.15}$$

where $E_{\text{cantilever}}$ is the energy stored in the cantilever; therefore the energy loss of the freely oscillating cantilever per second (dissipated power) is given by

$$P_{\text{dissipation(free)}} = f_0 E_{\text{loss-per-cycle}} = \frac{2\pi f_0}{Q} E_{\text{cantilever}} = \frac{\omega_0}{Q} E_{\text{cantilever}} \ . \tag{17.16}$$

Since $E_{\text{cantilever}}$ is kept constant in the CA mode, we can measure the dissipated power $P_{\text{dissipation}}$ induced by the dissipative interaction forces by measuring the variation of the magnitude of the excitation signal. If the amplitude of the excitation signal A_{exc} is increased by ΔA_{exc}, $P_{\text{dissipation}}$ is given by

$$P_{\text{dissipation}} = \frac{\omega_0}{Q}\left(1 + \frac{\Delta A_{\text{exc}}}{A_{\text{exc}}}\right) E_{\text{cantilever}} . \qquad (17.17)$$

17.2.4
Frequency Modulation Detector

In the following, we describe some basics on the FM detector design. The most commonly used FM detectors is a phase-locked loop (PLL) detector [7–10]. A quadrature detector was also used in the early days after the development of the FM detection method. The principles of the PLL detector and the quadrature detector are both based on the phase comparison between the input signal and the reference signal. Considering the practical applications in DFM, the working frequency range is one of the most important factors to design the FM detector.

In the quadrature detector, the input signal is phase-shifted by a passive phase shifter and used as the reference signal. This phase shifter should have a steep phase sensitivity on a slight change in the input frequency; therefore the working frequency range is very narrow. Since the resonance frequency of the cantilever varies from one to another, one must tune the parameters of the phase shifter each time the cantilever is changed unless a frequency conversion (heterodyne) technique, a well-known technique in radio communication technology where the input frequency is converted to intermediate frequency, is utilized.

In the PLL detector, the phase of the input signal is compared with that of the signal from the voltage-controlled oscillator (VCO) using a phase detector. The output of the phase detector, a measure of the difference of the phase of the two signals, controls the oscillation frequency of the VCO so that the difference of the phase is always constant. In this situation, the phase of the VCO output locks to that of the input signal. The control voltage of the VCO is a measure of the frequency of the input signal. The response time of the PLL, which is related to a demodulation bandwidth, is determined by the parameters of a loop filter. In the case of the PLL detector, the working frequency range is determined by the tuning range of a VCO. Thus, it is very easy to implement the PLL detector that can be used for a wide variety of force sensors with any resonance frequencies without changing circuit parameters by utilizing a VCO with a wide tuning range. An interesting feature of the PLL detector for the FM detection method is that the output signal of the VCO can be used as an excitation signal for the cantilever. Further details of the PLL detector for the FM detection method are described in the following subsection.

17.2.5
Phase-Locked-Loop Frequency Modulation Detector

Figure 17.4 shows schematics of various PLL detectors. In the early days, the PLL detector implemented by a PLL IC, as shown in Fig. 17.4a, was often used for the FM

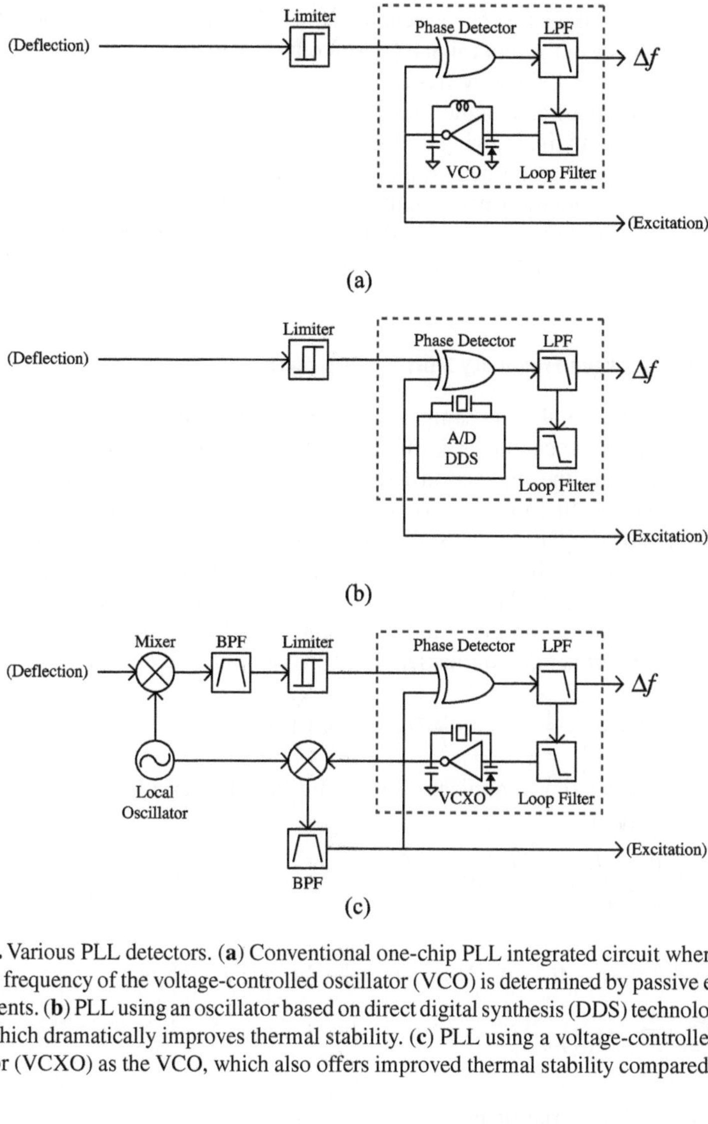

Fig. 17.4. Various PLL detectors. (**a**) Conventional one-chip PLL integrated circuit where the oscillation frequency of the voltage-controlled oscillator (VCO) is determined by passive electrical components. (**b**) PLL using an oscillator based on direct digital synthesis (DDS) technology as the VCO, which dramatically improves thermal stability. (**c**) PLL using a voltage-controlled crystal oscillator (VCXO) as the VCO, which also offers improved thermal stability compared with (**a**)

detection method; however, the VCOs in such ICs often suffer from a large thermal drift of the oscillator frequency. Such a large thermal drift is due to the fact that the output frequency of the VCO is determined by passive electrical components. A solution to improve thermal stability is to use a quartz crystal oscillator as a frequency standard. For example, using an oscillator based on direct digital synthesis (DDS) technology or a numerically controlled oscillator (NCO) as the VCO dramatically improves thermal stability [8, 9]. A schematic illustrating such PLLs is shown in Fig. 17.4b. The other solution is to employ a voltage-controlled crystal oscillator (VCXO) as the VCO [10]. This VCXO-PLL detector is used together with a frequency conversion (heterodyne) circuit, allowing the VCXO-PLL detector to be used for a wide variety of cantilevers as schematically shown in Fig. 17.4c. For this purpose, the frequency of the input signal (f_0) is converted to the intermediate frequency, which is 4.5 MHz ± 450 Hz by frequency mixing with the output signal of a local oscillator. The output signal of the VCXO can be used to make an excitation signal by another frequency mixing. Owing to fast analog circuit technology, the VCXO-PLL detector offers a very wide demodulation bandwidth, which is more than 10 kHz. Such a wide demodulation bandwidth is very important not only for high-speed imaging with the FM detection method, but also for Kelvin probe force microscopy (KFM) based on the FM detection, which will be described in Sect. 17.5.1.1.

17.2.6 Relationship Between Frequency Shift and Interaction Force

As mentioned before, the tip–sample interaction force is detected as the resonance frequency shift of the cantilever in the FM detection. The relationship between the frequency shift and the interaction force was first derived by Giessibl [11] using a Hamiltonian–Jacobi formalism (first-order perturbation theory). Similar results were also obtained by using the Krilov–Bogoliubov–Mitropolsky method, which is a kind of perturbation theory [12], or by using a least-action principle [13]. In this subsection, we follow the latter method using the least-action principle.

We assume that the motion of the cantilever tip end as the oscillatory motion can be expressed in terms of a Fourier series. The action integral can be written as

$$S = \int_0^{T_0} \left(\frac{1}{2}\dot{s}^2 - \frac{1}{2}\omega_0^2 s^2 - \frac{\omega_0^2}{k} U_{\mathrm{ts}}\left(s + s_0\right) \right) \mathrm{d}t = 0 \,. \tag{17.18}$$

The correct orbital $s(t)$ minimizes S, meaning that the variation δS must be zero; therefore we obtain

$$\delta S = \left[\frac{\mathrm{d}}{\mathrm{d}t}\left(\frac{\partial L}{\partial \dot{s}}\right) - \frac{\partial L}{\partial s} \right] \delta s = \int_0^{T} \left(\ddot{s} + \omega_0^2 s - \frac{\omega_0^2}{k} F_{\mathrm{ts}}\left(s + s_0\right) \right) \delta s \, \mathrm{d}t = 0 \,, \tag{17.19}$$

by using $F_{\mathrm{ts}}(z) = -\mathrm{d}U_{\mathrm{ts}}/\mathrm{d}z$. Now we make the periodic ansatz

$$s(t) = \sum_{n=1}^{\infty} a_n \cos n\omega t \,. \tag{17.20}$$

The corresponding variation is

$$\delta s(t) = \sum_{n=1}^{\infty} \frac{\partial s}{\partial a_n} \delta a_n = \sum_{n=1}^{\infty} \cos n\omega t \delta a_n \,, \tag{17.21}$$

where a_n defines the set of variational parameters. Inserting (17.20) and (17.21) into (17.19), we obtain

$$\delta S = \frac{\pi}{\omega} \sum_{n=1}^{\infty} \left(-n^2\omega^2 + \omega_0^2\right) a_n \delta a_n - \int_0^{T} \left(\frac{\omega_0^2}{k} F_{\mathrm{ts}}\left(s + s_0\right) \sum_{n=1}^{\infty} \cos n\omega t \delta a_n \right) \mathrm{d}t \,. \tag{17.22}$$

Now we suppose $a_n = 0$ for $n \geq 2$ for simplicity. The action integral S is minimized when δS vanishes for variation of a_1,

$$\frac{\partial S}{\partial a_1} = \frac{\pi}{\omega} a_1 \left(\omega_0^2 - \omega^2\right) - \int_0^T \left(\frac{\omega_0^2}{k} F_{ts}\left(s + s_0\right) \cos \omega t\right) \mathrm{d}t = 0 . \tag{17.23}$$

Therefore, we obtain

$$\omega = \omega_0 \sqrt{1 - \frac{1}{a_1 \pi k} \int_0^T \left(F_{\mathrm{ts}}\left(s + s_0\right) \omega \cos \omega t\right) \mathrm{d}t} . \tag{17.24}$$

Substituting $\cos \omega t$ with $u = \cos \omega t$, we can write (17.24) as

$$\omega = \omega_0 \sqrt{1 - \frac{2}{a_1 \pi k} \int_{-1}^1 F_{\mathrm{ts}}\left(s + s_0\right) \frac{u}{\sqrt{1 - u^2}} \mathrm{d}u} . \tag{17.25}$$

If the shift of the angular resonance frequency $\Delta\omega = \omega - \omega_0$ is very small compared with ω_0, the angular frequency shift $\Delta\omega$ is approximated as

$$\Delta\omega = \omega - \omega_0 = -\frac{\omega_0}{a_1 \pi k} \int_{-1}^1 F_{\mathrm{ts}}\left(s + s_0\right) \frac{u}{\sqrt{1 - u^2}} \mathrm{d}u . \tag{17.26}$$

As shown in Fig. 17.5, a_1 is equal to oscillation amplitude A_0, and s_0 is equal to $d_0 + A_0$, where d_0 is the closest distance between the tip from the sample surface. Therefore, the frequency shift is finally obtained as

$$\Delta f = -\frac{f_0}{\pi k A_0} \int_{-1}^1 F_{\mathrm{ts}}\left(d_0 + A_0\left(1 + u\right)\right) \frac{u}{\sqrt{1 - u^2}} \mathrm{d}u . \tag{17.27}$$

For the case when the oscillation amplitude is small compared with the range of interaction, namely, $A_0 \ll \lambda$, where λ is a scaling parameter for the range of interaction, $F_{\mathrm{ts}}\left[d_0 + A_0\left(1 + u\right)\right]$ in the integrand is nonzero only for the small

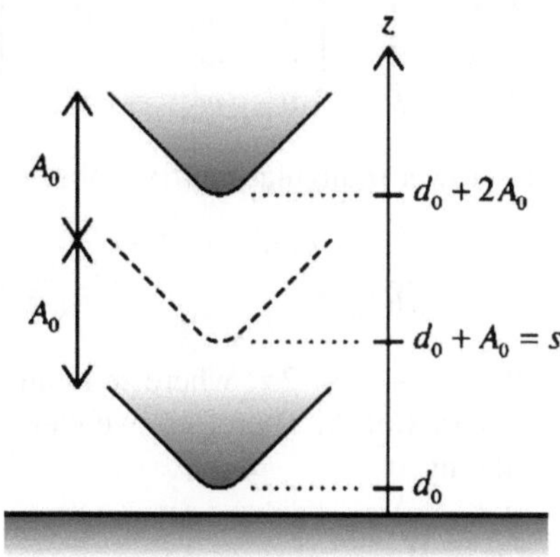

Fig. 17.5. The oscillatory motion of the cantilever

range around $u = -1$. Substituting u with $u = \cos(\pi - \xi) = -1 + \xi^2/2$, we can write (17.27) as

$$\Delta f = \frac{f_0}{\pi k A_0} \int_0^{\infty} F_{ts} \left[d_0 + A_0 \left(\xi^2/2 \right) \right] d\xi \, . \tag{17.28}$$

Furthermore, substituting ξ with $\xi^2/2 = (\lambda / A_0)\, y^2$, we obtain

$$\Delta f = \frac{\sqrt{2} f_0 d^{1/2}}{\pi k A_0^{3/2}} \int_0^{\infty} F_{ts} \left(d_0 + \lambda y^2 \right) dy \, . \tag{17.29}$$

From (17.29), we learn that the frequency shift Δf is proportional to $f_0/kA_0^{3/2}$ when the oscillation amplitude is relatively large ($A_0 \gg \lambda$). Giessibl [11] proposed a "normalized frequency shift", γ, as the parameter independent of the experimental parameters,

$$\gamma = \frac{\Delta f}{f_0} k A_0^{3/2} \, . \tag{17.30}$$

17.2.7 Inversion of Measured Frequency Shift to Interaction Force

In the last subsection, the relationship between the frequency shift and the interaction force was derived as (17.27). In order to calculate the interaction force from the measured frequency shift versus distance curve which was recorded experimentally, (17.27) should be inverted. Giessibl [14] proposed a practical method to invert (17.27) to recover the tip–sample interaction force. Suppose that the frequency shift versus distance curve composed of a set of N data points is measured by decreasing the distance d with a step of Δ as $d_i = d_1 - (i-1)\Delta$ for $i \geq 1$. Each frequency shift datum (Δf_i) corresponds to the frequency shift measured when the distance of the base of the cantilever from the surface is $d_i + A_0$. Equation (17.27) can be written as a linear equation, $\boldsymbol{f} = \boldsymbol{W}\boldsymbol{F}_{ts}$,

$$\begin{pmatrix} \Delta f_1 \\ \Delta f_2 \\ \Delta \cdots \\ \Delta f_N \end{pmatrix} = \begin{pmatrix} w'_{11} & 0 & \cdots 0 \\ w'_{21} & w'_{22} & \cdots 0 \\ \cdots & \cdots & \cdots\cdots \\ w'_{N1} & w'_{N1} & \cdots\, w'_{NN} \end{pmatrix} \begin{pmatrix} F_{ts1} \\ F_{ts2} \\ \cdots \\ F_{tsN} \end{pmatrix} . \tag{17.31}$$

$\boldsymbol{W}$ is a lower triangular matrix whose elements are given by

$$w'_{ij} = \frac{f_0}{\pi k A_0} \int_{1-2(i-j+1)/(2\alpha+1)}^{1-2(i-j)/(2\alpha+1)} \frac{\tau}{\sqrt{1-\tau^2}} d\tau \, , \tag{17.32}$$

for $0 \leq i - j \leq 2\alpha$, where α is an integer expressing the vibration amplitude A_0 in terms of Δ. Therefore, we can obtain the interaction force at each point by calculating

$$\boldsymbol{F}_{ts} = \boldsymbol{W}^{-1} \boldsymbol{f} \, . \tag{17.33}$$

Since the number of elements in the matrix $\boldsymbol{W}$ is as large as N^2, the recovery of the force is not easy.

Recently, Sader and Jarvis [15] succeeded in inverting (17.27) analytically. They obtained

$$F(d_0) \cong \frac{2k}{f_0} \int_{s_0}^{\infty} \left[\left(1 + \frac{A_0^{1/2}}{8\sqrt{\pi (z - s_0)}} \right) \Delta f(z) - \frac{A_0^{3/2}}{\sqrt{2 (z - s_0)}} \frac{\mathrm{d}\Delta f(z)}{\mathrm{d}z} \right] \mathrm{d}z \,, \tag{17.34}$$

which is valid irrespective for the vibration amplitude (A_0). This equation asymptotically approaches the following two equations, for the small-amplitude case ($A_0 \ll \lambda$),

$$F_{\mathrm{small}}(d_0) \cong \frac{2k}{f_0} \int_{s_0}^{\infty} \Delta f(z) \, \mathrm{d}z \,, \tag{17.35}$$

and for the large-amplitude case ($A_0 \gg \lambda$),

$$F_{\mathrm{large}}(d_0) \cong -\frac{\sqrt{2} k A_0^{3/2}}{f_0} \int_{s_0}^{\infty} \left(\frac{1}{\sqrt{z - s_0}} \frac{\mathrm{d}\Delta f(z)}{\mathrm{d}z} \right) \mathrm{d}z \,, \tag{17.36}$$

which was derived by the inversion of Abel's integral equation [16]. Although calculations of (17.33) and (17.34) give almost similar results, the latter is easier to implement.

17.3 Noise in Frequency Modulation Atomic Force Microscopy

17.3.1 Thermal Noise Drive

The minimum detectable force achieved by DFM has been discussed in the literature since its invention. Here we summarize the derivation of some important equations to show that the minimum detectable force is essentially the same independent of the detection methods, the AM detection and the FM detection. This is because the thermal displacement noise of the cantilever governs the minimum detectable force in all cases. Prior to discussion of the minimum detectable force, we calculate the thermal noise drive, which is an origin of the random displacement of the cantilever end.

If we suppose that the thermal noise drive is "white," the magnitude is described as

$$\Psi_{\mathrm{th}}(\omega) = I_{\mathrm{th}} \,. \tag{17.37}$$

We can calculate the mean-square displacement of the cantilever end, $\langle z_{\mathrm{th}}^2 \rangle$, which is excited thermally by the thermal noise drive by performing integration of the displacement noise density over the entire bandwidth using $|G_{\mathrm{cantilever}}(\omega)|$ given in (17.5) as

$$\left\langle z_{\mathrm{th}}^2 \right\rangle = \frac{1}{2\pi} \int_0^{\infty} N_{\mathrm{th}}(\omega) \, \mathrm{d}\omega = \frac{1}{2\pi} \int_0^{\infty} |G(\omega)|^2 \, I_{\mathrm{th}} \, \mathrm{d}\omega = \frac{I_{\mathrm{th}} Q}{4km^* \omega_0} \,, \tag{17.38}$$

where $N_{\mathrm{th}}(\omega)$ is the spectral noise density of the thermal displacement noise of the cantilever end. On the other hand, $\langle z_{\mathrm{th}}^2 \rangle$ is given by the equipartition theorem as

$$\left\langle z_{\mathrm{th}}^2 \right\rangle = k_{\mathrm{B}}T/k \,. \tag{17.39}$$

Therefore, we can obtain the thermal noise drive as

$$I_{\mathrm{th}} = 4m^* \omega_0 k_{\mathrm{B}} T/Q \,. \tag{17.40}$$

17.3.2 Minimum Detectable Force in Static Mode

In static mode AFM, where the cantilever is not intentionally vibrated, the interaction force is directly measured by monitoring the static deflection of the cantilever end. In this case, the minimum detectable force is equal to the product of the thermal displacement of the cantilever end at a given bandwidth B and the spring constant of the cantilever (k). The mean-square thermal displacement of the cantilever end is calculated by integrating the thermal noise density for the bandwidth ranging from 0 to B as

$$\left\langle z_{\mathrm{th}}^2 \right\rangle^{\mathrm{static}} = \frac{1}{2\pi} \left| G(0) \right|^2 \frac{4m^* \omega_0 k_{\mathrm{B}} T}{Q} (2\pi B) = \frac{4k_{\mathrm{B}} TB}{k\omega_0 Q} \,. \tag{17.41}$$

Therefore, the minimum detectable force is

$$F_{\mathrm{min}}^{\mathrm{static}} = k\sqrt{\left\langle z_{\mathrm{th}}^2 \right\rangle^{\mathrm{static}}} = \sqrt{\frac{4kk_{\mathrm{B}} TB}{\omega_0 Q}} \,. \tag{17.42}$$

17.3.3 Minimum Detectable Force Using the Amplitude Modulation Detection Method

For DFM using the AM detection method, the cyclic force is detected with the force sensitivity enhanced at the resonance frequency by the factor of Q; therefore, the detected force is $1/Q$ times of the product of the thermal displacement of the cantilever end at a given bandwidth B and the spring constant of the cantilever (k). The mean-square thermal displacement of the cantilever end is calculated by integrating the thermal noise density for the bandwidth B around the resonance frequency. For simplicity, we assume that the bandwidth B is so narrow that the thermal noise spectral density is almost constant within this range; therefore, the mean-square thermal displacement of the cantilever end is given as

$$\left\langle z_{\mathrm{th}}^2 \right\rangle^{\mathrm{AM}} = \frac{1}{2\pi} \left| G(\omega_0) \right|^2 \frac{4m^* \omega_0 k_{\mathrm{B}} T}{Q} (2\pi B) = \frac{4k_{\mathrm{B}} TBQ}{k\omega_0} \,. \tag{17.43}$$

Therefore, the minimum detectable force is

$$F_{\mathrm{min}}^{\mathrm{AM}} = k\sqrt{\left\langle z_{\mathrm{th}}^2 \right\rangle^{\mathrm{AM}}} \Big/ Q = \sqrt{\frac{4kk_{\mathrm{B}} TB}{\omega_0 Q}} \,, \tag{17.44}$$

which is exactly the same as (17.42). When the oscillation amplitude is small, the minimum detectable force gradient is also calculated using the approximation [17]

$$F'^{\mathrm{AM}}_{\min} = F^{\mathrm{AM}}_{\min} \Big/ \sqrt{\langle z^2_{\mathrm{osc}} \rangle}\,, \tag{17.45}$$

where $\langle z^2_{\mathrm{osc}} \rangle$ is the mean-square vibration amplitude given by $\langle z^2_{\mathrm{osc}} \rangle = A_0^2/2$. Therefore, we obtain the minimum detectable force gradient for the AM detection method as

$$F'^{\mathrm{AM}}_{\min} = \sqrt{\frac{4kk_{\mathrm{B}}TB}{\omega_0 Q \langle z^2_{\mathrm{osc}} \rangle}}\,. \tag{17.46}$$

17.3.4 Minimum Detectable Force Using the Frequency Modulation Detection Method

For DFM using the FM detection method, one should first calculate the minimum detectable frequency shift [4]. The minimum detectable frequency shift is equal to the frequency noise of the cantilever oscillator at a given bandwidth B. The mean-square frequency shift is calculated as

$$\left\langle (\delta f_{\mathrm{th}})^2 \right\rangle = \int_B \left[\delta f_{\mathrm{th}}\left(f_{\mathrm{mod}}\right)\right]^2 \mathrm{d}f_{\mathrm{mod}}\,, \tag{17.47}$$

where $\delta f_{\mathrm{th}}(f_{\mathrm{mod}})$ is the frequency noise density at a modulation frequency f_{mod}. If we assume that the dominant noise source is the thermal displacement noise of the cantilever end, the frequency noise of the oscillator at the modulation frequency f_{mod} arises from the thermal displacement of the cantilever end at the frequencies of $f_0 \pm f_{\mathrm{mod}}$; therefore, $\delta f_{\mathrm{th}}(f_{\mathrm{mod}})$ is given by [18]

$$\delta f_{\mathrm{th}}\left(f_{\mathrm{mod}}\right) = \sqrt{\frac{2E_{\mathrm{np(th)}}\left(f_{\mathrm{mod}}\right)}{E_{\mathrm{c}}}}\, f_{\mathrm{mod}} = \sqrt{\frac{E_{\mathrm{n(th)}}\left(f_{\mathrm{mod}}\right)}{E_{\mathrm{c}}}}\, f_{\mathrm{mod}}\,. \tag{17.48}$$

$E_{\mathrm{np(th)}}\left(f_{\mathrm{mod}}\right)$ and $E_{\mathrm{n(th)}}\left(f_{\mathrm{mod}}\right)$ are the thermal phase noise energy and the thermal oscillator energy at the modulation frequency f_{mod}, respectively. E_{c} is the oscillator energy given by $E_{\mathrm{c}} = k\langle z_{\mathrm{osc}} \rangle^2$. Since $E_{\mathrm{n(th)}}\left(f_{\mathrm{mod}}\right)$ is given by $E_{\mathrm{n(th)}}\left(f_{\mathrm{mod}}\right) = kN_{\mathrm{th}}(\omega_{\mathrm{mod}})$, $\delta f_{\mathrm{th}}(f_{\mathrm{mod}})$ becomes

$$\delta f_{\mathrm{th}}\left(f_{\mathrm{mod}}\right) = \sqrt{\frac{N_{\mathrm{th}}\left(f_{\mathrm{mod}}\right)}{\langle z_{\mathrm{osc}} \rangle^2}}\, f_{\mathrm{mod}}\,. \tag{17.49}$$

In the FM detection method, the spectral width of the oscillator is decreased with increasing the vibration amplitude. In this situation, we can ignore noise components with the modulation frequency on the order of the oscillator linewidth and less. Using an approximation of $|G_{\mathrm{cantilever}}(\omega)|$ in (17.7), we obtain

$$N_{\mathrm{th}}\left(f_{\mathrm{mod}}\right) = N_{\mathrm{th}}\left(\omega_{\mathrm{mod}}\right) = \left|G_{\mathrm{cantilever}}\left(\omega_0 \pm \omega_{\mathrm{mod}}\right)\right|^2 \Psi_{\mathrm{th}} = \frac{k_{\mathrm{B}}T}{m^* \omega_0 Q \omega^2_{\mathrm{mod}}}\,. \tag{17.50}$$

Equation (17.49) becomes

$$\delta f_{\rm th}\left(f_{\rm mod}\right)=\sqrt{\frac{k_{\rm B}T}{m^*\omega_0 Q\omega_{\rm mod}^2\left\langle z_{\rm osc}\right\rangle^2}}f_{\rm mod}=\frac{1}{2\pi}\sqrt{\frac{\omega_0 k_{\rm B}T}{kQ\left\langle z_{\rm osc}\right\rangle^2}}\ ; \tag{17.51}$$

therefore the mean-square frequency noise at the given bandwidth B is

$$\left\langle(\delta f_{\rm th})^2\right\rangle=\int_{\rm B}\frac{\omega_0 k_{\rm B}T}{4\pi^2 kQ\left\langle z_{\rm osc}\right\rangle^2}{\rm d}f_{\rm mod}=\frac{\omega_0 k_{\rm B}TB}{4\pi^2 kQ\left\langle z_{\rm osc}\right\rangle^2}\ . \tag{17.52}$$

The root-mean-square frequency noise is

$$\sqrt{\left\langle(\delta f_{\rm th})^2\right\rangle}=\frac{1}{2\pi}\sqrt{\frac{\omega_0 k_{\rm B}TB}{kQ\left\langle z_{\rm osc}\right\rangle^2}}. \tag{17.53}$$

If the oscillation amplitude is small, the minimum detectable force gradient becomes

$$F_{\rm min}^{\prime\rm FM}=2k\frac{\sqrt{\left\langle(\delta f_{\rm th})^2\right\rangle}}{f_0}=\sqrt{\frac{4kk_{\rm B}TB}{\omega_0 Q\left\langle z_{\rm osc}\right\rangle^2}}\ , \tag{17.54}$$

which becomes the same expression as (17.46), the minimum detectable force gradient for the AM detection method. Furthermore, if we assume

$$F_{\rm min}^{\rm FM}=F_{\rm min}^{\prime\rm FM}\sqrt{\left\langle z_{\rm osc}^2\right\rangle} \tag{17.55}$$

as in (17.45), we obtain the following expression, which is exactly the same as (17.42) and (17.44):

$$F_{\rm min}^{\rm FM}=\sqrt{\frac{4kk_{\rm B}TB}{\omega_0 Q}}\ . \tag{17.56}$$

17.3.5 Effect of Displacement-Sensing Noise on Minimum Detectable Force

In the previous subsections, we discussed the minimum detectable force obtained in the static mode and dynamic modes, using the AM detection method and the FM detection method. In the discussion, we suppose that the thermal displacement noise of the cantilever end is the dominant noise source. However, in the practical instruments, the noise of the displacement-sensing system may also become dominant. In this subsection, we discuss the effect of the displacement-sensor noise on the minimum detectable force assuming the displacement-sensing system has a "white" displacement-sensor noise density of $n_{\rm ds}$.

In static mode AFM, the minimum detectable displacement at the given noise bandwidth increases up to

$$\sqrt{\left\langle z_{\rm th}^2\right\rangle+n_{\rm ds}^2 B}\ ; \tag{17.57}$$

therefore, the minimum detectable force becomes

$$F_{\min}^{\text{static (with sensor noise)}} = k\sqrt{\left\langle z_{\text{th}}^2 \right\rangle + n_{\text{ds}}^2 B} = \sqrt{\frac{4kk_{\text{B}}TB}{\omega_0 Q} + k^2 n_{\text{ds}}^2 B} \,. \tag{17.58}$$

The sensor noise also affects the minimum detectable force gradient for DFM using the AM detection method in the same manner. From (17.44), we obtain

$$F_{\min}^{\text{AM (with sensor noise)}} = \frac{k\sqrt{\left\langle z_{\text{th}}^2 \right\rangle + n_{\text{ds}}^2 B}}{Q} = \sqrt{\frac{4kk_{\text{B}}TB}{\omega_0 Q} + \frac{k^2 n_{\text{ds}}^2 B}{Q^2}} \,. \tag{17.59}$$

For the minimum detectable force gradient, we obtain

$$F_{\min}^{\prime\text{AM (with sensor noise)}} = \frac{k\sqrt{\left\langle z_{\text{th}}^2 \right\rangle + n_{\text{ds}}^2 B}}{Q\sqrt{\left\langle z_{\text{osc}}^2 \right\rangle}} = \sqrt{\frac{4kk_{\text{B}}TB}{\omega_0 Q \left\langle z_{\text{osc}}^2 \right\rangle} + \frac{k^2 n_{\text{ds}}^2 B}{Q^2 \left\langle z_{\text{osc}}^2 \right\rangle}} \,. \tag{17.60}$$

For DFM using the FM detection method, the displacement-sensor noise increases the minimum detectable frequency shift. The frequency noise density due to the displacement-sensor noise at a modulation frequency f_{mod}, $\delta f_{\text{ds}}(f_{\text{mod}})$, is given by [18]:

$$\delta f_{\text{ds}}\left(f_{\text{mod}}\right) = \sqrt{\frac{E_{\text{n(ds)}}\left(f_{\text{mod}}\right)}{E_{\text{c}}}} f_{\text{mod}} \,, \tag{17.61}$$

which is similar to (17.48). Therefore, $E_{\text{n(ds)}}(f_{\text{mod}})$ is the apparent oscillator energy at the modulation frequency f_{mod}, which is equal to kn_{ds}^2. The frequency noise density due to the displacement-sensor noise becomes

$$\delta f_{\text{ds}}\left(f_{\text{mod}}\right) = \sqrt{\frac{n_{\text{ds}}^2}{\left\langle z_{\text{osc}}^2 \right\rangle}} f_{\text{mod}} = \frac{n_{\text{ds}}}{\sqrt{\left\langle z_{\text{osc}}^2 \right\rangle}} f_{\text{mod}} = \sqrt{2}\frac{n_{\text{ds}}}{A_0} f_{\text{mod}} \,. \tag{17.62}$$

The square summation of the frequency noise density due to the thermal displacement of the cantilever end and that due to the displacement sensor gives the total frequency noise density at the modulation frequency f_{mod} as

$$\delta f\left(f_{\text{mod}}\right) = \sqrt{\left[\delta f_{\text{th}}\left(f_{\text{mod}}\right)\right]^2 + \left[\delta f_{\text{ds}}\left(f_{\text{mod}}\right)\right]^2} \,. \tag{17.63}$$

Figure 17.7 shows plots of $\delta f(f_{\text{mod}})$ using practical experimental parameters. A cantilever with the spring constant of 30 N/m, the resonance frequency of 300 kHz and the Q-factor of 10,000 is oscillated at its resonance frequency with a vibration amplitude (A_0) of 5 nm at a temperature of 300 K. The frequency noise density due to the thermal noise, $\delta f_{\text{th}}(f_{\text{mod}})$, is constant as described in (17.51). The total frequency noise density $\delta f(f_{\text{mod}})$ increases with increasing sensor noise density. In this figure, we show plots of $\delta f(f_{\text{mod}})$ for the cases when the sensor noise densities, n_{ds}, are 0.1, 0.2, and 0.3 pm/$\sqrt{\text{Hz}}$.

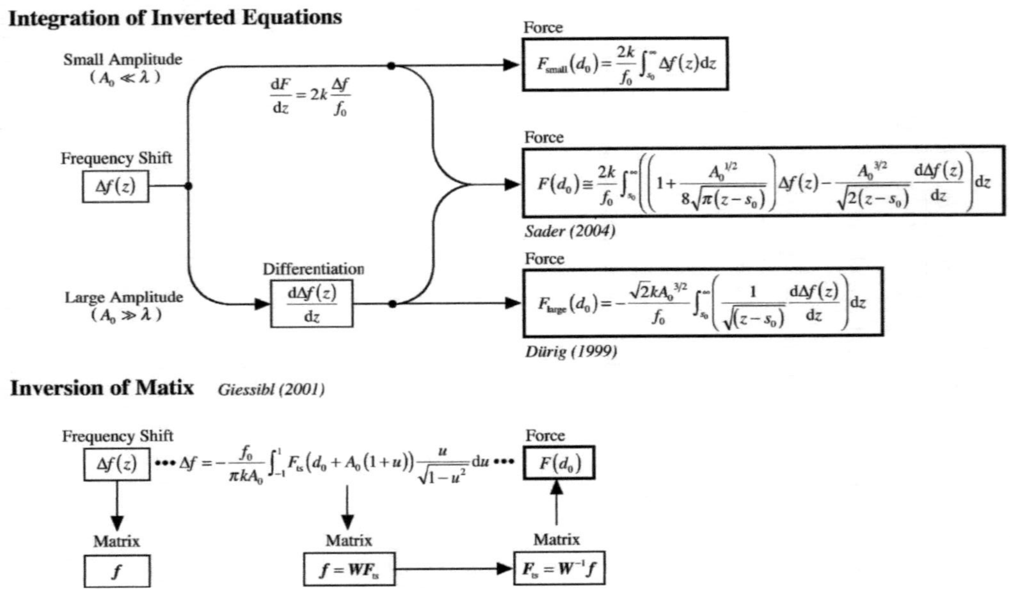

Fig. 17.6. Various methods for recovery of interaction forces from measured frequency shifts. One can recover force independent of the vibration amplitude of the cantilever by integration of the inverted equation found by Sader and Jarvis, or by the matrix inverse

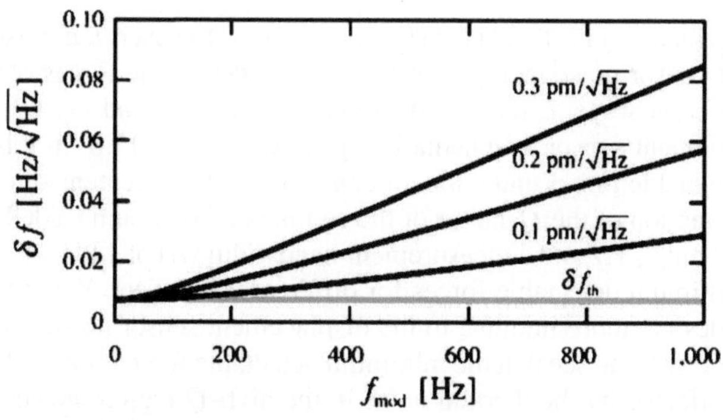

Fig. 17.7. Total frequency-noise densities $\delta f(f_{\text{mod}})$ as a function of the modulation frequency f_{mod} using practical experimental parameters. The cantilever with parameters $k = 30\,\text{N/m}$, $f_0 = 300\,\text{kHz}$, and $Q = 10{,}000$, is oscillated with $A_0 = 5\,\text{nm}$ at $T = 300\,\text{K}$. $\delta f(f_{\text{mod}})$ increases gradually when the sensor-noise density, n_{ds}, is increased up to $0.3\,\text{pm}/\sqrt{\text{Hz}}$

We obtain the mean-square frequency noise at the given bandwidth B by integrating (17.63) as

$$\left\langle(\delta f)^2\right\rangle = \int_B \left(\frac{f_0 k_{\text{B}} T}{2\pi k Q \left\langle z_{\text{osc}}\right\rangle^2} + \frac{n_{\text{ds}}^2 f_{\text{mod}}^2}{\left\langle z_{\text{osc}}^2\right\rangle} \right) \mathrm{d}f_{\text{mod}} = \frac{f_0 k_{\text{B}} T B}{2\pi k Q \left\langle z_{\text{osc}}\right\rangle^2} + \frac{n_{\text{ds}}^2 B^3}{3\left\langle z_{\text{osc}}^2\right\rangle} . \tag{17.64}$$

Therefore the root-mean-square frequency noise becomes

$$\sqrt{\left\langle(\delta f)^2\right\rangle} = \sqrt{\frac{f_0 k_{\text{B}} T B}{2\pi k Q \left\langle z_{\text{osc}}\right\rangle^2} + \frac{n_{\text{ds}}^2 B^3}{3\left\langle z_{\text{osc}}^2\right\rangle}}, \tag{17.65}$$

and the minimum detectable force gradient for the case when the oscillation amplitude is small, given by (17.54), increases up to

$$F'^{\text{FM (with sensor noise)}}_{\min} = 2k\frac{\sqrt{\left\langle(\delta f)^2\right\rangle}}{f_0} = \sqrt{\frac{4kk_{\text{B}}TB}{\omega_0 Q \left\langle z_{\text{osc}}\right\rangle^2} + \frac{16\pi^2 k^2}{3\omega_0^2}\frac{n_{\text{ds}}^2 B^3}{\left\langle z_{\text{osc}}^2\right\rangle}} . \tag{17.66}$$

The minimum detectable force also increases up to

$$F^{\text{FM (with sensor noise)}}_{\min} = \sqrt{\frac{4kk_{\text{B}}TB}{\omega_0 Q} + \frac{16\pi^2 k^2}{3\omega_0^2} n_{\text{ds}}^2 B^3} . \tag{17.67}$$

17.3.6 Comparison of Minimum Detectable Force for Static Mode and Dynamic Modes

In summary, the minimum detectable forces obtained in the static mode and dynamic modes, using the AM detection method and the FM detection method, are all the same as described in (17.42), (17.44), and (17.56). However, the existence of the displacement-sensor noise increases the minimum detectable forces obtained in all modes in different ways as described in (17.58), (17.59), and (17.67). The effect of the displacement sensor is schematically shown in Fig. 17.8. In this figure, the minimum detectable forces under the influence of the displacement-sensor noise are plotted as a function of the Q-factor of the cantilever. We assume a deflection noise density of $0.3\,\mathrm{pm}/\sqrt{\mathrm{Hz}}$ and a measurement bandwidth (B) of 1 kHz. Figure 17.8a,b shows the minimum detectable forces for different cantilevers. We can see that the dynamic modes are more immune to the displacement sensor noise compared with the static mode. We can see that the minimum detectable force for the AM detectable force is less affected by the thermal noise in the high-Q region; however, it should be noted that the AM detection is not practical in the high-Q environment because of the large response time, as mentioned in the previous section. On the other hand, we can see that the minimum detectable force for the FM detectable force is less affected by the thermal noise; however, we should note that the minimum detectable frequency shift for the FM detection method was calculated assuming that the origin of the frequency noise of the oscillator is from the thermal displacement of the cantilever end, and that the frequency noise due to the displacement-sensor noise is

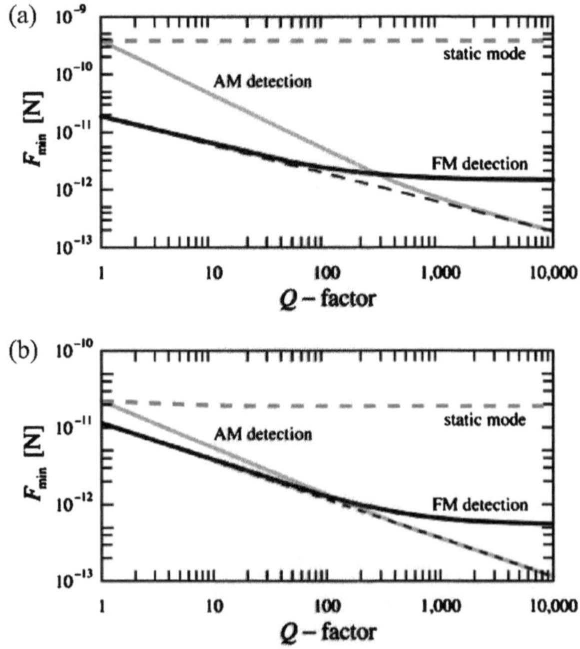

Fig. 17.8. The minimum detectable forces F_{min} obtained in the static mode and the dynamic modes as a function of the Q-factor of the cantilever for the measurement bandwidth of 1 kHz at a temperature of 300 K. The *dark dashed line* shows F_{min} achieved for all modes when the displacement-sensor noise is not taken into account. For the other plots, n_{ds} of $0.3\,\mathrm{pm}/\sqrt{\mathrm{Hz}}$ is assumed. (**a**) F_{min} for the cantilever with parameters $k = 2\,\mathrm{N/m}$ and $f_0 = 40\,\mathrm{kHz}$. (**b**) for the cantilever with parameters $k = 30\,\mathrm{N/m}$ and $f_0 = 300\,\mathrm{kHz}$

added to it. In reality, the displacement-sensor noise may affect the frequency noise of the oscillator since the displacement sensor is included in the self-oscillation loop.

17.4 High-Resolution Imaging of Organic Molecules in Various Environments

Since atomic-resolution imaging by FM-AFM has been well established, the technique has been widely used as a powerful analysis tool to investigate surface structures of various materials, including insulators on a nanometer scale. No special treatment for imaging samples, such as staining or metal coating, which are common techniques in electron microscopy imaging, is required. Because of these distinctive advantages over other analysis methods, FM-AFM is essentially important especially for the study of organic materials as well as a wide variety of practical samples, which often have poor conductivity. In this section, the present status of high-resolution imaging of organic molecules is described.

17.4.1 Alkanethiol Self-Assembled Monolayers

Alkanethiol [CH_3–$(CH_2)_{n-1}$–SH:C_n] molecules are spontaneously arranged on a Au substrate and formed into self-assembled monolayers (SAMs). Their highly ordered structures as well as their unique self-organization character have attracted a wide variety of interest in terms of both basic surface science and practical applications. In addition, since the thiol end group is often used as an electrical junction of an organic molecule and a gold electrode in molecular electronics, nanoscale investigations of electrical properties at the interface between the alkanethiol molecule and the gold substrate are essential.

Molecular conformations and packing arrangements in the monolayer films have been intensively studied by a variety of techniques [19–25]. Transmission electron diffraction measurements showed that the molecules are hexagonally packed to form a $(\sqrt{3} \times \sqrt{3})R\,30°$ overlayer of the Au(111) lattice [22,23]. Furthermore, a helium atom diffraction study revealed the existence of a larger unit cell composed of four distinct molecules corresponding to the c(4 × 2) superlattice with respect to the $(\sqrt{3} \times \sqrt{3})R\,30°$ structure [24]. Although the origin of the c(4 × 2) unit cell has been explained by analogy to bulk n-alkane crystals [24] or by the *gauche* defects caused by the dimerization of the molecules [25], it is still under discussion.

Surface structures of the c(4 × 2) superlattice have been directly visualized by scanning tunneling microscopy (STM), which revealed some different packing arrangements even in a c(4 × 2) unit cell. Delamarche et al. [26] presented molecularly resolved STM images of four different packing arrangements and referred to them as α-, β-, γ-, and δ-phases, as shown in Fig. 17.9a–d [26]. The STM contrasts they presented clearly confirmed the structural models predicted by Camillone et al. [24]. On the other hand, some research groups presented clear STM images revealing the existence of three different molecular contrasts in a c(4×2) unit cell [27,28]. The structure is schematically depicted in Fig. 17.9e, and is hereafter referred to as the ε-phase.

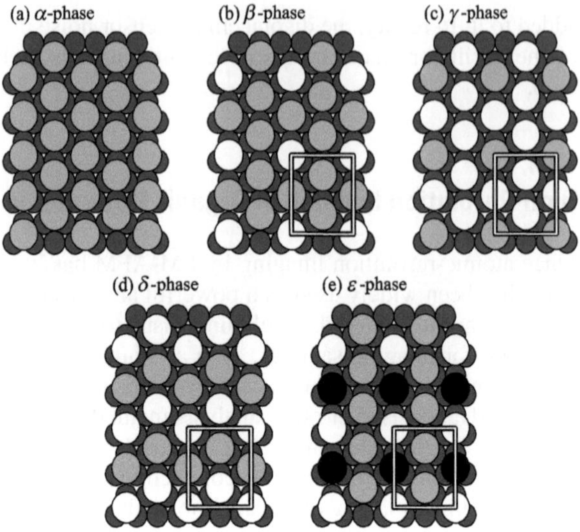

Fig. 17.9. Models of alkanethiol self-assembled monolayers (SAMs) on Au(111) surfaces proposed from scanning tunneling microscopy images. (**a**) Hexagonally packed $(\sqrt{3} \times \sqrt{3})R30°$ structure. (**b**)–(**e**) c(4 × 2) superlattice structures with a rectangular unit cell composed of four distinct molecules

High-resolution noncontact AFM (NC-AFM) imaging has been used for the structural analysis of SAMs. Uchihashi et al. [29] first reported molecularly resolved NC-AFM images of nonanethiol (C_9) monolayer films. Fukuma et al. [30] demonstrated molecular-resolution NC-AFM images of long-chain alkanethiol (C_{16}) SAMs, which were hard to image by STM. They also studied c(4 × 2) superlattice structures of dodecanethiol SAMs (C_{12}).

Figure 17.10a is an NC-AFM image of two different C_{12} domains separated by a gold step [31]. Since the monolayer was prepared at elevated temperature, the film has only some molecular-scale defects and no depressions that are usually formed by gold etching during the self-assembly process [32]. Figure 17.10b,c shows NC-AFM images taken on the left and right gold terraces seen in Fig. 17.10a, respectively. These molecularly resolved NC-AFM images clearly show two different contrast patterns of c(4 × 2) superlattice structures. The C_{12} monolayer formed on the left terrace is composed of zigzag-shaped molecular rows corresponding to the δ-phase model shown in Fig. 17.9d, while the one formed on the right terrace has some protruded molecules forming a rectangularly shaped unit cell (ε-phase, Fig. 17.9e). Higher-resolution NC-AFM images of δ- and ε-phases were obtained with larger Δf values, as shown in Fig. 17.11a,b, respectively. Note that these images were taken on different domains from those shown in Fig. 17.10, so the orientation of the molecular rows is different. In this experiment, we could not find the other three contrast patterns of c(4 × 2) superlattice structures (α-, β-, and γ-phases, Fig. 17.9a–c).

Although high-resolution STM images of C_{12} monolayers prepared with the same procedure as used in our experiment were presented by Bumm et al. [32], no detailed discussions on the surface structure have been presented. The result obtained in our experiment suggested that the annealing treatment during film formation reduces not only defect density, but also the number of domains corresponding to $(\sqrt{3} \times \sqrt{3})R30°$ structures (Fig. 17.9a). Instead, the defect-reduced SAMs are composed predominantly of c(4 × 2) superlattice structures, especially in δ- and ε-

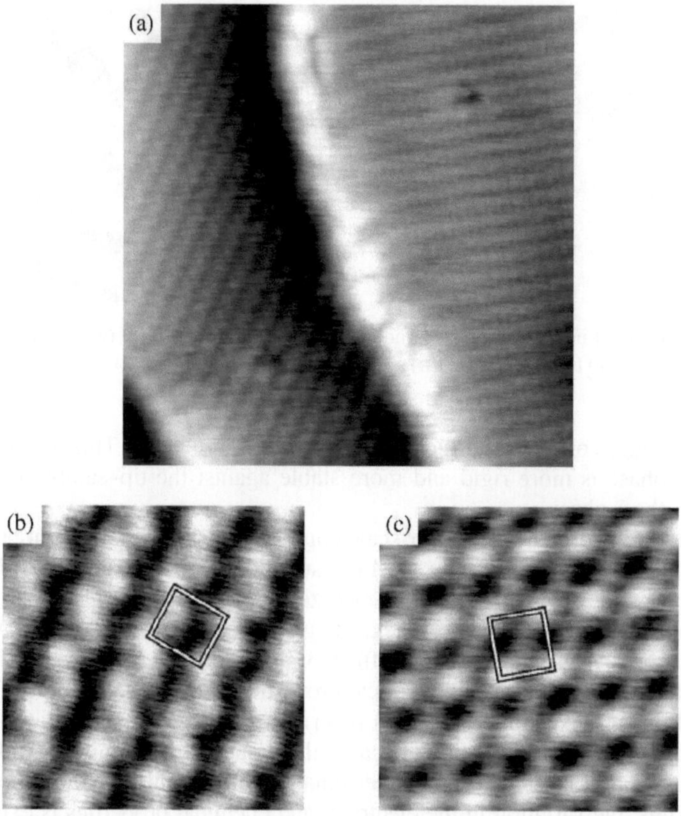

Fig. 17.10. Noncontact atomic force microscopy (NC-AFM) images of a C_{12} monolayer on an Au(111) surface. (**a**) 20 nm × 20 nm, $\Delta f = -70$ Hz, $A = 5$ nm. (**b**) 4.5 nm × 4.5 nm, $\Delta f = -70$ Hz, $A = 5$ nm. (**c**) 4.5 nm × 4.5 nm, $\Delta f = -50$ Hz, $A = 5$ nm

phases. In general, an annealing treatment during film formation often brings the film structure to a thermodynamic equilibrium. The result indicated that δ- and ε-phases might be thermodynamically more stable than the other packing arrangements.

While the upper and the lower parts in Fig. 17.11a exhibit molecular-scale features of the δ-phase, the middle part of the image shows a distorted contrast owing to unstable imaging conditions. Namely, as the tip was brought close to the surface, the imaging conditions, such as frequency shift and the cantilever oscillation amplitude, became so unstable that high resolution was hard to achieve on δ-phase domains. On the other hand, the conditions during the NC-AFM imaging of the ε-phase were stable enough for us to obtain high-resolution NC-AFM images of the ε-phase as shown in Fig. 17.11b.

In order to obtain high-resolution NC-AFM images of organic thin films, it is essential that the film structure is "rigid" enough to withstand the tip–sample interaction force [30]. If the film rigidity is not high enough, molecules can be displaced by the tip–sample interaction force. Consequently, stable operation becomes difficult. In this experiment, it was found to be more difficult to obtain high-resolution

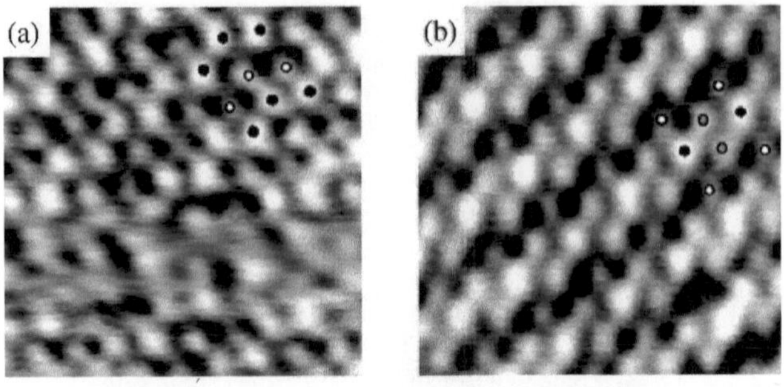

Fig. 17.11. NC-AFM images of a C_{12} monolayer on an Au(111) surface. (**a**) δ-phase (4.5 nm × 4.5 nm, $\Delta f = -280$ Hz, $A = 5$ nm). (**b**) ε-phase (4.5 nm × 4.5 nm, $\Delta f = -260$ Hz, $A = 5$ nm)

NC-AFM images of δ-phase structures than ε-phase structures. This result suggests that the ε-phase is more rigid and more stable against the tip–sample interaction force than the δ-phase.

So far, three different molecular-scale contrast features with different brightness have been found in the ε-phase by STM investigations [27,28]. The NC-AFM image shown in Fig. 17.11b also supports the existence of three molecular-scale contrast features as indicated by the white, black, and gray circles in the image [31]. Although contrast in NC-AFM images can be affected not only by the topography itself but also by electronic structure and chemical properties of the surface, molecular-scale NC-AFM images taken on the c(4 × 2) superlattice structures are likely to represent the true surface topography. This is because the monolayer is composed of the same molecular species and the surface is terminated with chemically inert methyl end groups. Thus, the variation in the electrical and chemical properties is expected to be relatively small.

17.4.2 Submolecular-Scale Contrast in Copper Phthalocyanines

Short-range chemical interactions between a tip front atom and a surface atom play essential roles in the formation of the AFM imaging contrast. Taking them into account is indispensable for the interpretation of submolecular-scale contrast in molecule images. Copper phthalocyanine (CuPc) thin films on MoS_2 surfaces were investigated by NC-AFM for this purpose [33]. CuPc is a symmetrical macrocyclic compound which consists of four iminoisoindoline units with a copper ion accommodated in the central cavity of the Pc ring. When CuPcs are deposited on a MoS_2 surface, they form a closely packed structure with their molecular planes almost parallel to the substrate [34–37]. Since CuPc has widely delocalized π-electron orbitals sticking out of the molecular plane, relatively strong tip–sample chemical interactions are expected in NC-AFM imaging of flat-lying CuPcs.

Figure 17.12a,b shows topographic and dissipation images of a CuPc monolayer on the MoS_2 surface, respectively [33]. These images were processed with a tilt compensation filter and a smoothing filter. The topographic image shows individual molecules in the closely packed structure. On the other hand, the dissipation image

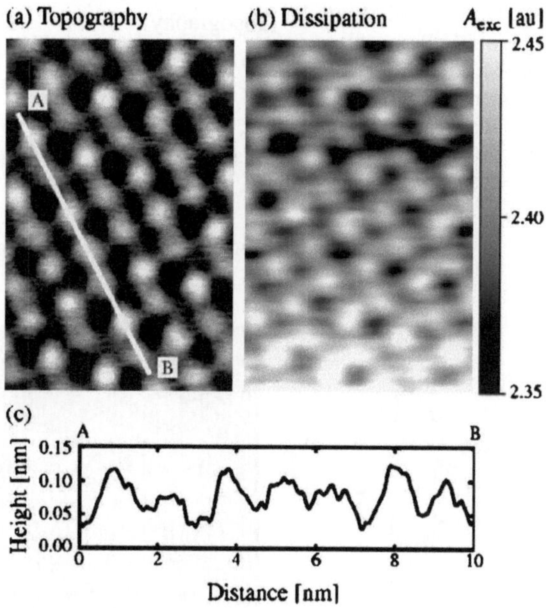

Fig. 17.12. NC-AFM images of a copper phthalocyanine (CuPc) monolayer on a MoS_2 surface. (**a**) Topographic and (**b**) dissipation images (8 nm × 12 nm, $\Delta f = -40$ Hz, $A = 5.5$ nm). (**c**) Cross-sectional plot measured along the *bright line A–B* shown in (**a**)

shows inverted contrast with respect to the topographic image. Figure 17.12c shows a cross-sectional plot measured along the bright line A–B indicated in Fig. 17.12a, revealing the existence of a molecular height variation. The plot shows that the magnitude of height variation is about 0.05 nm. No long-range regularity is confirmed in the molecular height variation.

Figure 17.13a,b shows topographic and dissipation images taken on the same sample, respectively. These images were also processed with a tilt compensation filter and a smoothing filter. The topographic image shows submolecular-scale contrast, revealing the four-leaf structure of the CuPcs. The image also shows an asymmetric feature inside the molecule. The central part of the CuPcs is imaged as an "apparent hole," which is about 50–60 pm lower than the average molecular plane. The dissipation image also shows that submolecular-scale dissipation contrasts are observed even with a relatively fast scanning speed and a relatively large time constant of tip–sample distance regulation. Thus, the contrasts are not likely to be caused by the topographic artifacts. The magnitude of energy dissipation was drastically changed at the lower part of the image, suggesting an atomic-scale tip change [38].

The molecular tilt angle of CuPcs on the MoS_2 surface was studied using angle-resolved UV photoelectron spectroscopy (ARUPS) by Okudaira et al. [37]. The ARUPS result showed the best agreement with a molecular tilt angle of 6°, showing that the molecules do not always lie completely flat on the surface. The asymmetric feature inside the molecules observed in this experiment can be explained by the inclination of the molecular planes.

Since a copper ion is small enough to be fully accommodated in the central cavity of a Pc ring, CuPc has a completely planar structure. In addition, van der Waals radii for carbon, nitrogen, and copper atoms are 170, 155, and 140 pm, respectively. Accordingly, the apparent holes (50–60 pm in depth) found in the NC-AFM image

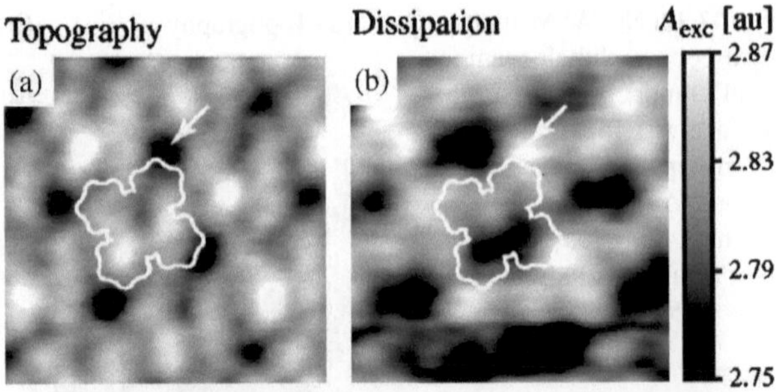

Fig. 17.13. NC-AFM images of a CuPc monolayer on a MoS_2 surface. (**a**) Topographic and (**b**) dissipation images (4 nm × 4 nm, $\Delta f = -45$ Hz, $A = 5.5$ nm)

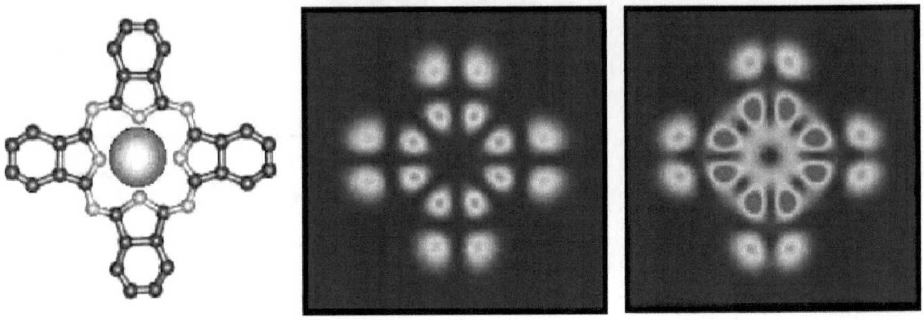

Fig. 17.14. Molecule structure of CuPc (*left*). Calculated electron densities of the highest occupied molecular orbital (*center*) and the lowest unoccupied molecular orbital (*right*)

cannot be explained simply by the molecular structure. Such apparent holes have also been found in previously reported STM images of CuPcs on various surfaces such as MoS_2 [36] graphite [36], Au(111) [39,40], and Cu(100) [41]. The formation mechanism of these STM contrasts has been explained by taking account of the spatial distribution of molecular orbitals [39–41]. CuPc has lower charge densities of the highest occupied molecular orbital (HOMO) and the lowest unoccupied molecular orbital (LUMO) at the center of the molecule [41], as shown in Fig. 17.14. Thus, the strong dependence of the tunneling probability on HOMO and LUMO densities leads STM images to represent not only the geometrical structure but also the electronic structure of CuPcs.

Similarly, the short-range chemical interactions dominated by frontier orbitals such as HOMO and LUMO play an important role in the formation of atomic-scale NC-AFM contrasts; hence, the apparent holes found in the NC-AFM image reflect the spatial distribution of the frontier orbitals.

17.4.3 Atomic Force Microscopy Imaging in Liquids

Atomic-scale FM-AFM imaging of atomically flat samples in UHV environments is now becoming routine work; however, the imaging capability in liquid by FM-AFM is severely hindered by the extreme reduction of a cantilever Q-factor owing to the hydrodynamic interaction between the cantilever and the liquid. The Q-factors in UHV usually exceed 10,000, while those in liquid environments are smaller than 10.

In fact it was not likely that FM-AFM would work in such an environment with a low Q-factor because a high Q-factor of a cantilever was indispensable for the stable operation of FM-AFM having an electromechanical resonator for the self-oscillation circuit.

The use of the small-amplitude mode and the large noise reduction in the cantilever deflection sensor brought great progress in FM-AFM imaging in liquids [42, 43]. The force sensitivity can be increased by FM detection with small-amplitude oscillation because of the increase in the duration of the proximity interactions. Note that the small-amplitude mode can be used only when the noise in the deflection sensor (laser beam deflection method) is sufficiently reduced down to the level of the thermal fluctuation of the cantilever. The phase noise in the FM detector is proportional to the ratio of the measurement noise to the oscillation amplitude.

Dominant noises in the laser beam deflection sensor are "optical feedback noise" and "optical interference noise", both of which come from the high coherence of the laser light. The modulation of the laser power with a high frequency signal whose frequency is typically 300–500 MHz is effective to reduce the coherence and hence the noise [44]. This technique was applied to the deflection sensor. Figure 17.15a,b shows the frequency spectra of cantilever Brownian motion measured using the improved deflection sensor in air and water, respectively [44]. The solid lines show experimentally measured values using the improved deflection sensor, while the dotted curves show the theoretical values calculated using the following equation.

$$\begin{aligned} z_{\mathrm{th}}(f) &= \sqrt{N_{\mathrm{th}}(f)} = \sqrt{|G(\omega)|^2 I_{\mathrm{th}}} \\ &= \sqrt{\frac{1}{\left[1-(f/f_0)^2\right]^2 + [f/(f_0 Q)]^2} \frac{2k_{\mathrm{B}}T}{\pi f_0 k Q}}\,, \end{aligned} \tag{17.68}$$

where f, f_0, and k are the vibration frequency, the resonance frequency, and the spring constant of a cantilever, respectively.

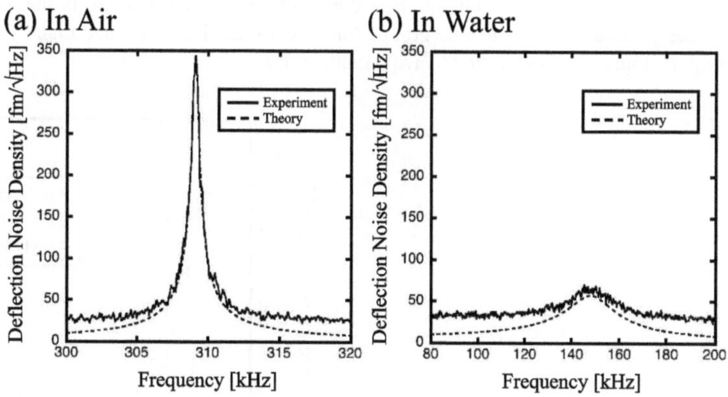

Fig. 17.15. Frequency spectra of cantilever Brownian motion measured in (**a**) air ($Q = 574$, $f_0 = 309.1$ kHz) and (**b**) water ($Q = 7$, $f_0 = 148.4$ kHz). The *solid lines* show experimentally measured values, while the *dotted lines* show theoretically calculated values with (17.68). A Si cantilever with an Al backside coating (Nanosensors, NCHR) was used ($k = 40$ N/m)

The peaks found in the spectra correspond to the Brownian vibration at the cantilever resonance, while the background white noise comes from the deflection-sensor noise. The results reveal that the deflection-noise densities arising from our deflection sensor were 25 $\text{fm}/\sqrt{\text{Hz}}$ in air and 30 $\text{fm}/\sqrt{\text{Hz}}$ in liquid. These values are much smaller than the deflection-noise densities obtained with a deflection sensor in a commercially available atomic force microscope (typically 100–1000 $\text{fm}/\sqrt{\text{Hz}}$ in air). Since the bandwidth of the FM detector (B_{FM}) is usually less than 1 kHz, the deflection-noise components which induce the frequency noises should be in the frequency range from f_0 to $f_0 + 1$ kHz. In this frequency region, the experimentally measured values are nearly the same as the theoretically calculated values in both air and liquid environments.

Consequently, true atomic-resolution imaging in liquids was successfully achieved by FM-AFM [42, 43]. A cleaved surface of a polydiacetylene [poly(2,4-hexadiyne-1,6-diol bis(p-toluene sulfonate)), poly-PTS] single crystal was imaged in pure water by FM-AFM as shown in Fig. 17.16. The herringbone structures of the (p-toluene sulfonate) side groups were clearly detected. Figure 17.17 shows frequency shift versus distance curves obtained for this sample. Each curve is clearly

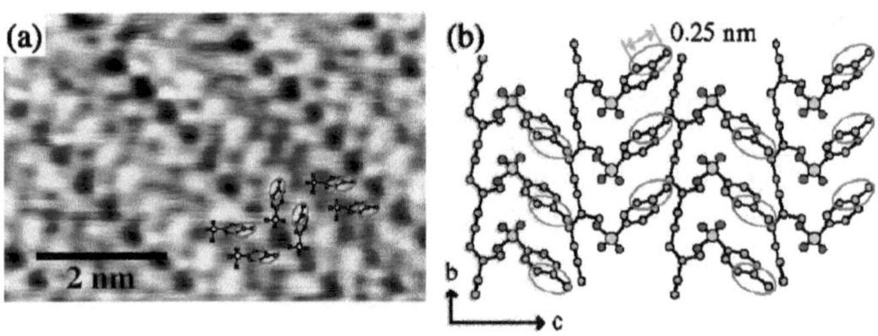

Fig. 17.16. (**a**) FM-AFM image of the *bc* plane of a poly(2,4-hexadine-1,6-diol bis(p-toluene sulfonate)) (poly-PTS) single crystal taken in pure water (**a**). Scan area: 6 nm × 4 nm, $\Delta f = +290$ Hz. ($f_0 = 140$ kHz, $k = 42$ N/m). (**b**) Crystal structure of the *bc* plane. The lattice constants of the crystal are $a = 1.449$ nm, $b = 0.491$ nm, and $c = 1.494$ nm. In the *bc* plane, one side of the p-toluene sulfonate side groups (omitted for clarity) is located under the other side. Hydrogen atoms are also omitted to avoid complexity

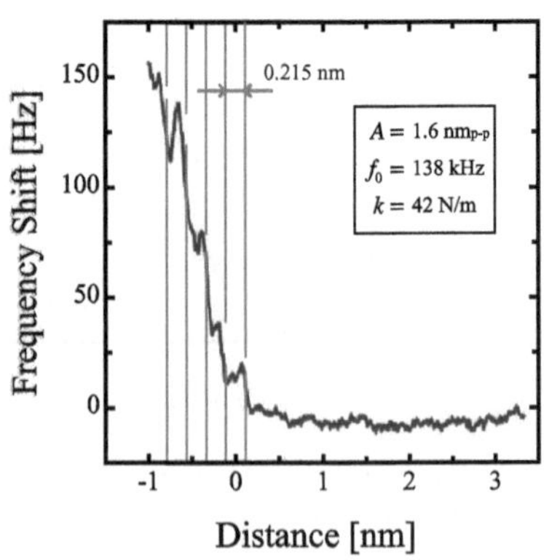

Fig. 17.17. Frequency shift–distance curve measured on the *bc* plane of a poly-PTS single crystal in water ($A = 0.8$ nm, $f_0 = 138$ kHz, $k = 42$ N/m)

modulated with a period of about 0.2 nm in the positive frequency shift region, which is probably caused by a hydration shell structure.

Achievements of true atomic-resolution imaging in liquids by dynamic mode AFM open up a wide variety of application fields. In particular, it is a remarkably powerful tool for the study of molecular-scale biology.

17.5 Investigations of Molecular Properties

17.5.1 Surface Potential Measurements

17.5.1.1 Kelvin Probe Force Microscopy

Electrostatic forces between an atomic force microscope tip and organic molecules are often related to the electronic structures of the molecules deposited on a solid substrate; thus, the measurements of the electrostatic forces in AFM are essentially important in terms of the following points:

1. AFM topographic imaging is greatly affected by the electrostatic interaction forces, which strongly depend on molecular species and materials. Topographic contrast can be heavily modified without compensation for electrostatic forces.
2. Local electronic structures at the interface between organic semiconductors and metal electrodes play a crucial role in molecular electronics, especially in carrier injection processes. Electrostatic force study by AFM can provide useful information on the electrical properties of nanoscale electrical junction regions.

KFM [45], which has been a common method for studying electrical properties in the nanometer-scale area, is a dynamic mode AFM technique combined with

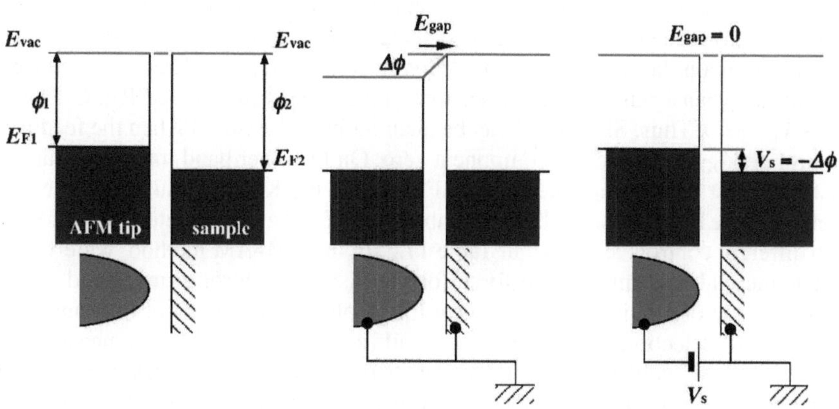

Fig. 17.18. Energy diagrams of an atomic force microscope tip and a sample. They are electrically isolated (*left*). Both are grounded (*center*). Both vacuum levels are equalized by an external bias voltage (*right*)

a conventional Kelvin probe method used to measure macroscopic contact potential differences. It allows us to map the electron affinity or the work function on a nanometer scale.

We consider a simple two-conductor system consisting of a metallic probe tip and a flat metallic sample, the Fermi energies of which are E_{F1} and E_{F2}, respectively. Assume that their vacuum levels are equal when either conductor is not electrically fixed. The configuration of the Fermi and the vacuum levels in this case is shown in Fig. 17.18 (left). When both are electrically shorted (the Fermi levels are equalized), an electric field between the two conductors is produced because of the work function difference between the two materials. The direct origin of this electric field is the charges induced on the surfaces of the atomic force microscope tip and the sample, which works as an electric double layer. If there exist molecules in the gap, the polarization caused by the molecules must be also considered.

When an alternating current (AC) voltage V_{AC} with an angular frequency ω_m is applied between the probe tip and the sample, the tip vibrates with the frequency ω_m owing to the electrostatic force. Using the capacitance between the tip and the sample C, which is a function of the tip-to-sample distance z, we can describe the electrostatic force F_z^{el} as

$$\begin{aligned} F_z^{el} &= \frac{1}{2}\frac{\partial C}{\partial z}\left(V_S + V_{DC} + V_{AC}\cos\omega_m t\right)^2 \\ &= \frac{1}{2}\frac{\partial C}{\partial z}\left[\left(V_S + V_{DC}\right)^2 + 2\left(V_S + V_{DC}\right)V_{AC}\cos\omega_m t + V_{AC}^2\cos^2\omega_m t\right], \end{aligned} \tag{17.69}$$

where V_{DC} and V_S are an externally applied bias voltage to the tip and a surface potential (SP) or a contact potential difference (CPD), respectively. Since electrostatic force is proportional to the applied voltage squared, the vibration contains DC, ω_m, and $2\omega_m$ components, which are a static attractive force between the electrodes (the tip and the sample) making up the capacitor, a force caused by the AC electric field acting on the previously described charges, and a force between the capacitor electrodes again, but induced by the AC voltage, respectively. The ω_m component disappears when a bias voltage applied to the tip compensates for the SP or CPD, i.e., $V_S + V_{DC} = 0$. Thus, SP or CPD can be quantitatively measured when the feedback control of V_{DC} keeps the ω_m component zero. On the other hand, the tip-to-sample distance must be also regulated as in AFM during the KFM measurement. Several methods have been classified by the combination of FM and AM detections for these two different controls as shown in Table 17.2. In the AM–AM method, widely used in commercial instruments, the force for the distance control is measured by the change in the oscillation amplitude near the resonance frequency of the cantilever, namely, AM detection or *tapping mode*, while an electrostatic force is measured by applying an AC voltage with a nonresonant frequency, which is relatively low in many cases. However, this method has a disadvantage that the sensitivity in electrostatic force detection is low because of its nonresonant detection. The sensitivity is improved in *lift mode* [46], where the topographic and electrostatic signals are alternately measured by two line scans. After topographic information has been ac-

Table 17.2. Various methods for detecting electrostatic force in electric field modulation (EFM)

EFM method	Tip-to-sample distance control method	Cantilever vibration frequency	Electrostatic force detection method	EFM frequency
AM–AM	AM (tapping mode)	$\sim f_{\mathrm{res}}$	AM	Nonresonant frequency
Lift-mode AM–AM	AM (tapping mode)	$\sim f_{\mathrm{res}}$	AM	$\sim f_{\mathrm{res}}$
FM–AM	FM	f_{res}	AM	$f_{\mathrm{res}}^{(2)} (\sim 6.3 f_{\mathrm{res}})$
FM–FM	FM	f_{res}	FM	f_{m} (within the bandwidth in FM detection)

f_{res} fundamental resonance frequency of cantilever
$f_{\mathrm{res}}^{(2)}$ second resonance frequency of cantilever
AM amplitude modulation detection
FM frequency modulation detection

quired during the first scan, the tip-to-sample distance is regulated during the second scan on the basis of the information already obtained. During the second scan, the tip is vibrated not by a mechanical vibrator but by an AC bias voltage modulated at the resonance frequency and thus the vibration is purely caused by the electrostatic force. It should be noted that the cantilever resonance is used for both detections of topographic and electrostatic information, which means that both signals can be sensitively measured.

Although the mutual perturbation between topographic and electrostatic signals can be remarkably suppressed in lift mode, a thermal or a mechanical drift of the tip-to-sample distance and some mechanical noises may cause an error in the potential measurement.

There are two methods for measuring electrostatic force in NC-AFM, which are shown as FM–AM and FM–FM methods in Table 17.2. The FM–AM method was first developed for the improvement of the response time in KFM measurements in a vacuum, where the Q-factor of the cantilever is tremendously high (more than 10,000), causing the response time of the vibration amplitude to be extremely long. The response is improved by the use of a high modulation frequency, the second resonance of the cantilever, for the AC modulation of the bias voltage. The use of the second resonance enables us to achieve high sensitivity and high-speed detection of electrostatic force [47]. In this method, electrostatic forces are measured using AM detection, while FM detection is used to maintain the tip-to-sample distance. On the other hand, in the FM–FM method the frequency of the AC modulation voltage is set at a frequency within the bandwidth of FM detection, which is usually less than 10 kHz. The resultant modulation of the electrostatic force causes the FM in the cantilever mechanical oscillation, which is finally detected as a modulated frequency shift by the FM detector [48]. Note that the electrostatic force is measured as an oscillating frequency shift of the resonance even if the modulation frequency of the voltage applied to the tip is not resonant. The electrostatic force induced by nonresonant ω_{m} causes the frequency modulation of resonant frequency at ω_{m}. Since the frequency shift measurement by FM detection is a highly sensitive method, the response time of which is not limited by the Q-factor, the FM–FM method has an excellent advantage in the measurement sensitivity and response time.

17.5.1.2
Surface Potential Measurements of Molecular Films

When molecules are deposited on a solid surface, the interfacial polarization is caused mainly by charge transfer between the films and the substrate and/or the sum of the electric dipoles of the molecules.

SP measurement is a most common method for investigation of molecular electrical properties. There have been several studies on the SP of alkanethiol SAMs. Evans and Ulman [49] investigated the SP of alkanethiol SAMs using the classical Kelvin method. They reported that the SP of the SAMs was positive with respect to the substrate and that the value linearly increased by 9.3 mV per CH_2 unit as the alkyl chain length was increased.

Lü et al. [50] first applied KFM combined with intermittent contact AFM to the local SP measurement of a patterned film with two different thiol SAMs which were prepared by microcontact printing. They measured the SP dependence of alkanethiol SAMs on the chain length, which was 14.1 ± 3.1 mV per CH_2 unit. They suggested that the relative difference of SP values between two different thiol SAMs is measured with high accuracy. The relative SP measurement by KFM against the potential of a reference material increases reliability because surface condition changes of the probe tip greatly affect the measurements.

Ichii et al. [51] used KFM combined with an NC-AFM setup for investigating the local SP of phase-separated SAMs (PS-SAMs) containing two species of thiol

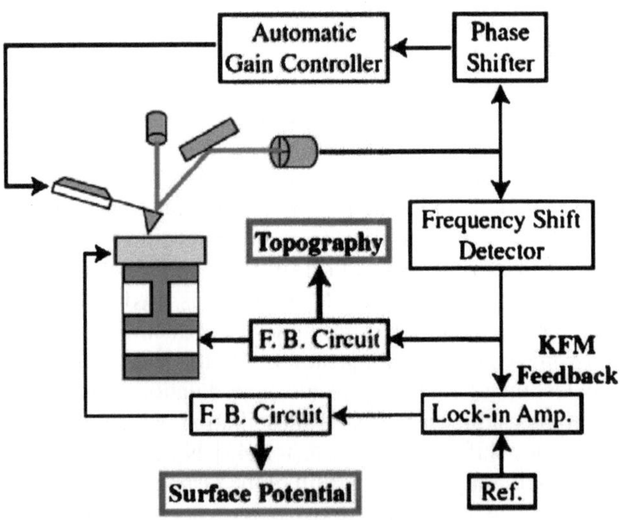

Fig. 17.19. The experimental setup for Kelvin probe force microscopy (KFM) combined with NC-AFM

molecules on a nanometer scale. Their experimental setup is shown in Fig. 17.19. One of the thiol SAMs in the PS-SAMs can be used as a reference in the relative SP measurement as mentioned before. The PS-SAM film is a more suitable sample for the SP study on a nanometer scale because it is self-assembled on an atomically flat Au(111) surface where the variation of the work function is negligible. They prepared three kinds of PS-SAMs, each of which was composed of two species with different chain-length alkanethiols, i.e., C_8/C_{12}, C_{10}/C_{12}, and C_{12}/C_{14}, respectively. The chain-length dependence of the SP was investigated and the origins of the SP distributions were discussed in terms of the molecular arrangements and the dipole moment.

Figure 17.20a shows a topographic image of a C_8/C_{12} PS-SAM film. The image clearly exhibits the feature of the phase separation. The bright areas correspond to the domains composed of C_{12} molecules, while the dark domains are composed of C_8 as shown in a schematic illustration in Fig. 17.20b. From the cross-sectional plots (not shown here), the height difference was about 0.5 nm.

This value agrees well with the difference (0.44 nm) in film thickness between C_8 SAMs and C_{12} SAMs, where the molecules are tilted 30° from the surface normal. It should be noted that the C_8 domains contain a depression called an etch pit, while no etch pits can be seen in the C_{12} domains. Bumm et al. [52] showed that the thermal treatment in heated ethanol produces a decrease in the number of etch pits as well as creating defect regions in SAM films. In our experiment, the C_{12} domains were heated in ethanol though the C_8 domains were not. This difference in the preparation between C_{12} and C_8 domains probably affected the number of etch pits.

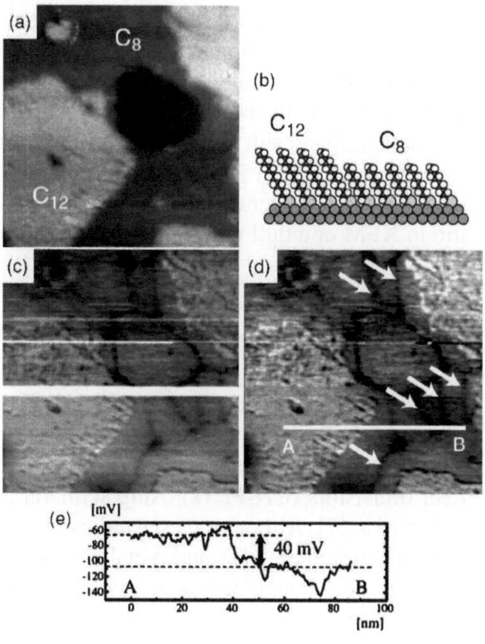

Fig. 17.20. (**a**) NC-AFM image of a C_8/C_{12} phase-separated SAM (PS-SAM) (100 nm × 100 nm, $\Delta f = -120$ Hz). (**b**) Models of possible structures of the C_8/C_{12} PS-SAM. (**c**) Surface potential image simultaneously obtained with (**a**). (**d**) Filtered surface potential image reduced the effect of "tip changes" from (**c**). (**e**) Cross-sectional plots taken on the unfiltered surface potential image

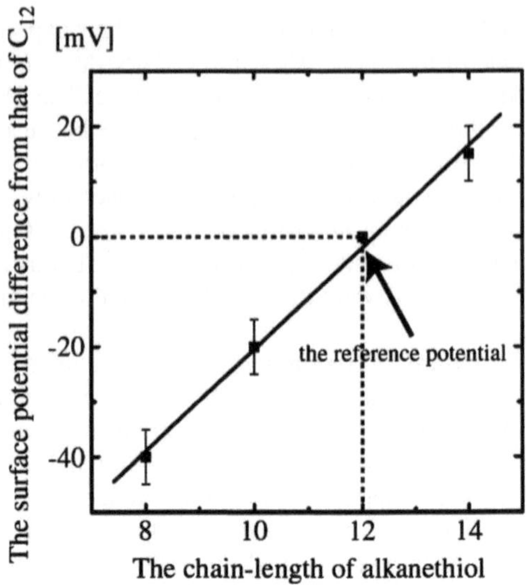

Fig. 17.21. Relative surface potential against the potential of C_{12} domains as a function of the chain length. The *error bars* were estimated from the noise in the experiment

Figure 17.20c shows the SP image of the C_8/C_{12} PS-SAM simultaneously taken with the topographic image. The filtered SP image in Fig. 17.20d to reduces the effect of the tip changes from the raw data. The SP of C_8 is lower than that of C_{12}. From the cross-sectional plots shown in Fig. 17.20e, which were plotted in the horizontal direction on the unfiltered SP image, the SP difference was about 40 mV.

Figure 17.21 shows plots of the difference in SP between the C_{12} and the C_n domains as a function of the chain length. The error bars were estimated from the noise in the experiment. The plots show that the SP increases linearly as the alkyl chain length is increased. The gradient is about 9 mV per CH_2 unit. Although the domain sizes in the PS-SAMs investigated in this study were several tens of nanometers, this result agrees well with the previous macroscopic study. However, in the case of C_{12}/C_{14} PS-SAMs, we could also find regions where the measured SP difference was smaller than expected. This was probably due to a low signal-to-noise ratio in KFM or a dull tip.

17.5.1.3
Carbon Nanotube Field Effect Transistor

Single-walled carbon nanotubes (SWNTs) as quantum wires are promising candidates for building blocks of next-generation electronic devices because of their unique electrical and mechanical properties [53]. Recently, carbon nanotube field effect transistors (CN-FETs) using semiconducting SWNTs have been intensively studied [54–57]. Although these studies indicate that the Schottky barriers at the interfaces between SWNTs and metal electrodes play important roles in the CN-FET operating mechanism, only a small number of studies have included local electrical measurements at the interfaces [58]. Miyato et al. [59] used KFM to investigate the

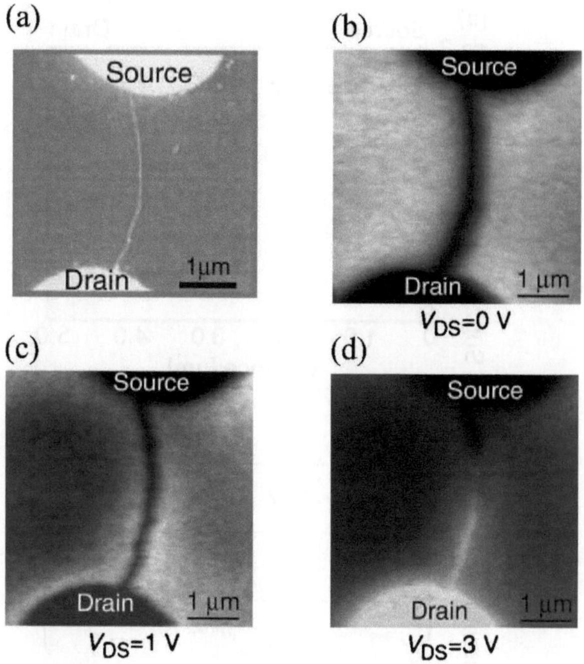

Fig. 17.22. (**a**) Topographic image of SWNTs connected to Ti electrodes, the gap distance between which is about 4 μm. A single bundle of SWNTs extending between the electrodes can be seen. (**b**)–(**d**) Surface potential images of the carbon nanotube field effect transistor sample shown in (**a**) obtained at drain voltages of 0, 1, and 3 V, respectively, while the gate voltage was fixed at 3 V. The brightness of the bundle of SWNTs is increased as the drain voltage is increased

local SP distribution along the SWNT channel between two electrodes in an FET device.

Figure 17.22a shows a topographic image of a CN-FET sample with Ti electrodes. The SWNT extended almost perpendicular to the edge of the electrodes and formed a gate channel. The gate channel in this sample was a single bundle with an average height of 4 nm.

The SP was measured by KFM when the bias voltage was changed. Figure 17.22b–d shows SP images of the CN-FET, the topographic image of which is shown in Fig. 17.22a, at three different drain voltages. It clearly shows that the contrast of the SWNT channel is varied according to the drain voltage change. The SP profiles along the bundle of SWNTs from the source to the drain were obtained from the SP images at different V_{DS}.

Figure 17.23 shows the potential profiles of the SWNTs in contact with the Ti electrodes. The shaded regions on the left and right sides correspond to the source and drain electrodes. We can clearly see that the SP drops at the drain edge are greatly modulated by the gate voltages,. The potential at a distance of 4 μm in Fig. 17.23 changes by about 1 V when the gate voltage is increased from −2 to 2 V. This agrees with the fact that the drain current is decreased as the gate voltage is increased in the gate voltage range where hole conduction is dominant. These SP changes at the interface between the Ti drain electrode and the SWNTs probably originate from the Schottky barrier change caused by gate bias modulation [54–56].

In addition, some potential drops exist in the middle of the SWNT bundle as indicated by the arrows in Fig. 17.23. As the drain bias voltage is increased,

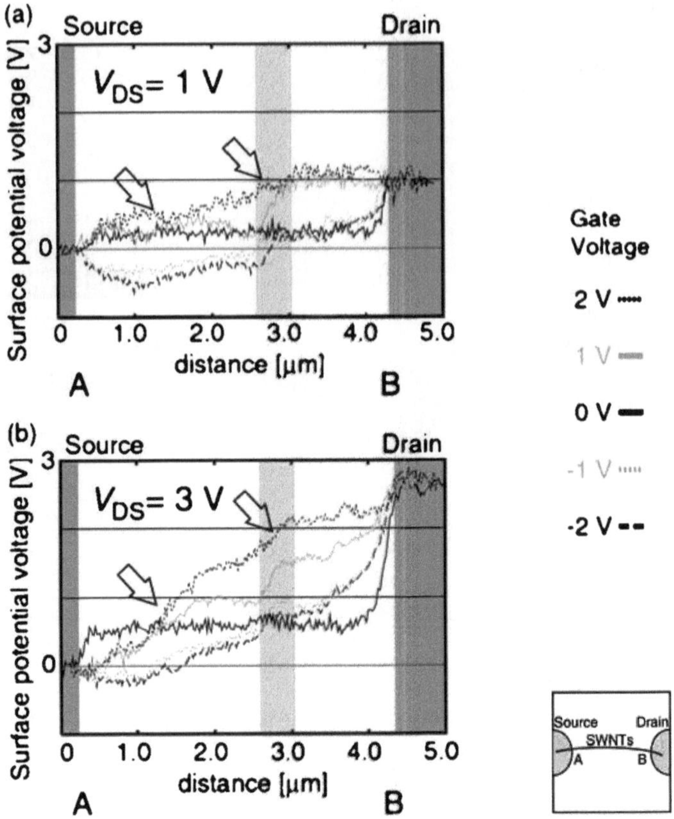

Fig. 17.23. Surface potential profiles along the bundle of SWNTs from the Ti source to the Ti drain for various gate voltages ranging from −2 to 2 V at a fixed drain voltage. The *dark shaded areas* correspond to the electrodes. (**a**) $V_{DS} = 1$ V. (**b**) $V_{DS} = 3$ V

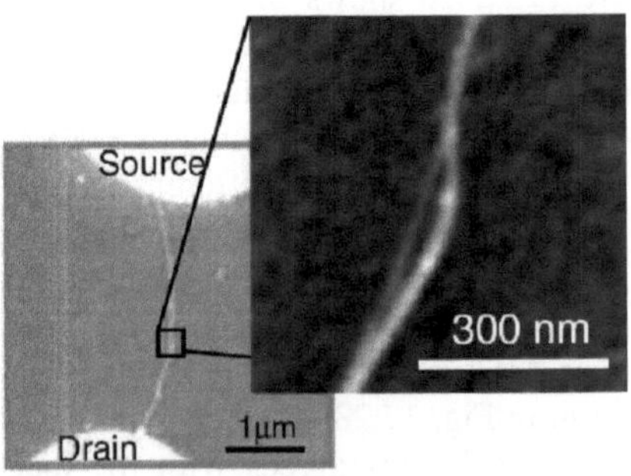

Fig. 17.24. The topographic image at a position where the potential was sharply changed (light shaded region in Fig. 17.23) is magnified. The SWNT bundle was untwisted around this point

these drops become larger. These are probably caused by some structural defects in the nanotubes such as untwisting points, breaking points, or connecting points. Figure 17.24 shows a magnified topographic image corresponding to the shaded region where the potential drops sharply in Fig. 17.23.

17.5.2 Energy Dissipation Measurements

The cantilever oscillates at the resonance frequency $(1/T)$ in FM-AFM. When the tip approaches the sample surface, the oscillation amplitude A is sharply reduced even before the start of the tip–surface contact. This is because the oscillation energy of the cantilever $(kA^2/2)$ is dissipated through tip–surface interactions (F_{ts}). The dissipation energy per unit time $P_{dissipation}$ is given by

$$P_{dissipation} = \frac{1}{2}kA^2\frac{\omega_0}{Q} + \frac{A\omega}{T}\int_0^T F_{ts}\sin\omega t\,dt\,. \tag{17.70}$$

The first term expresses the energy loss per unit time of a cantilever-free oscillation. The integration term has nonzero value when F_{ts} shows hysteresis behavior between the approaching and the retracting processes of the tip. The possible origins of the force hysteresis are various: changes in molecular configuration and/or conformation, lattice defects, tip changes. Besides, the displacement current between the tip and the sample creates the energy dissipation due to the Joule heat loss when some contact potential difference exists between them.

There have been a large number of studies on the energy dissipation in FM-AFM in both experimental and theoretical aspects [60,61]. Although experimental results on the local energy dissipation definitely contain useful nanoscale information of sample materials, the interpretation is quite complicated because of the existence of the various origins mentioned before.

Ichii et al. [62] studied the energy dissipation of phase-separated SAMs of thiol molecules. They prepared phase-separated SAMs containing 1-decanethiol [CH_3–

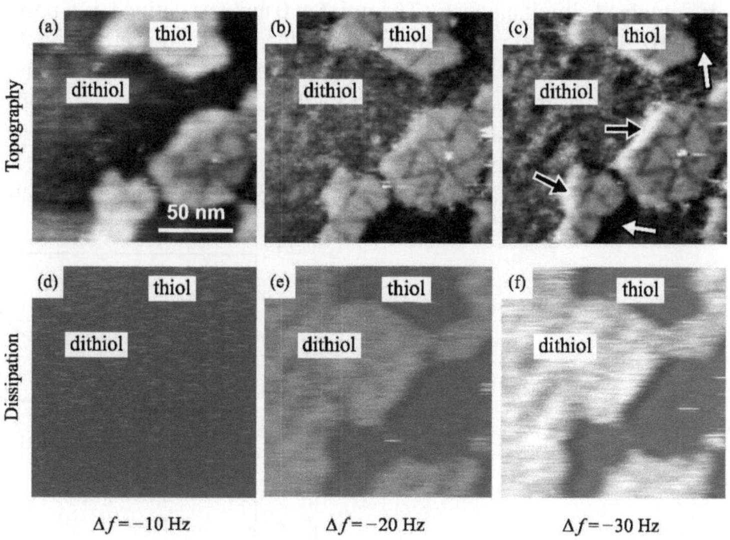

Fig. 17.25. (**a**)–(**c**) Topographic images and (**d**)–(**f**) energy dissipation images of PS-SAMs with different frequency shifts: $\Delta f = -10$ Hz for (**a**) and (**d**), −20 Hz for (**b**) and (**e**), and −30 Hz for (**c**) and (**f**). Scan area 150 nm × 150 nm

$(CH_2)_9$–SH; C_{10}] and 1,10-decanedithiol [HS–$(CH_2)_{10}$–SH; C_{10}D] on an Au(111) surface. Figure 17.25a–c shows topographic images of SAMs composed of C_{10} and C_{10}D, exhibiting the feature of phase separation clearly. In Fig. 17.25a the bright areas, which are higher than surrounding regions, are the domains of C_{10} molecules forming the closely packed domains such as $(\sqrt{3} \times \sqrt{3})R30°$ structures or c(4×2) structures. The lower regions probably correspond to C_{10}D molecules which are arranged on the gold surface with their molecular axes parallel to the substrate owing to the strong affinity between sulfur and gold atoms.

Figure 17.25d–f shows energy dissipation images taken simultaneously with those in Fig. 17.25a–c, respectively. The contrast in energy dissipation between C_{10}D domains and C_{10} domains increased as the absolute value of the frequency shift ($|\Delta f|$) was raised. In fact, this contrast enhancement is mainly ascribed to the increase of energy dissipation on C_{10}D domains as shown in Fig. 17.26. The energy dissipation on C_{10} domains increased only slightly.

Since the SP of C_{10}D domains was different from that of C_{10} domains, an electrostatic interaction might be related to the energy dissipation. However, the dissipation on C_{10}D domains was still very large even during the cancellation of the SP by KFM using a highly doped Si cantilever. Therefore, the origin of the large dissipation on C_{10}D domains is related to the perturbation of molecular arrangements of C_{10}D domains. They suggested that the structures of C_{10}D molecules were changed by the tip–sample interactions, and hence the energy dissipation on C_{10}D domains became large when $|\Delta f|$ was set to a large value.

They also investigated alkanethiol PS-SAMs with different head groups [63]. The sample used was a PS-SAMs composed of 1-dodecanethiol (C_{12})/11-mercapto-1-undecanol (C_{11}OH). Figure 17.27 shows plots of the tip-to-sample distance dependence of the frequency shift (Δf) and the energy dissipation, respectively. These plots were obtained simultaneously. The point at 0 nm was defined as a point where $\Delta f = -10$ Hz. In order to reduce the effect of the SP difference between C_{12} and C_{11}OH domains, these plots were taken with the KFM feedback on. When the tip-to-sample distance was relatively large, Δf on the C_{11}OH domain was almost equal to Δf on C_{12}. On the other hand, Δf on the C_{11}OH domain rapidly decreased when the tip-to-sample distance was smaller than −0.3 nm. The height difference between the C_{12} and C_{11}OH domains at $\Delta f = -30$ Hz is 0.3 nm in this figure, which agrees

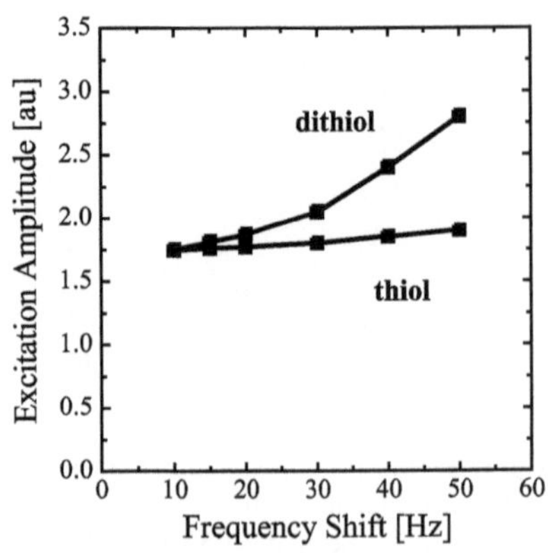

Fig. 17.26. Energy dissipation (excitation amplitude) as a function of the frequency shift (Δf) taken on the C_{10} domains and C_{10}D regions

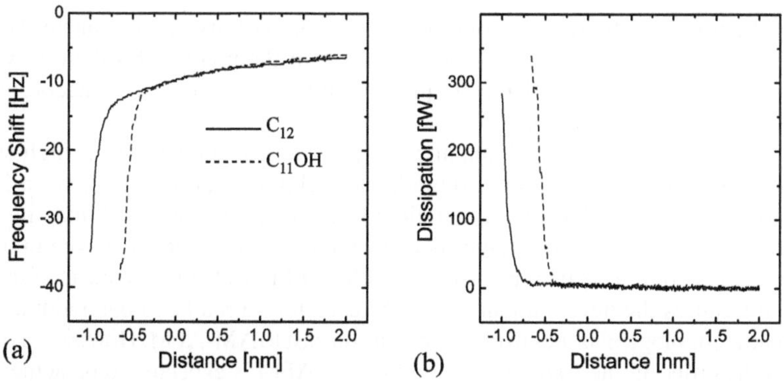

Fig. 17.27. Tip-to-sample distance dependences of the frequency shift (**a**) and the energy dissipation (**b**), obtained on the $C_{12}/C_{11}OH$ PS-SAMs

well with that in the corresponding topographic image. Since the theoretical molecular lengths of the two molecules are almost the same, it is natural that the observed height gap should come from the difference in tip–molecule interactions; however, this difference is so large that we cannot ascribe it to the short-range chemical forces between the tip and functional head groups of the thiol SAMs. What should be considered is the possible existence of water molecules on the $C_{11}OH$ domain, which was mentioned before. If this is the case, the dipole layer of the water film can cause a strong electrostatic force. When the tip-to-sample distance is relatively large, the electrostatic force is compensated by KFM feedback electronics; however, as the tip approaches the surface, some local variation of the dipole on the surface which cannot be thoroughly compensated by the KFM feedback probably affects the topographic image.

The energy dissipation also rapidly increases on the $C_{11}OH$ domain when the tip is close to the surface as shown in Fig. 17.27b. This can be also explained by the fluctuation of the water molecules as described before.

17.6 Summary and Outlook

This chapter has focused on recent progress in FM-AFM especially for molecular-scale imaging and analysis of organic materials. The basics and instrumentation of FM-AFM has been also described. The success in atomic-scale FM-AFM imaging in various environments, including in liquids, which was achieved by the fundamental noise reduction in the instruments, is a quantum leap toward a wide variety of practical applications. However, there are still several problems to be solved for practical imaging applications.

In FM-AFM imaging of a sample having a large height difference in surface structures, the cantilever oscillation tends to be unstable or stops in the worst case. When the oscillation stops, the AFM imaging signal is inevitably lost; besides the tip can crash anytime into the sample. By setting the frequency shift at a small value (the tip is positioned far from the sample surface), we can eliminate the risk of the oscillation stopping at the severe sacrifice of the imaging resolution. A new intelligent control method of the frequency shift, flexibly changing according to sample topography, is required.

A big concern in FM-AFM imaging toward practical applications is high-speed imaging. In fact there has been a remarkable technological development in high-speed dynamic mode AFM imaging in liquids. High-speed AM-AFM imaging with a frame rate of more than ten frames per second has already been achieved in liquid environments [64]. The imaging speed of FM-AFM is also expected to improve. Since it requires the development of an FM detector with a large bandwidth, there are some technological issues to overcome in contrast to AM-AFM. However, a frame rate on the order of one frame per second in FM-AFM is expected to be achievable in the near future.

References

1. Binnig G, Quate CF, Gerber C (1986) Phys Rev Lett 56:930
2. Hansma PK, Cleveland JP, Radmacher M, Walters DA, Hillner PE, Bezanilla M, Fritz M, Vie D, Hansma HG, Prater CB, Massie J, Fukunaga L, Gurley J, Elings V (1994) Appl Phys Lett 64:1738
3. Martin Y, Williams CC, Wickramasinghe HK (1987) J Appl Phys 61:4723
4. Albrecht TR, Grütter P, Horne D, Ruger D (1991) J Appl Phys 69:668
5. García R, San Paulo A (1999) Phys Rev B 60:4961
6. Yamada H (2002) In: Non-Contact Atomic Force Microscopy edited by Morita S, Wiesendanger R, Meyer E, p 193–212, Springer, Berlin, Heidelberg, New York
7. Dürig U, Steinauer HR, Blanc N (1997) J Appl Phys 82:3641
8. Edwards H, Taylor L, Duncan W, Melmed AJ (1997) J Appl Phys 82:980
9. Loppacher C, Bammerlin M, Battiston F, Guggisberg M, Müller D, Hidber HR, Lüthi R, Meyer E, Güntherodt HJ (1998) Appl Phys A 66:S215
10. Kobayashi K, Yamada H, Itoh H, Horiuchi T, Matsushige K (2001) Rev Sci Instrum 72:4383
11. Giessibl F (1997) Phys Rev B 56:16010
12. Sasaki N, Tsukada M (1998) Jpn J Appl Phys 37:L533
13. Dürig U (1999) Surf Interface Anal 27:467
14. Giessibl F (2001) Appl Phys Lett 78:123
15. Sader J, Jarvis SP (2004) Appl Phys Lett 84:1801
16. Dürig U (1999) Appl Phys Lett 75:433
17. Smith DPE (1995) Rev Sci Instrum 66:3191
18. Robins WP, in Phase Noise in Signal Sources (Peregrinus, London, 1982)
19. Porter MD, Bright TB, Allara DL, Chidsey CED (1987) J Am Chem Soc 109:3559
20. Nuzzo RG, Korenic EM, Dubois LH (1990) J Chem Phys 93:767
21. Nuzzo RG, Dubois LH, Allara DL (1990) J Am Chem Soc 112:558
22. Strong L, Whitesides GM (1988) Langmuir 4:546
23. Chidsey CED, Liu G-Y, Scoles G, Wang J (1990) Langmuir 6:1804
24. Camillone N, Chidsey CED, Liu G-Y, Scoles G (1993) J Chem Phys 98:3503
25. Fenter P, Eberhardt A, Eisenberger P (1994) Science 266:1216
26. Delamarche E, Michel B, Biebuyck HA, Gerber C (1996) Adv Mater 8:719
27. Kobayashi K, Yamada H, Horiuchi T, Matsushige K (1998) Jpn J Appl Phys 37:6183
28. Noh J, Hara M (2001) Langmuir 17:7280

29. Uchihashi T, Ishida T, Komiyama M, Ashino M, Sugawara Y, Mizutani W, Yokoyama K, Morita S, Tokumoto H, Ishikawa M (2000) Appl Surf Sci 157:244
30. Fukuma T, Kobayashi K, Horiuchi T, Yamada H, Matsushige K (2001) Appl Phys A 72:S109
31. Fukuma T, Kobayashi K, Yamada H, Matsushige K (2004) J Appl Phys 95:1222
32. Bumm LA, Arnold JJ, Charles LF, Dunbar TD, Allara DL, Weiss PS (1999) J Am Chem Soc 121:8017
33. Fukuma T, Kobayashi K, Yamada H, Matsushige K (2004) J Appl Phys 95:4742
34. Hara M, Sasabe H, Yamada A, Garito AF (1989) Jpn J Appl Phys 28:L306
35. England CD, Collins CE, Schuerlein TJ, Armstrong NR (1994) Langmuir 10:2748
36. Ludwig C, Strohmaier R, Petersen J, Gompf B, Eisenmenger W (1994) Vac J Sci Technol B 12:1963
37. Okudaira KK, Hasegawa S, Ishii H, Seki K, Harada Y, Ueno N (1999) J Appl Phys 85:6453
38. Bennewitz R, Foster AS, Kantrovich LN, Bammerlin M, Loppacher C, Schär S, Guggisberg M, Meyer E (2000) Phys Rev B 62:2074
39. Lu X, Hipps KW, Wang XD, Mazur U (1996) J Am Chem Soc 118:7197
40. Hipps KW, Lu X, Wang XD, Mazur U (1996) Phys J Chem 100:11207
41. Lippel PH, Wilson RJ, Miller MD, Wöll C, Chiang S (1989) Phys Rev Lett 62:171
42. Fukuma T, Kobayashi K, Matsushige K, Yamada H (2005) Appl Phys Lett 86:193108
43. Fukuma T, Kobayashi K, Matsushige K, Yamada H (2005) Appl Phys Lett 87:034101
44. Fukuma T, Kimura M, Kobayashi K, Matsushige K, Yamada H (2005) Rev Sci Instrum 76:053704
45. Nonnenmacher M, O'Boyle MP, Wickramasigh HK (1991) Appl Phys Lett 58:1921
46. Jacobs HO, Knapp HF, Müller S, Stemmer A (1997) Ultramicroscopy 69:239
47. Kikukawa A, Hosaka S, Imura R (1995) Appl Phys Lett 66:3510
48. Kitamura S, Iwatsuki M (1998) Appl Phys Lett 72:3154
49. Evans SD, Ulman A (1990) Chem Phys Lett 170:462
50. Lü J, Delamarche E, Eng L, Bennewitz R, Meyer E, Güntherodt H-J (1999) Langmuir 15:8184
51. Ichii T, Fukuma T, Kobayashi K, Yamada H, Matsushige K (2004) Nanotechnology 15:S30
52. Bumm LA, Arnold JJ, Charles LF, Dunbar TD, Allara DL, Weiss PS (1999) J Am Chem Soc 121:8017
53. Dekker C (1999) Phys Today 52:22
54. Heinze S, Tersoff J, Martel R, Derycke V, Appenzeller J, Avouris P (2002) Phys Rev Lett 89:106801
55. Appenzeller J, Knoch J, Derycke V, Martel R, Wind S, Avouris P (2002) Phys Rev Lett 89:126801
56. Radosavljevi M, Heinze S, Tersoff J, Avouris P (2003) Appl Phys Lett 83:2435
57. Javey A, Guo J, Wang Q, Lundstrom M, Dai H (2003) Nature 424:654
58. Cui X, Freitag M, Martel R, Brus L, Avouris P (2003) Nano Lett 3:783
59. Miyato Y, Kobayashi K, Matsushige K, Yamada H (2005) Jpn J Appl Phys 44:1633
60. Gauthier M, Tsukada M (1999) Phys Rev B 60:11716
61. Loppacher C, Bennewitz R, Pfeiffer O, Guggisberg M, Bammerlin M, Schär S, Barwich V, Baratoff A, Meyer E (2000) Phys Rev B 62:13674
62. Ichii T, Fukuma T, Kobayashi K, Horiuchi T, Yamada H, Matsushige K (2003) Appl Surf Sci 210:99
63. Ichii T, Fukuma T, Kobayashi K, Yamada H, Matsushige K (2004) Jpn J Appl Phys 43:4545
64. Ando T, Kodera N, Takai E, Maruyama D, Saito K, Toda A (2001) Proc Natl Acad Sci USA 98:12468

18 Noncontact Atomic Force Microscopy

Yasuhiro Sugawara

18.1 Introduction

Atomic force microscopy (AFM) is a novel technique for high-resolution imaging of conducting as well as nonconducting surfaces. The physical property sensed in AFM is the interaction force between the sample surface and a sharp probing tip. In 1995, noncontact AFM (NC-AFM) using the frequency modulation (FM) detection method [1] achieved true atomic resolution, in which the probing tip is not in contact with the sample surface, and the surface structure is obtained from the change of the resonant frequency of the oscillating cantilever. At present, various surfaces such as clean semiconductors, ionic crystals, metal oxide, metal-deposited semiconductors, pure metals and layered materials have been observed successively with atomic resolution. Now, NC-AFM is expected to become a powerful scientific tool for resolving the atomic features in a variety of fields such as materials and biological sciences. Here, we overview the capability of NC-AFM as an atomic-resolution microscopy.

18.2 NC-AFM System the Using FM Detection Method

In order to detect very weak atomic forces acting between the probing tip and the surface and to avoid the destruction of the very sharp tip apex and the surface, the FM detection method is used in ultrahigh vacuum (UHV) [1]. In the FM detection method, the cantilever serves as the frequency-determining element of an oscillator. Force acting on the cantilever causes instantaneous frequency modulation of the oscillator output, which is demodulated with a FM detector. The FM technique offers increased sensitivity through increased Q without restricting system bandwidth. In Fig. 18.1, we show a schematic diagram of NC-AFM using the FM detection method. The cantilever is driven at the resonant frequency by the positive feedback system with a variable gain amplifier, an automatic gain control (AGC) circuit and a phase shifter. The AGC circuit is used to maintain the vibration amplitude of the oscillating cantilever at a constant level. The phase shifter is used to adjust the phase of the feedback system. The frequency shift Δf of the cantilever induced by force interaction is detected by the FM demodulator. In AFM imaging, the distance between the probing tip and the sample surface is controlled to keep a constant frequency shift Δf by the feedback controller.

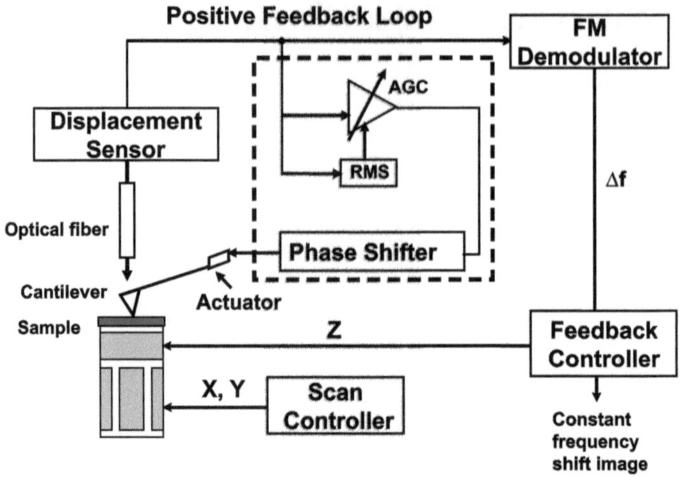

Fig. 18.1. Frequency modulation noncontact atomic force microscopy (NC-AFM)

In the experiments, we used two types of NC-AFM systems: one is a noncontact atomic force microscope operating at room temperature; the other is a noncontact atomic force microscope operating at low temperatures. At room temperature, some atomically resolved imagings are reported on various surfaces. On the other hand, at low temperatures, there are only a few reports. There are several advantages of the NC-AFM measurement in a low-temperature environment: (1) frequency noise of the cantilever caused by the thermal fluctuation decreases, and a high-sensitivity measurement is expected; (2) investigations of some special phenomena such as phase transitions are expected. So, we present the design of the low-temperature NC-AFM system [2].

As shown in Fig. 18.2, the low-temperature NC-AFM system consists of an observation chamber, a preparation chamber and a load lock chamber. The observation chamber has a UHV-compatible cryostat and an AFM unit. Deflection of the cantilever is detected by a fiber-optic interferometer. In order to avoid long-distance translation of the atomic force microscope unit while protecting the fragile optical fiber and to reduce outgassing in the bottom bath cryostat, the "top bath" cryostat is used as a cryostat. The top bath cryostat consists of a liquid helium Dewar (4.6 l) and a liquid nitrogen Dewar (3.8 l), which are suspended by three pipes from the top of the observation chamber. In order to avoid the temperature increase of the AFM unit by the radiation, we use three radiation shields (namely, inner, middle and outer shields), which are made from silver-plated copper. The inner shield is connected to the bottom of the helium Dewar. The middle shield is connected to the top of the helium Dewar and cooled by evaporated helium gas. The outer shield is connected to the bottom of the liquid nitrogen Dewar. The three shields have openings used for sample and cantilever exchange and for observing. The openings have two shutters to suppress thermal radiation and gas adsorption onto a sample surface. The two shutters are placed between each of the three shields. An inner shutter is in contact with the inner shield. An outer shutter is in contact with the outer shield. The AFM unit is cooled down to 5 K for 14 h with 4.6 l liquid helium. The quick

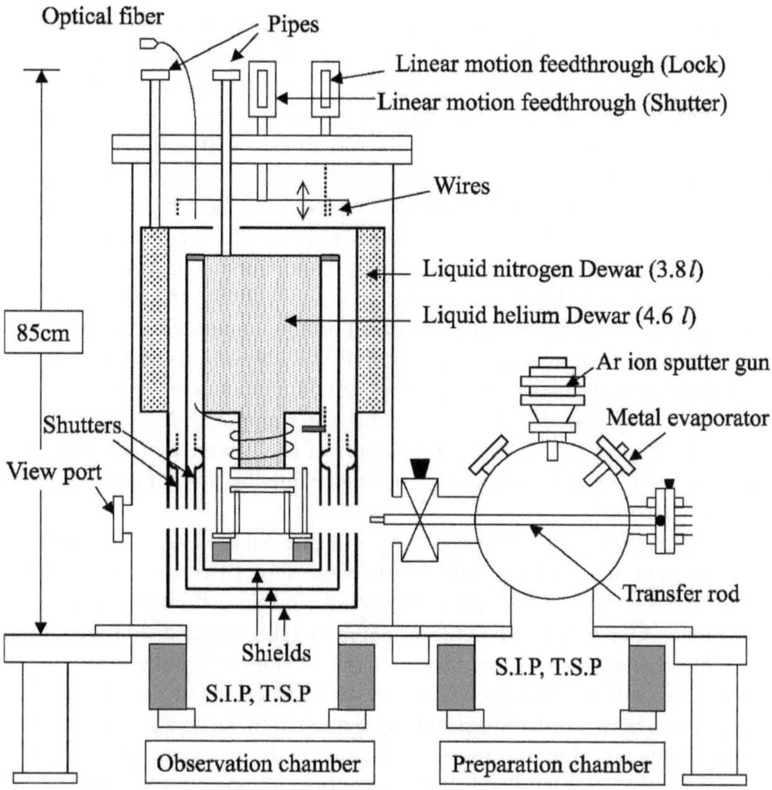

Fig. 18.2. Side view of the observation and the preparation chambers for a low-temperature NC-AFM system

sample and cantilever exchange is performed even at low temperatures. The optimal and reproducible positioning of the optical fiber with respect to a cantilever can be performed with a specially designed three-dimensional micropositioner within 10 min.

As the force sensor, we used a conductive silicon cantilever with a sharpened tip. The spring constant and mechanical resonant frequency were 40–60 N/m and 150–170 kHz, respectively. The nominal radii of curvature for the tip apex were 5–10 nm. The silicon tip was cleaned by sputtering with Ar ions. There are the dangling bonds out of the silicon tip apex. The NC-AFM image was obtained under the constant frequency shift.

18.3 Identification of Subsurface Atom Species

NC-AFM has the capability to identify or recognize atom species on a sample surface, if we can control the atomic species at the tip apex. That is, we succeed in identification of Si and Ge atoms by imaging the Si/Ge intermixing Si(111) surface using the Si tip, where both Si and Ge atoms are group IV elements [3]. We further investigate the capability of NC-AFM to identify the atom species under the surface.

It is well known that B atoms (group III elements) easily diffuse in the Si bulk crystal at high temperature owing to its small atomic radius. By using this nature of the B atom, we used the Si(111) $\sqrt{3} \times 5\sqrt{3}$-B surface with different atom species under the surface. As shown in Fig. 18.3, on the Si(111) $\sqrt{3}\sqrt{\times 3}$-B surface, there are two types of structures. One is the Si-T_4 structure, where both the topmost and the subsurface atoms are Si atoms; the other is B-S_5 structure, where the topmost and the subsurface atoms are Si and B atoms, respectively. The topmost Si atom on the Si-S_5 structure is 0.02 nm higher than that on the B-S_5 structure. On the Si-T_4 structure, a dangling bond orbital is localized on the topmost Si atom. On the other hand, on the B-S_5 structure, an empty orbital is localized on the topmost Si atom owing to the charge transfer from the topmost Si atom to the subsurface B atom. In this experiment, first, Si(111)7 × 7 substrate was prepared by in situ thermal treatments of a Si sample. Several monolayers of B_2O_3 were then deposited on the sample at 843 K from a hot-filament evaporator. Finally, this sample was annealed at 1073 K for 10 min.

Figure 18.4 shows an NC-AFM image on the Si(111) $\sqrt{3} \times 5\sqrt{3}$-B surface measured at room temperature. The AFM image shows two types of spots: bright and dim spots. The height difference between the bright and dim spots is estimated to be 0.03 nm, which is larger than the structural height difference of 0.02 nm. This is due to the fact that tip–sample force interaction is dominated by chemical interaction.

A question is raised as to which spot is the B site under the surface. In the chemical force interaction between the Si tip and the Si adatom in the Si-T_4 structure, the level of the dangling bond orbital for the Si tip is equal to that of the dangling bond orbital for Si adatoms. On the other hand, in the force interaction between the Si tip and the Si adatom in the B-S_5 structure, the level of the dangling bond orbital differs from the level of the empty orbital. In this case, the energy shift for the dangling bond and the dangling bond orbital interaction is larger than the energy shift for the dangling bond and the empty orbital interaction. As a result, the force interaction between the dangling bond orbital and the dangling bond orbital is stronger than that between

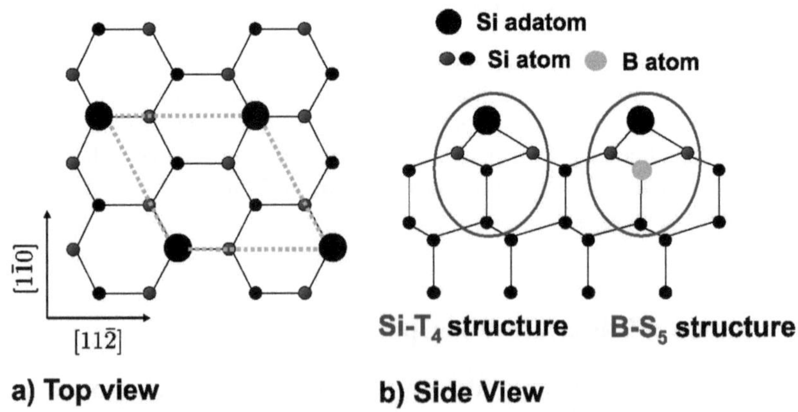

Fig. 18.3. Structure of Si(111) $\sqrt{3} \times \sqrt{3}$-B surface with different atom species under the surface

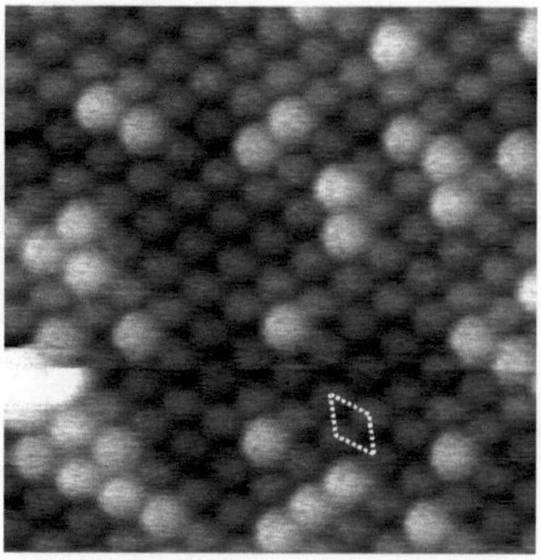

Fig. 18.4. NC-AFM image of Si(111)$\sqrt{3} \times \sqrt{3}$-B surface. The *bright spots* and *dim spots* correspond to Si-T_4 and B-S_5 structures, respectively. The scan size is 7.7 nm × 7.7 nm

the dangling bond orbital and the empty orbital. This suggests that the bright spots correspond to the Si-T_4 structure, while dim spots correspond to the B-S_5 structure. This experimental result suggests that the contrast of the NC-AFM images changes depending on the atom species under the surface. Thus, we found that NC-AFM has the capability to identify or recognize atom species under a sample surface.

18.4 Tip-Induced Structural Change on a Si(001) Surface at 5 K

The Si(001) surface has attracted much attention because of its practical importance in most large scale integration (LSI) devices that are fabricated on this surface. However, recent scanning tunneling microscopy (STM) studies performed at low temperatures questioned whether the c(4 × 2) phase is the most stable phase. For instance, the p(2 × 1) structure was observed in [4, 5]. It was suggested in [4] that the p(2 × 1) structure is due to the time averaging of the rapid flip-flop motion at 5 K. At the same time, the symmetric p(2 × 1) structure observed in [2] at 20 K was attributed to static symmetric dimers stabilized by the "antiferromagnetic" dimer–dimer interaction. Thus, up to now, the appearance of the p(2 × 1) phase in STM images at low temperature remains a controversial issue. Here, we investigate the tip-induced structural change on a Si(001) surface by NC-AFM at 5 K.

Figure 18.5a shows the NC-AFM topographic image of the Si(001) surface taken at $\Delta f = -10$ Hz, where we can assume that the tip–surface interaction is rather weak. The vibration amplitude of the oscillating cantilever was 13.5 nm. Here, the c(4 × 2) phase with a zigzag pattern is clearly observed. This result is in good agreement with a previous AFM study [6]. In Fig. 18.5b, we show a stable topographic image taken at a larger $\Delta f = -22$ Hz, where the tip–surface interaction is much stronger. We find that a surface phase is observed with bright lines along dimer rows and flicker noise. It is noted that NC-AFM images with flicker noise were observed for the first time, although the same phenomenon has been confirmed by STM studies [4]. The appearance of the flicker noise strongly suggests that surface dimers are frequently flipped by the tip–surface interaction. Notice that the p(2 × 1) phase was not observed at this frequency shift. Figure 18.5c shows a topographic

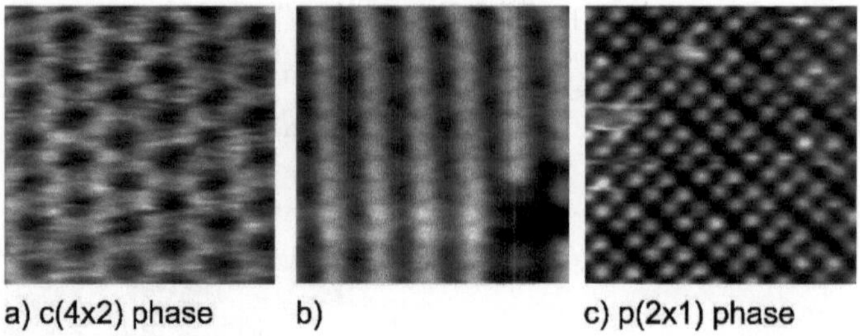

Fig. 18.5. NC-AFM topographic images of the Si(001) surface at 5 K observed at (**a**) $\Delta f = -10$ Hz, (**b**) -22 Hz and (**c**) -30 Hz. The vibration amplitude of the oscillating cantilever was 13.5 nm. In all images the scan area is 3.0 nm × 3.0 nm

image taken at an even larger $\Delta f = -30$ Hz, in which case a very strong tip–surface interaction is expected. We find that the p(2 × 1) phase with symmetric dimers is observed at these conditions. For the first time, the appearance of the p(2 × 1) phase was confirmed by NC-AFM at 5 K. The dimer does not flip back and remains in the flipped position during subsequent tip oscillations since there is not enough thermal energy in the system at 5 K. When during the lateral scan the tip eventually moves to the other atom of the same dimer, the dimer flips again. This means that each atom of the dimer is imaged equally as the upper atom, and results in the p(2 × 1) phase with symmetric dimers observed, with the increased apparent distance between the corresponding bright spots as seen in Fig. 18.5c.

The stable p(2 × 1) phase is caused by the continuous change of the buckling direction during the scan in which the tip acts as a manipulation ("writing") device [7]. Using a special scanning protocol at high-frequency shifts, one can create a particular dimer pattern on the surface. At low-frequency shifts, the tip performs as a probe that does not damage the surface, "reading" the actual pattern of buckled dimers; thus the c(4 × 2) phase is observed. This regime can thus be used to image the surface before and after the intrusive scan to check the surface structure change.

18.5 Influence of Surface Stress on Phase Change in the Si(001) Step at 5 K

In NC-AFM measurement on a Si(001) surface at 5 K, the asymmetric c(4 × 2) phase changed into the symmetric p(2 × 1) phase under the strong tip–sample force interaction [7]. The strong tip–sample interaction affects the buckling direction of asymmetric dimers of the c(4 × 2) phase, and the stable p(2 × 1) phase is caused by the continuous change of the buckling direction during the scan. Namely, the tip-induced dimer flipping events. The dimer is locked in the excited state at 5 K until during the scan the tip comes close to the other dimer atom and flips the dimer again.

How do the phenomena described above occur near the S_A step on the Si(001) surface? It is known that when structural reconstruction of the phase occurs on the Si(001) surface the surface stress concentrates on the step, and a big distortion is detected at the S_A step. At the upper side of the S_A step, line stress works and pulls long the direction of a dimer row, while at the lower side of the S_A step there is

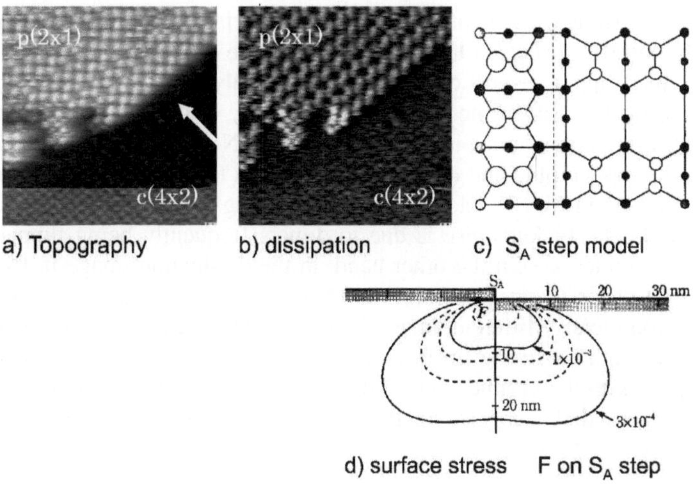

Fig. 18.6. (**a**) NC-AFM topographic and (**b**) dissipation images on a Si(001) surface with a step at 5 K. In all images the scan area is 9.2 nm × 9.2 nm. The dimer row in the [110] direction is indicated by the *arrow*. (**c**) S_A step model and (**d**) distribution of the surface stress on the S_A step. Note that the c(4 × 2) phase exists on the upper S_A step, while the p(2 × 1) phase exists on the lower S_A step

compression stress in the direction of a dimer row (Fig. 18.6d). Thus, line stress F is in a direction to the upper side of the S_A step, and the atom of the step edge receives relief. So, we investigated the phase change on the S_A step by the tip-induced dimer flipping events.

Figure 18.6 shows the topographic and dissipation images measured at the strong tip–surface interaction. For the first time, we found that the c(4 × 2) phase was observed at the upper side of the S_A step, while the p(2 × 1) phase was observed at the lower side of the S_A step. This means that the tip-induced dimer flipping events do not occur at the upper side of the S_A step, but they occurred at the lower side of the S_A step. This result originates from the influence of the surface stress on the flipping events.

18.6 Origin of Anomalous Dissipation Contrast on a Si(001) Surface at 5 K

The tip–sample interaction in NC-AFM leads to energy dissipation, which has been used as another imaging mechanism of surface topography. The macroscopic dissipation mechanism due to long-range electrostatic damping by Joule dissipation of charge carriers is well established. However, the origins of atomic-scale dissipation, leading to atomically resolved contrast, in NC-AFM are still under debate. Essentially, three models have been suggested to explain the energy dissipation in NC-AFM: a stochastic friction force mechanism (thermal fluctuation process), an adhesion hysteresis mechanism (stick–slip process) and an artifact (a soft coupling between the electronics and the nonlinearity of the interaction).

In order to study the role of the relaxation of surface atoms in the energy dissipation, the Si(001) surface at low temperature is one of the most suitable surfaces, because dimer flipping events on the surface are easily induced by the atomic force microscope tip [7]. So, in the present experiments, we investigated the origin of the energy dissipation on a Si(001) surface using NC-AFM at 5 K. In Fig. 18.7, we show the topographic and dissipation images on the Si(001) surface. In the topographic image in Fig. 18.7a, the bright lines along dimer rows and flicker noise are observed. The flicker noise is due to dimers frequently being flipped by the tip–surface interaction. On the other hand, in the dissipation image in Fig. 18.7b, the c(4 × 2) phase is clearly observed. Thus, for the first time, we succeeded in obtaining completely different image contrasts. This difference is explained by the fact that, under a strong tip force interaction, if the dimer is flipped by a tip, it will flip back to the ground state of the c(4 × 2) phase. This experimental result strongly suggests that energy dissipation can be explained by the adhesion hysteresis mechanism.

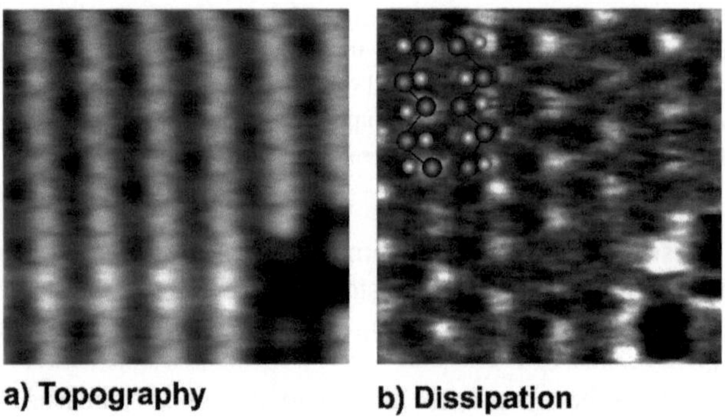

Fig. 18.7. (**a**) Topographic and (**b**) dissipation images on a Si(001) surface simultaneously measured using NC-AFM at 5 K. Scan size 3 nm × 3 nm

18.7 Summary

We demonstrated that NC-AFM has the capability to identify or recognize atom species under a sample surface. The controversial issue of the origin of the p(2 × 1) reconstruction of the Si(001) surface observed in low temperature STM experiments was clarified using low-temperature NC-AFM at 5 K. The stable p(2 × 1) phase was observed and it was caused by the surface dimer flipping during the scan induced by the interaction force with the tip. Using an appropriate scanning protocol, one can modify the dimers' buckling pattern on the surface at will. Note that this result shows the significance of tip-induced structure modifications and manipulation to nanotechnology by NC-AFM for the first time. Finally, we showed that the tip–sample interaction in NC-AFM leads to atomic-scale energy dissipation, which can be explained by the adhesion hysteresis mechanism.

References

1. Albrecht T, Grütter P, Horne D, Rugar D (1991) J Appl Phys 69:668
2. Suehira N, Tomiyoshi Y, Sugawara Y, Morita S et al. (2001) Rev Sci Instrum 72:2971
3. Morita S et al. (2005) Scanning Probe Microscopy: Characterization, Nanofabrication and Device Application of Functional Materials. Kluwer, Dordrecht, p 173
4. Yokoyama T, Takayanagi K (2000) Phys Rev B 61:R5078
5. Kondo Y, Amakusa T, Iwatsuki M, Tokumoto H (2000) Surf Sci 453:L318
6. Uozumi T, Tomiyoshi Y, Suehira N, Sugawara Y, Morita S (2002) Appl Surf Sci 188:279
7. Li YJ, Nomura H, Ozaki N, Naitoh Y, Kageshima M, Sugawara Y, Hobbs C, Kantorovich L (2006) Phys Rev Lett 96:106104

19 Tip-Enhanced Spectroscopy for Nano Investigation of Molecular Vibrations

Norihiko Hayazawa · Yuika Saito

19.1 Introduction

An optical microscope that is capable of showing images of molecules in nanometer scale has been a dream of scientists, of physicists, chemists, and biologists. However, it has been thought to be difficult to achieve owing to the strict limitation of a spatial resolution because of the wave nature of light [1]. While there have been attempts to overcome the diffraction limit by using the nonlinear response of materials [2, 3], near-field optics could provide better detecting accuracy [4–6]. In 1928, Synge [7] proposed a method, near-field scanning optical microscopy (NSOM), which can overcome the limited resolution of a conventional optical microscope (far-field microscope) where the resolution is limited by the so-called diffraction limit formulated by Abbe [8] in 1873. Since the first proposal of Synge until the invention of the scanning tunneling microscope (STM) [9], NSOM has not almost been utilized experimentally especially in the visible light region because of the difficulty in controlling the probe close enough to a sample surface (typically less than 10 nm). Triggered by the STM invention, the scanning probe microscope technique was focused upon, and in 1984 Pohl et al. [4] reported the fundamental technique of NSOM [4], where they used an aperture-type probe coated with an opaque metal film. Following this, NSOM has been exploited for many applications, for example, spectroscopy and optical memory, by using an aperture-type probe (typically, a tapered optical fiber coated with opaque aluminum). Thus, the resolution of an aperture-type probe can go beyond the diffraction limit; however, it is still limited by the aperture size (approximately 100 nm). In recent years, apertureless-type probes such as a cantilever of an atomic force microscope (AFM) [10], a metallic probe of a STM [11], and a small metal particle [12] have been exploited for the achievement of much higher spatial resolution. Because, in principle, the signals obtained with apertureless-type probes are the scattered light at the tip apex, the resolution is strongly dependent on the tip diameter, which led to a high resolution with a smaller diameter of the tip.

In terms of spectroscopy by NSOM, it can allow us to obtain the optical properties of samples beyond the diffraction limit of the light. One of the attractive points is to achieve single-molecule detections. So far, fluorescence spectroscopy by using an aperture-type probe has been widely developed and near-field fluorescence microscopy has been thoroughly investigated as a tool to observe the molecular distribution of a sample. Fluorescence imaging of single molecules using a near-field scanning optical microscope was first reported by Betzig and Chichester [13], and the dynamics, such as fluorescence lifetime of a single molecule, was studied by Xie and Dunn [14] and Ambrose et al. [15]. Aside from fluorescence spectroscopy, Raman spectroscopy [16–19] and IR absorption spectroscopy [20–25] using an aperture probe were recently reported by some researchers. These two spectroscopic techniques are what we called vibrational spectroscopy and, compared with fluorescence spectroscopy, vibrational spectroscopy can provide precise information on the vibrations of molecules. However, the cross sections of Raman scattering (approximately 10^{-30} cm^{-2}) and IR absorption (approximately 10^{-19} cm^{-2}) are very small compared with that of fluorescence (approximately 10^{-16} cm^{-2}). For example, a very long exposure time was required owing to the low throughput of the aperture. From this point of view, enhancement of the cross sections is a very important factor. In order to apply NSOM to the vibrational spectroscopic sensitivity, the surface-enhancement effect of small metallic structures, for example, metal tip, metal film, and metal particle, is very powerful tool to compensate for these small cross sections, for example, surface-enhanced Raman scattering (SERS) for Raman spectroscopy [26–28] and surface-enhanced IR absorption (SEIRA) for IR absorption spectroscopy [29, 30]. A metallic probe tip as a small metallic nano structure for a surface enhancer is a promising technique for near-field enhanced Raman spectroscopy and makes it realistic. As a virtue of the surface-enhancement effect, such an enhancement effect can also be applicable to one-photon-excited [31–35] or two-photon-excited [36] fluorescence spectroscopy or enhanced SEIRA spectroscopy [37] by using metallic probe tip near-field scanning optical microscopes.

In this chapter, we present nanoimaging of molecular distributions colored by Raman-scattering spectral shifting, which is probed with a metallic tip. The metallic probe tip has been used to enhance the optical field only in the vicinity of the probe tip [6, 11, 12, 31–41]. The effect is similar to the one seen in the detection of molecules on the metal-island film, known as SERS [42], while in this case a single metallic tip works for the field enhancement in nanometer scale. This technique was recently termed tip-enhanced Raman spectroscopy (TERS).

In Sect. 19.2, we introduce the concept of TERS using a metallic probe tip in reflection and transmission modes. In Sect. 19.3, how to fabricate the metallic probe tips is introduced. In Sect. 19.4, some of the experimental results of the tip-enhanced Raman imaging showing nanometer-scale molecular distribution are demonstrated using a transmission mode. For the detailed chemical analysis of materials, polarization-controlled TERS is explained and demonstrated for single-walled carbon nanotubes in Sect. 19.5. In Sect. 19.6, TERS based on a reflection mode is applied for opaque samples such as strained silicon. Finally, in Sect. 19.7, we discuss some of trends toward much higher spatial resolution.

19.2 TERS (Reflection and Transmission Modes)

19.2.1 Experimental Configuration of TERS

The experimental setup for TERS can be classified into two families: transmission mode (Fig. 19.1a) and reflection mode (Fig. 19.1b).

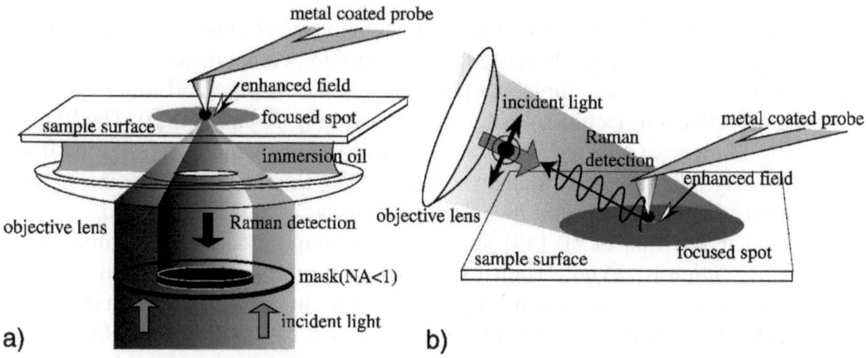

Fig. 19.1. Tip-enhanced Raman spectroscopy (TERS) configurations: (**a**) transmission mode; (**b**) reflection mode. NA numerical aperture

Transmission mode has been widely used in TERS because in this mode the metallic tip is illuminated through the sample plane by an inverted microscope objective lens, which is highly efficient in both illumination and collection especially when using a high numerical aperture (NA) objective lens [43–45]. The advantage is enormous for Raman spectroscopy, where the scattering cross section is small (Sect. 19.1). Above all, evanescent illumination is possible to reduce the background signals by rejecting incident NA components smaller than 1 [31]. However, transmission-mode TERS is not applicable to opaque or thick samples.

In a reflection mode, the tip is illuminated directly from the same side of the tip [46–50], which is advantageous and promising for observing opaque samples. In addition, an important requirement for attaining efficient enhancement under the tip is the polarization of the electric field of the incident light along the tip axis (Sect. 19.5). This condition is difficult to achieve in the transmission-mode geometry but can easily be achieved in the reflection-mode geometry. The problem of the reflection mode is the interference of an AFM scanner, which prevents the access of an objective lens. Measurements with small NA suffer from low efficiency both in illumination and in collection. Accordingly, most of the works in reflection mode were confined to Rayleigh scattering imaging [10, 11, 47] and a few reports on TERS [49, 50].

In the following, we will review some examples of transmission-mode and reflection-mode setups.

19.2.2 Transmission Mode

Hayazawa et al. [43, 51] and Stöckle et al. [44] first reported transmission-mode setups. Here we review the system reported in [43], which has been a basic experimental setup of many subsequent works on TERS. The system was constructed on the basis of an AFM using a silver-coated cantilever set on an inverted microscope system. The incident laser beam (continuous wave, 532 nm) was expanded 20 times with a beam expander and an evanescent mask was inserted in the beam path at the conjugated position of the pupil plane of an objective lens to achieve an evanescent illumination.

This configuration realized a strong longitudinal field at the focused spot and reduced undesirable background signals caused by transmission components (NA < 1) [31]. A cover slip, on which the sample was put, was set on an inverted oil-immersion microscope objective lens ($\times 60$, 1.4 NA, 250-μm working distance, WD). The Raman signal was efficiently collected by the same objective lens, which was separate from the incident light, by a half or a dichroic mirror. The Raman signal was introduced into a spectrometer and detected by a liquid-nitrogen-cooled CCD camera.

In 2003, Hartschuh et al. [45] succeeded in imaging carbon nanotubes with a spatial resolution of 25 nm. Their experimental setup was based on an inverted optical microscope with an xy scan stage for raster scanning. A laser beam (633 nm) was reflected by a dichroic beam splitter and focused by an objective (NA $= 1.4$) on the sample surface without the evanescent mask described earlier. A silver tip was positioned near the focus of the beam by means of a sensitive shear-force feedback mechanism. Raman scattered light was collected with the same objective and detected through a narrow band pass filter followed by a single-photon-counting avalanche photodiode. The details will be discussed in Sect. 19.4.

19.2.3 Reflection Mode

Anderson et al. [46] demonstrated an early stage of work in 2002. The instrument combined a multimode AFM and a near-IR Raman microscope (785 nm). The modified sample stage for rough positioning allowed AFM operation under a Raman microscope objective ($\times 10$) and realized side illumination of the tip–sample interface. They reported the point SERS spectra of diamond particles.

The reflection TERS that was reported by Sun and Shen [48] in 2003 employed a long bent cantilever, which was made from tungsten wire through electrochemical etching and then coating with silver. This system was integrated on the platform of a commercial near-field system (AFM combined with micro Raman spectroscopy). The compact design allowed the scanning head under the microscope objective ($\times 50$, 0.45 NA, 13.8-μm WD). An argon ion laser (514 nm) was focused onto the metal tip from the top of the cantilever. They reported the one dimensional TERS scanning of a silicon grating.

In 2005, Mehtani et al. [49] constructed a side-illumination system. The nearly horizontal confocal microscope uses a long WD objective ($\times 50$, 0.42 NA, 20.5-μm WD) to facilitate the optical coupling of the Raman spectrometer with the AFM stage. The objective lens was fixed on an XY stage controlled by step motors with an accuracy of approximately 40 nm to position the objective with respect to the tip. The 514-nm line of argon ion laser was used. A TERS signal was observed for methylene blue, poly(3,4-ethylene dioxythiophene)/poly(styrene sulfonate) polymer, CdS, and carbon nanotubes at arbitrary positions of the film. Enhancement factor of 10^3–10^4 were achieved with both silver- and gold-coated silicon nitride tips.

The aforementioned setups demonstrate the promise of reflection TERS for the analysis of various materials. However, reflection-mode TERS has been done at the cost of AFM performances and Raman collection efficiency to allow the access of an objective lens to the tip apex and is still in the stage of development. The most recent reflection-mode results applied for strained silicon will be reviewed in Sect. 19.6.

19.3 How to Fabricate the Tips?

In TERS, the very weak signal detected by Raman scattering from a small number of molecules can be significantly enhanced by coating the tip with noble metals such as silver or gold to increase the scattering cross section by generating local surface plasmons (SERS effect; Sect. 19.1). Since the spatial resolution and sensitivity of TERS experiments strongly depend on the performance of the metallized tip, the development of a near-field tip exhibiting strong and stable enhancement, as well as good spatial resolution, is a necessity.

Silver colloids are ideal for local surface plasmon generation since the imaginary part of the dielectric constant in silver is quite small and their absorption lies in the visible wavelength range. An ideal tip for localized plasmon excitation should be covered with silver particles whose size can be readily controlled at 20–100 nm [43, 51–55]. A sharp conical silver tip is another candidate for SERS, and can concentrate the electric field at the apex [45, 56]. In this section, we will briefly review methods to fabricate metallized tips.

19.3.1 Vacuum Evaporation and Sputtering Technique

Vacuum evaporation is widely used to form metallized tips [53–55]. The shapes of the metal islands that are produced by this method are not spherical but rather half hemispheres, even if a high evaporation rate is used. To increase the aspect ratio of the island in evaporated films, a subsequent annealing process is required [53]; however, the annealing process often causes damage on the tip.

Another method to increase the aspect ratio is to deposit another material under the silver layer. This procedure modifies the surface characteristics of the substrate and thereby changes the morphology of upper silver surfaces. An example of this fabrication method is performed as follows [54]. First, gold palladium alloy (AuPd) was deposited on a silicon cantilever tip to modify the surface of the silicon tip with a thickness of 5 nm. Then, silver was evaporated with a thickness of 40 nm at the rate of $0.5\,\text{Å}\,\text{s}^{-1}$. Owing to the precoated AuPd, a silver sphere was formed at the tip apex. Figure 19.2 shows a photograph of a metallized tip taken with a scanning electron microscope (SEM). A particle of around 40 nm in diameter can be seen at the apex.

Sputtering is also a convenient method for forming metallized tips, but this confines grain sizes to a rather small range -usually less than approximately 40 nm and is not controllable [55].

19.3.2 Electroless Plating

An alternative method for tip metallization is an electroless plating method [57–62]. In this section we will shows an example of the tip metallization by the chemical method known as a mirror reaction.

Fig. 19.2. Scanning electron microscope (SEM) image of silver-coated probe made by vacuum evaporation, ×100,000, 5 kV, 7 μA

The mirror reaction forms a silver surface of well-packed particles on various kinds of substrate, such as glass, silicon, silicon nitride, mica, or polymers. The details of the preparation of silver surfaces, which involves the reduction of silver nitrate by glucose, were given in [58]. By varying the reaction conditions, principally by varying the concentration of silver nitrate solution, one can control the colloidal particle size. To demonstrate this, three probes of different particle sizes have been prepared.

Figure 19.3 shows SEM images of the silver metallized tip made with different silver nitrate concentrations. The main lateral grain sizes obtained were 82 ± 14, 41 ± 6, and 23 ± 3 nm for surfaces A–C respectively. Surfaces A–C show almost complete surface coverage with good packing at the tip apex, and the particle shapes are nearly spherical. TERS were measured using the tip A–C. The particle size from 80 to 40 nm showed good SERS and the efficiency decreased for particle sizes as small as 20 nm. This tendency was, to some extent, explained by the particle size dependence in the absorption spectra calculated by Mie theory. This chemical method will be useful not only for near-field Raman imaging, but also for quantitative understanding of the SERS effect.

19.3.3 Etching of Metal Wires Followed by Focused Ion Beam Milling

When a tuning fork base setup is employed, metallized tips can be made directly from metal wires by chemical etching [45,63,64]. For further precise control of the tip apex, a subsequent focused ion beam (FIB) milling process is used. The ion beam milling is relatively fast and convenient for a small batch of tips.

Vasile et al. [64] and Sánchez [36] reported a tip fabrication method by FIB. A blunt tip having radii of curvature of approximately 1 μm was formed by electrochemical etching in NaOH solution. The tip holder was mounted in an FIB apparatus so that the axes of the tip were collinear with the ion beam. At the first stage of milling by a gallium ion beam, an annular beam deflection pattern was used. To remove the inevitable conical shoulder, a larger annular pattern to sputter the shoulder was applied. A single, final cut was made after the shoulder had been cleaned, resulting in the final tip profile with a radius of 10–15 nm.

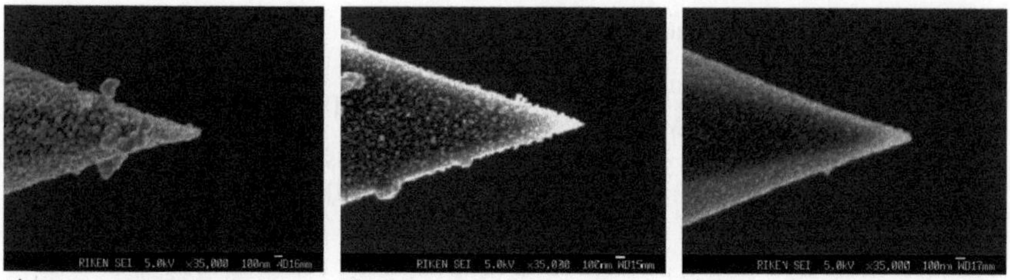
a) b) c)

Fig. 19.3. SEM image of silver-coated probe made by electroless plating: (**a**) average particle size approximately 82 nm; (**b**) approximately 41 nm; (**c**) approximately 23 nm. The *scale bar* indicates 100 nm. ×70,000, 5 kV, 12 μA

19.3.4 Other Methods

Other methods for fabrication of SERS active silver surfaces are reviewed. These methods have not been tried for tip metallization, but are highly expected to apply for this purpose.

Some groups are fabricating metal films for SERS by attaching preformed colloidal particles to a substrate [65, 66]. These surfaces are prepared by self-assembly carried out in solution. Cleaned substrates were immersed in solutions of organosilane, rinsed, and subsequently immersed in aqueous colloidal silver or gold solutions. Since silver or gold colloids have affinity to –CN, –SH, or $-NH_2$ groups, two-dimensional arrays form spontaneously.

One problem of this approach may be the stability of the film for AFM scanning. This instability also restricts the rigor with which such surfaces can be cleaned. Inadequate cleaning can be a problem for SERS, in which characteristically one observes peaks in the spectra due to organic contamination of the substrate.

Jia et al. [67] recently reported a method by photoreduction based on surface-catalyzed reduction of silver nitrate by cyanate under visible light irradiation. Particle size and shape were controlled by the irradiation time. If, by chance, a single seed of silver crystal is formed at the very end of a tip, the shape of a probe apex will be easily controlled. One should be concerned about the coverage of the silver at the very end of the tip apex. The curvature of a tip may cause problems for silver coating in these methods.

19.4 Tip-Enhanced Raman Imaging

In this section, visualization of molecular specific vibrations will be demonstrated by tip-enhanced spectroscopy in transmission mode. The basic system configuration is based on a combination of an AFM and an inverted optical microscope as illustrated in Fig. 19.4. An expanded and collimated light field from a visible laser enters into the epi-illumination optics of an inverted optical microscope. As the metallic tip is moved closer to the focused spot that is generated by a high NA (1.4) objective

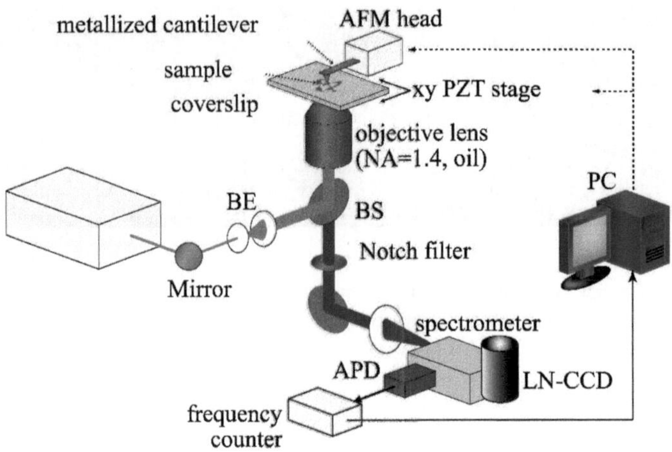

Fig. 19.4. System configuration of tip-enhanced spectroscopy. BE beam expander, BS beam splitter, AFM atomic force microscope, APD avalanche photodiode, LN-CCD liquid-nitrogen-cooled CCD camera, PC computer, PZT piezoelectric transducer

lens, a localized enhanced electric field is generated at the tip apex [6,68,69]. The localized enhanced electric field at the tip is scattered inelastically by the Raman-active molecules, which corresponds to tip-enhanced Raman scattering. The Raman scattering is collected by the same objective lens, and is directed to a spectrometer that is equipped with a liquid-nitrogen-cooled CCD camera for Raman spectra measurement and with an avalanche photodiode for Raman imaging. The avalanche photodiode is located after the exit slit of the spectrometer so that specific Raman-shifted lines can be detected. Excitation light or Rayleigh scattering is sufficiently rejected by a notch filter before the spectrometer. The distance between the sample and the metallic tip is regulated by AFM operation either in contact or in noncontact mode, and the sample is scanned with piezoelectric transducers in the X–Y plane. Scanning the XY piezoelectric transducer sample stage while simultaneously detecting the Raman signal with the avalanche photodiode, one can perform tip-enhanced Raman imaging at the specific Raman-shifted line.

19.4.1 Selective Detection of Different Organic Molecules

Aggregates of rhodamine 6G (R6G) and crystal violet (CV) molecules were used as a sample for TERS and tip-enhanced Raman imaging [70]. The samples were prepared by casting an ethanol solution of R6G and were dried on a cover slip covered with an 8-nm-thick silver film. The distribution of molecules for both samples was fairly inhomogeneous and was set to have a 1-nm average thickness of the layer of molecules. A silver film is one of most popular surface enhancers of the SERS effect, and is effective not only in enhancement of Raman scattering cross sections but also in the reduction of the fluorescence due to the fluorescence energy transfer from the molecules to metal [71]. Because both R6G and CV have absorption at the excitation wavelength of 532 nm and have fluorescence overlapped with Raman peaks, the observed Raman spectra are consequently due to surface-enhanced resonant Raman scattering (SERRS). Figure 19.5 shows the tip-enhanced Raman spectra of R6G, CV, and a complex of R6G and CV. Several Raman-shifted lines of R6G and CV molecules are observed, and all these peaks are in good agreement with the former works by other authors [72–75]. And, Watanabe et al. [76] recently reported the detailed analysis of the vibration modes of R6G molecules in TERS. To investigate the spatial resolving power of TERS and to identify molecular vibration distributions among different kinds of molecules, tip-enhanced Raman images of an aggregated sample of R6G and CV molecules were obtained at characteristic Raman peaks of each molecule. Figure 19.6 shows tip-enhanced Raman images at the same area of the sample. Figure 19.6a was obtained at 607 cm^{-1}, which corresponds to the Raman-shifted peak of the C–C–C in-plane bending vibration mode of R6G indicated by the arrow at "A" in Fig. 19.5. Figure 19.6b was obtained at 908 cm^{-1}, which corresponds to the Raman-shifted peak of the C–H out-of-plane bending vibration mode of CV indicated by the arrow at "B" in Fig. 19.5. Now we can selectively obtain the distributions of each vibration mode that we cannot distinguish in the topographic image shown in Fig. 19.6c. The distributions of each vibration mode are quite different, and show complicated structures reflecting the inhomogeneous distributions of both molecules. According to Fig. 19.6a, R6G molecules are mainly localized at the lower right of the area. On the other hand, Fig. 19.6b shows that CV molecules are randomly dispersed in the scanned area. In Fig. 19.6c, the island structures of the silver film are observed in the topographic image because the average thickness of the aggregated molecular layer is approximately 1 nm and is much thinner than the silver film (average thickness 8 nm). Accordingly, the distributions of both molecules are not clearly seen in the topographic image that reflects the pancake structure of the silver grains (30–50 nm in diameter and 8 nm in thickness). Note that without a metallic tip (far-field detection), it is impossible to obtain such high-resolution images because the far-field signal is averaged inside the diffraction-limited focused spot. In the case of the mixed sample used for imaging, both R6G and CV molecules are randomly dispersed on the silver-coated cover slip, so the far-field Raman signal from the focused spot is averaged to almost constant value. Figure 19.6 shows that the tip-enhanced Raman images give the molecular vibration distributions with a high sensitivity and high spatial resolution, even if the thickness of the molecular layer is 1 nm. Here, organic dye molecules adsorbed on silver films were used. While the silver films are necessary for dye molecules to quench the strong fluorescence, this configuration can work also as the so-called gap mode [77, 78] for the higher enhancement effect. With this sufficient tip-enhancement effect without fluorescence, we can carry out direct observation of molecular vibration without the aid of silver island films. In the next section, an example is discussed for carbon nanomaterials.

19.4.2 Observation of Single-Walled Carbon Nanotubes

One promising application of tip-enhanced Raman imaging is to analyze properties of nanomaterials, which are undetectable by a conventional microscope owing to

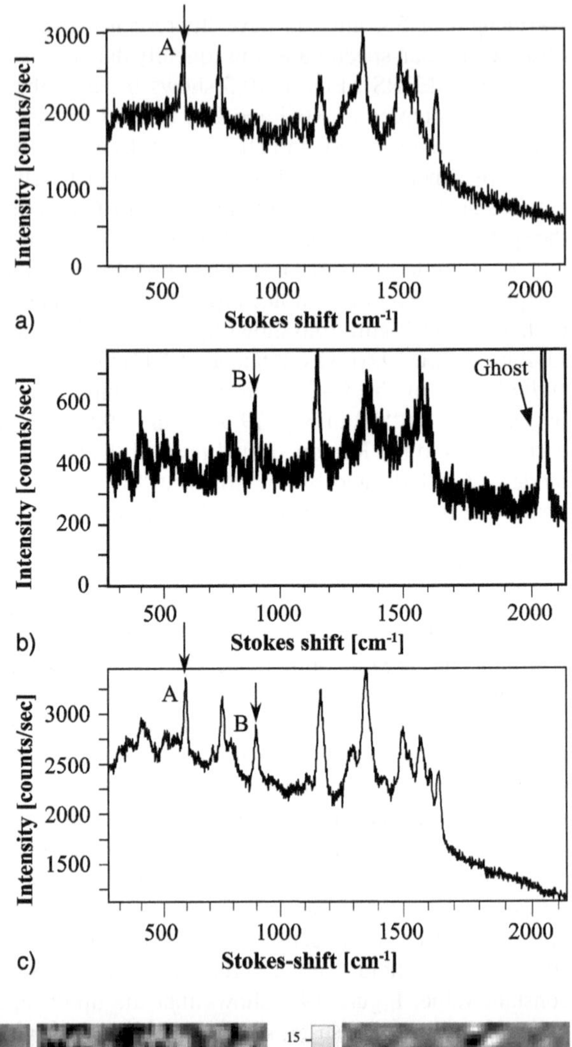

Fig. 19.5. Tip-enhanced Raman spectra of (**a**) rhodamine 6G (R6G), (**b**) crystal violet (CV), and (**c**) aggregates of R6G and CV. The Raman peaks at *A* and *B* were used for tip-enhanced Raman imaging in Fig. 19.6 [70]

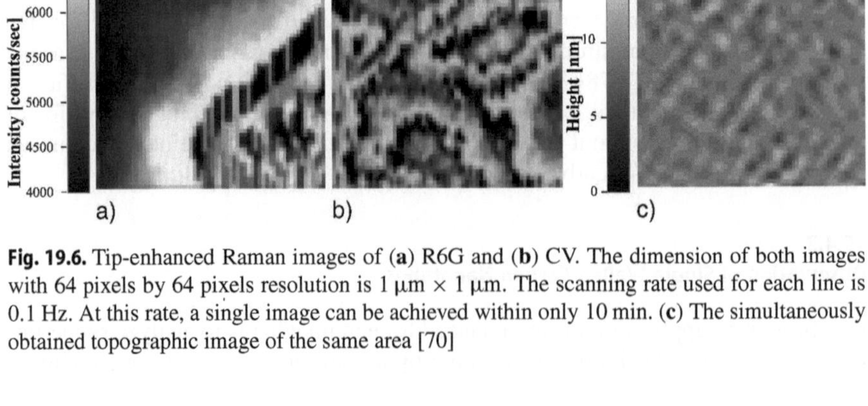

Fig. 19.6. Tip-enhanced Raman images of (**a**) R6G and (**b**) CV. The dimension of both images with 64 pixels by 64 pixels resolution is 1 μm × 1 μm. The scanning rate used for each line is 0.1 Hz. At this rate, a single image can be achieved within only 10 min. (**c**) The simultaneously obtained topographic image of the same area [70]

the diffraction limit. For example, TERS [54, 79] and tip-enhanced Raman imaging [45, 80, 81] has been realized as a very powerful tool to analyze single-walled carbon nanotubes (SWNTs). Depending on the chirality of each SWNT, the electric property of the tube is drastically changed and can be either conductive or semiconductive [82, 83]. Different chirality exhibits a different vibration mode, in which the representative vibration modes are the so called radial breathing mode (RBM approximately 150 cm^{-1}: strongly depending on the diameter), defect mode (D-band approximately 1300 cm^{-1}: caused by defects), graphite mode (G-band approximately 1600 cm^{-1}: split into several peaks depending on chirality), and G′-band (approximately 2600 cm^{-1}: overtone of D-band) [84]. However, these vibration modes are localized in nanometer scale. Accordingly, TERS has come into the spotlight. Figure 19.7 shows the tip-enhanced Raman imaging carried out on SWNTs [45] with the excitation wavelength at 633 nm. Because of no overlapping fluorescence and a high tip-enhancement effect, there are no silver island films underneath and the SWNTs are directly aligned on a cover slip. In this case, the G′-band was detected for the imaging while observing the surface topography simultaneously shown in Fig. 19.7. In the topographic image, aside from the SWNTs, humidity-related water on a cover slip was also observed and blurred the contrast of SWNTs. On the other hand, the tip-enhanced Raman image clearly visualizes the isolated SWNTs far beyond the diffraction limit of the light without the influence of the surface water. Diameter-selective visualizations of SWNTs were also demonstrated by detecting RBMs of each SWNT because different diameters exhibit different Raman shifts of the RBMs [85]. Figure 19.8 shows the tip-enhanced Raman images of a bundle of SWNTs where several different diameters are included [81]. Detected Raman shifts of the RBM are at 195 cm^{-1} (Fig. 19.8b), 244 cm^{-1} (Fig. 19.8c), and 278 cm^{-1} (Fig. 19.8d), and correspond to diameters of 1.23, 0.97, and 0.85 nm, respectively [86, 87]. Figure 19.8b reveals that the SWNTs having a diameter of 1.23 nm are localized at both edges of the bundle, Fig. 19.8c shows that the SWNTs with a diameter of 0.97 nm are prominently distributed towards the central and upper part of the bundle, and Fig. 19.8d indicates that the SWNTs with a diameter of 0.85 nm are mainly distributed towards the lower part of the bundle. Again, these

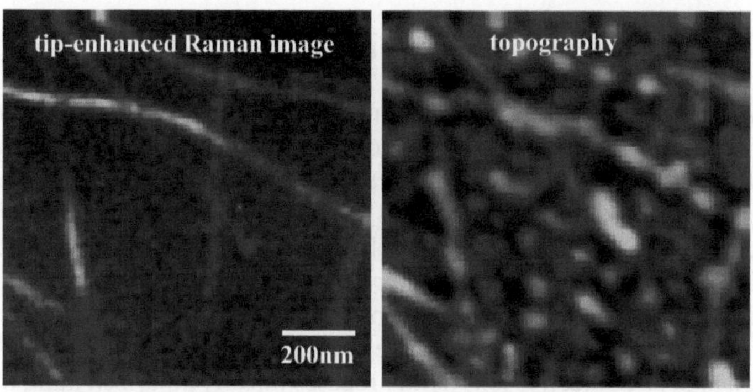

Fig. 19.7. (**a**) Tip-enhanced Raman image of single-walled carbon nanotubes (SWNTs) obtained at G′-band mode. (**b**) Simultaneously obtained topographic image of the same area [45]

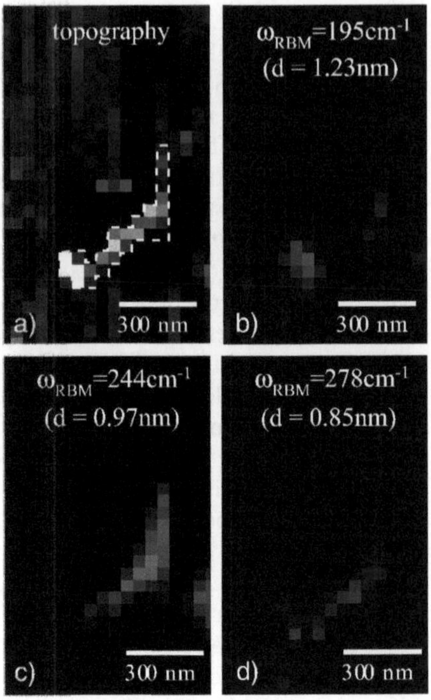

Fig. 19.8. (**a**) An AFM image of a SWNT bundle. Tip-enhanced Raman images were obtained at the frequencies of (**b**) 195 cm^{-1}, (**c**) 244 cm^{-1}, and (**d**) 278 cm^{-1} [81]

kinds of nanoimagings are not possible by conventional optical microscopes because all that nanoscale information is averaged within the diffraction-limited spatial resolution. For the detailed analysis of SWNTs, polarization-controlled TERS has been also developed and will be discussed in the next section.

19.5 Polarization-Controlled TERS

19.5.1 Polarization Measurement by Using a High NA Objective Lens

Polarization measurements provide unique structural information from which to extract a wealth of knowledge on orientations, intermolecular interactions, and symmetry deformations of molecules. Polarization-controlled TERS is a powerful tool for surface characterization, since it yields information about both chemical composition and assembled structures [54]. Unlike conventional Raman spectroscopy, the polarization direction in TERS is characterized by two axes, parallel (p-polarization) and perpendicular (s-polarization) to the probe tip axis. In this section, we will discuss the polarization treatments which are unique to TERS.

When a high NA objective lens tightly focuses linearly polarized light, the resulting polarization at the focal plane consists of both p- and s-components [88, 89]. Besides this polarization admixture, the field intensity distributions of each component cause a problem when the spatial resolution is down to nanometer scale. Figure 19.9

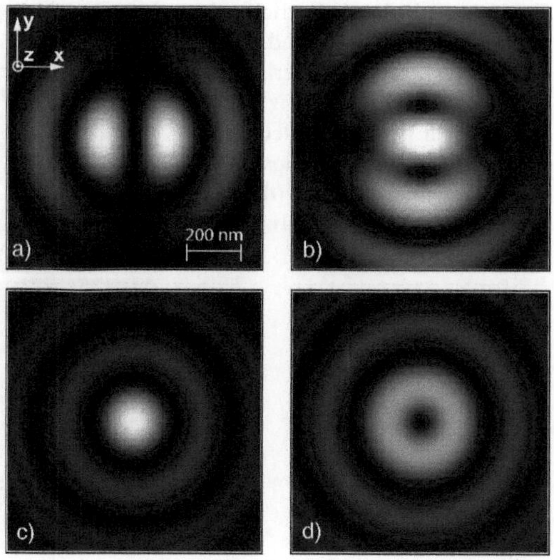

Fig. 19.9. Calculated field intensity distribution at a focused laser spot. (**a**) z-polarization component created by linear polarization (x-direction); (**b**), x- and y-polarization components; (**c**) z-polarization component excited by radial polarization; (**d**) x- and y-polarization components excited by azimuthal polarization

shows the calculated field intensity distribution of linearly polarized (x-polarized) light at the focal plane tightly focused by an objective lens (NA = 1.4), for p-polarization (Fig. 19.9a) and for s-polarization (Fig. 19.9b). The electric field of p-polarization has been canceled at the center of the spot, which creates alternately two side peaks. In TERS measurements, we carefully adjust the tip onto one of the two p-polarization field components [54, 89]. On the other hand, the field intensity of s-polarization shows a maximum at the center.

By controlling the incident polarizations, we can choose the polarization condition at the focal plane to be either p or s. Radially or azimuthally distributed polarizations form a p- or s-polarization. Figure 19.9c,d shows the calculated total electric field intensity distributions at a tightly focused spot using radial (p-polarization) and azimuthal (s-polarization). In the p-polarization configuration, the electric field intensity in the longitudinal direction (along the tip axis) is the strongest at the center. In the s-polarization configuration, the lateral field intensity is the strongest on the circumference. Note that both x-polarization-dominant and y-polarization-dominant areas are formed in the case of the s-polarization configuration.

19.5.2
Metallized Tips and Polarizations

The polarization efficiency in TERS strongly depends on the shape of a metallized tip. A p-polarized plane wave illumination onto the conically shaped tip can efficiently induce tip enhancement, while the s-polarized light cannot. It is generally accepted that tip enhancements require inducing p-polarization [36, 45, 68, 89]. However, the enhancement of the Raman signal also depends on the direction of the transition moment of molecules. Some vibrational modes couple well with s-polarization rather than with p-polarization. For example, to excite the Raman mode such as the G-band of carbon nanotubes, an s-polarization component is required [85, 90].

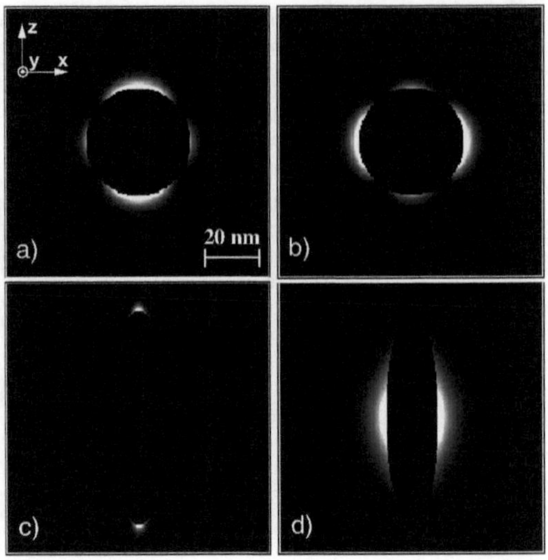

Fig. 19.10. Calculated electric field intensity distribution around a silver nanoparticle. (**a**) Spherical particle excited by y-polarization; (**b**) spherical particle excited by x-polarization; (**c**) 4:1 ellipsoid excited by y-polarization; (**d**) 4:1 ellipsoid excited by x-polarization

As we mentioned in Sect. 19.2, metallized tips prepared in a variety of ways may have different shapes, like nanospheres or sharp conical shapes. Any of these variations will affect the field intensity distribution around the tip apex, which will be discussed next.

A single nanosphere of size much smaller than the wavelength is considered as a dipole when a light field is incident on the sphere. The dipole generates an electric field perpendicular to the incident field as well as an electric field parallel to it. The calculation in Fig. 19.10 illustrates the dipole field intensity distribution formed around a silver sphere of approximately 40 nm diameter, excited by z-polarization (Fig. 19.10a) and x-polarization (Fig. 19.10b). Suppose the tip axis is along the z-direction, then Fig. 19.10a corresponds to the case when the tip is excited by p-polarization and Fig. 19.10b corresponds the case when the tip is excited by s-polarization. The amplitude of the former field, i.e., s-polarization field, is just half of that of the latter field, i.e., p-polarization field [52]. This means that the s-polarized field can be excited by using a tip having a metal nanosphere at the apex. Even though it may be rather inefficient compared with p-polarization, we also utilize the s-polarization, since the coupling of the incident field with the Raman transition moment of molecules is also important for tip enhancement.

In the case when the tip is sufficiently sharp, we simplify the apex as an ellipsoid. (long axis to short axis ratio of 4:1) Figure 19.10c and d shows the calculated electric field intensity when the incident field is p- or s-polarized, respectively [91]. Again, the tip axis is along the z-direction. While p-polarized excitation shows strong field concentration underneath the tip (Fig. 19.10c), only a negligible field is formed by s-polarized excitation (Fig. 19.10d). This figure reveals that only p-polarization is useful for field enhancement using a sharp metallic tip.

The field component perpendicular to the incident polarization has little intensity just underneath the tip compared with the parallel component, so it can be negligible in the case of TERS measurements.

19.5.3
Example of *p*- and *s*-Polarization Measurements in TERS

In this section, we show an example of the polarization measurements in TERS by applying both *p*- and *s*-polarizations. SWNTs, which have strong polarization dependence, were investigated with this polarization control. The RBM and the G-band, which belong to different vibrational symmetries, showed opposite polarization dependences.

The detailed experimental setup is shown in [54]. For polarization measurements, the combination of a half-wave plate, a polarizer and a wave plate was aligned in the incident beam path. Here, we show an example of a method to prepare quasi *p*- or *s*-polarizations by employing a spatial wave plate. The wave plate used consisted of four divided half-wave plates each with a different orientation of the slow axis. The plate provides both longitudinal (parallel to the tip axis) and lateral (perpendicular to the tip axis) polarization of the electric field on the sample plane by selecting the proper polarization of the incident field (Fig. 19.11).

We investigated the polarization dependence of near-field Raman scattering from SWNTs. Figure 19.12 shows a comparison of the near-field Raman signal intensities of the SWNTs measured under the *p*- and *s*-polarization conditions. "Near-field spectra" here means the subtraction of the spectra obtained without a tip from those obtained with a silver tip in contact with the sample. Black lines indicate the spectra measured under the *p*-polarization condition and gray lines indicate those measured under the *s*-polarization condition. Under the *p*-polarization condition, the RBM is efficiently enhanced but the G-band exhibits fewer enhancements. In contrast, undear the *s*-polarization condition, the G-band exhibits higher enhancement, while the RBM does not. The RBM and the G-band exhibited opposite enhancement behavior between the two polarizations. This figure indicates the selective enhancement of a particular vibrational mode by the near-field tip and these enhancements can be explained as follows.

The explanation is based on an assumption that the tip apex coated with silver particles excited by the incident laser field act as a single dipole and that the dipole provides sufficient field enhancement in either *p*- or *s*-polarization according to the incident polarization configurations. Under each polarization, the tip-enhanced field has either a longitudinal or a lateral polarization, which couples with the different

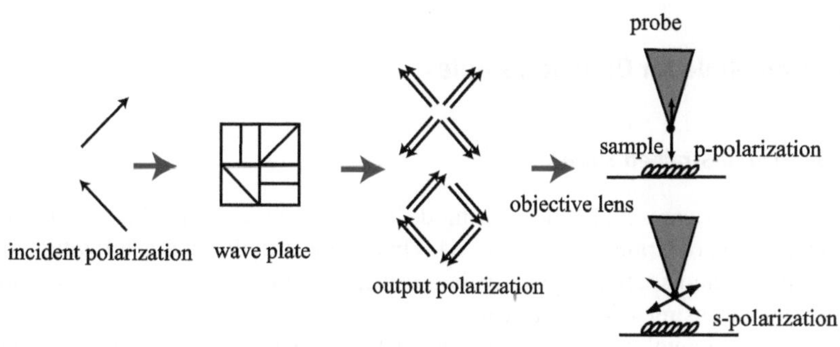

Fig. 19.11. The usage of the radial wave plate

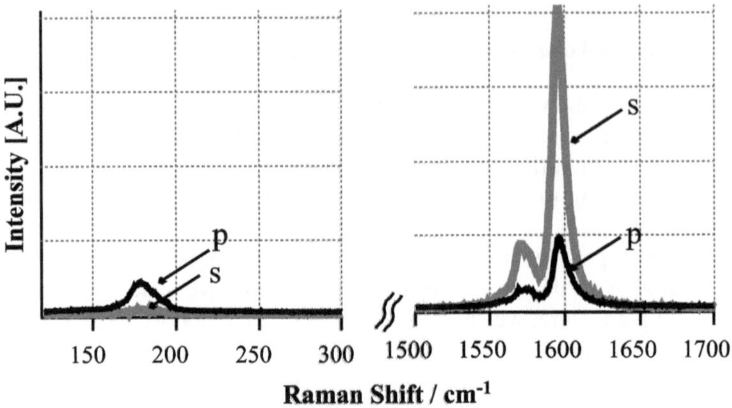

Fig. 19.12. Tip-enhanced near-field Raman spectra of SWNTs measured under *p*- and *s*-polarization conditions. Radial breathing mode and G-band show different polarization dependences. Excitation 532 nm, power 0.1 mW, accumulation time 60 s

electric resonance of the nanotubes. The vibrational transitions of carbon nanotubes of this exciting wavelength couple with electron transitions either parallel or perpendicular to the tube axis [90]. Since the G-band contains a vibrational transition moment along the tube axis, it is effectively excited by *s*-polarization. On the other hand, the vibrational transition moment of the RBM is located radially around the tube axis and is more efficiently excited by the *p*-polarization than by the *s*-polarization, since the entire excitation field in the *p*-polarization is perpendicular to the tube axis.

In this work, the polarization selectivity of the tip-enhanced efficiency within the G-band was also investigated. It has been reported that the G-band consists of several different symmetry components, A_1, E_1, and E_2 [84]. These symmetry species are expected to exhibit different polarization dependences. The results indicate that several symmetry species in the G-band have different enhancement efficiencies, and that these were successfully used to make symmetry assignments.

The work demonstrated that the silver-coated tip selectively enhanced the particular vibrational symmetry species. This method can be useful for TERS imaging especially when two vibrational bands lie in close proximity in the spectra.

19.6 Reflection Mode for Opaque Samples

19.6.1 TERS Spectra of Strained Silicon

In this section, we present a reflection-mode TERS successfully applied to nanoscale characterization of strained silicon (ε-Si). This nondestructive characterization technique with a nano spatial resolution is essential for fabricating high-performance electric devices using ε-Si substrates.

The experimental setup is shown in [50]. An incident continuous-wave laser (532 nm) is directed to the optical setup using a single-mode optical fiber. From the fiber output, the beam diameter of the laser is expanded (approximately 10 mm), while setting the polarization to p-polarization, which is parallel to the tip axis. This polarized light is focused on the sample using a long-working-distance (LWD) objective lens (×20, 0.28 NA, 30.5-mm WD). The apex of a tip is adjusted onto the focused spot using the same principle as a contact-mode operation of an AFM. The same LWD objective lens collects the tip-enhanced Raman signal. A dichroic mirror directs the Raman signal output via a multimode optical fiber to a spectrometer. The spectrum is detected using a liquid-nitrogen-cooled CCD camera. The ε-Si sample is prepared by forming a Ge-doped Si layer on a pure Si substrate. The concentration of Ge is gradually increased up to 25% to increase the lattice constant of Si–Si. The 30-nm-thick strained silicon layer is epitaxially grown on the buffer SiGe layer (thickness 1 μm) in which the concentration is kept at 25%.

Figure 19.13 shows the TERS spectra of ε-Si. The near-field spectrum in Fig. 19.13 is obtained when the silver tip is positioned close to the surface of the strained silicon. On the other hand, the far-field spectrum is obtained without the tip. We derive a background-corrected spectrum by taking the difference between the near-field and far-field spectra. The background-corrected spectrum contains nanometer-scale information of localized strains. In the far-field spectrum, the ε-Si peaks are recognized only as a shoulder of the strong background signal generated from the underlying SiGe layer. The near-field spectrum, however, shows a clear and distinct peak owing to this effect by tip enhancement.

Since the capped ε-Si layer is on the order of several tens of nanometers, a sufficient amount of incident light penetrates through the SiGe layer. The background

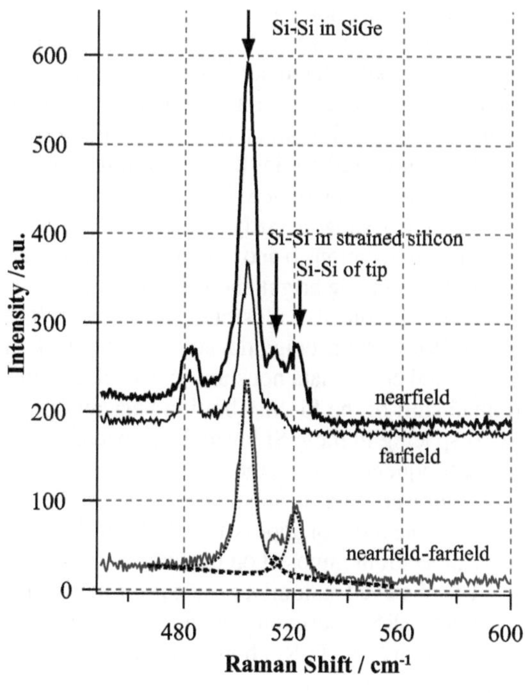

Fig. 19.13. Tip-enhanced near-field Raman spectra of strained silicon. The *bold line* and the *thin black line* correspond to near-field and far-field spectra, respectively. The *gray line* is the background-corrected spectrum derived by the subtraction of near-field and far-field spectra. *Dotted lines* are the Lorentzian curve fitting for the background-corrected spectrum

signal from the underlying SiGe buffer layer is still significant. In order to clarify the band peaks, the background-corrected spectrum is decomposed into three Lorentzian functions showing peaks at 503, 514, and 520 cm^{-1}. This corresponds to a Si–Si stretch from the underlying SiGe layer, the ε-Si layer and the tip, respectively. The background-corrected spectrum does not contain the unassigned peak at 481 cm^{-1}, which can be considered as stray light from the optical fiber. The displacement of the Si–Si Raman peaks of ε-Si from the inherent peak of Si (520 cm^{-1}) [92] provides us with plenty of information, such as local stress [92–95]. The estimated enhancement factor for ε-Si is 2.2×10^4 calculated based on the size of the diffraction-limited focused spot ($\phi = 3\,\mu m$) and the enhanced electric field corresponding to the tip diameter ($\phi = 40\,nm$).

19.6.2
Nanoscale Characterization of Strained Silicon

Lateral heterogeneity of strain in strained Si/SiGe structures is a serious problem when fabricating high-mobility electric device assemblies on ε-Si substrates. Here, we demonstrate the position dependence of near-field spectra for detailed characterization. Figure 19.14a shows the bright-field microscopic image of the ε-Si surface. The crosshatched patterns induced by lattice mismatch are seen on the epitaxial surface of the strained silicon. Figure 19.14b illustrates the contact AFM image ($3\,\mu m \times 3\,\mu m$) of the crosshatched pattern. The six crosses indicate positions where the tip-enhanced measurements were carried out. Figure 14c shows the background-corrected ε-Si peak extracted by Lorentzian curve fitting, as shown in Fig. 19.13. Based on the previous work [96], each ε-Si spectrum is curve-fitted with a 4-cm^{-1} spectral width. From Fig. 19.14c, one can observe that the position and the intensity of the ε-Si peaks fluctuate. The fluctuation is due to variations in the thickness of the ε-Si layer based on an underlying surface topography. The pattern in the underlying layer depends both on the Ge concentration of the relaxed material and on any residual, unrelaxed strain developed in the SiGe resulting from lattice-mismatched heteroepitaxial growth of SiGe on Si [94,95]. The bright areas indicated in locations a, d, e, and f show a fairly intense Raman signal compared with the dark areas indicated in locations b and c. From the cross-sectional view of Fig. 19.14b, the step height between the bright and the dark area is approximately 16 nm.

Figure 19.14c also illustrates the slight shift (approximately 2 cm^{-1}) of the ε-Si peaks. While the average deviation of the peak of the Si tip stays within the range $520 \pm 0.04\,cm^{-1}$ (data not shown), its value in the ε-Si spectrum is $514 \pm 0.8\,cm^{-1}$. Raman shift of the Si–Si vibration basically provides a measure of the interatomic Si–Si spacing in the ε-Si layer. This spacing causes fluctuations of channel strains, and subsequently affects the performance of the device. It has been reported that lateral strain variations is mostly associated with the crosshatched pattern and its variation is on the order of approximately 1 cm^{-1} [95]. Calculated from the shift in Raman peaks, the strain imposed on the capped ε-Si layer is about $(3 \pm 0.4) \times 10^9$ Pa [94].

The observed nanometer-scale information on localized strain cannot be detected by microSERS experiments owing to the averaging effect within a diffraction-limited focused spot [96]. TERS, however, has successfully revealed localized strain hidden in nanometer scale.

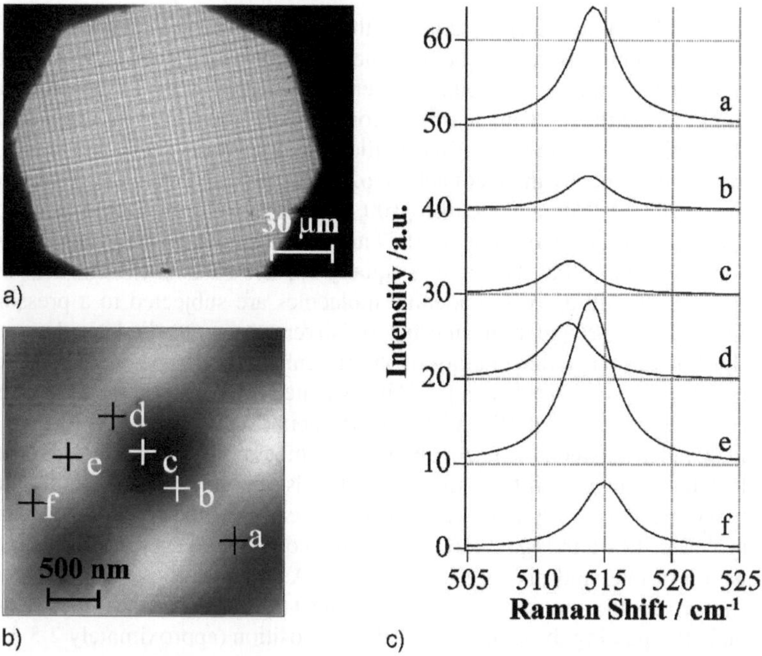

Fig. 19.14. (**a**) Bright-field microscopic image of the strained silicon surface. (**b**) AFM image of the strained silicon surface. *Crosses with letters* indicate the tip positions where TERS mapping was carried out. (**c**) Lorentzian curve-fitting results of position-dependent TERS spectra of ε-Si. *a–f* correspond to the positions shown in (**b**)

19.7 For Higher Spatial Resolution

19.7.1 Tip-Pressurized Effect

It was proved that a metallic tip worked as a surface enhancer for the SERS effect. However, the tip-enhanced electric field has the corresponding size of a tip diameter, such as approximately 30 nm, so the spatial resolution will be restricted by the tip diameter. To break through this limitation of spatial resolution, one of the ways is, of course, to fabricate a much sharper tip. The other possibility that is discussed here is to utilize the inherent feature of the existence of metallic atoms close to the molecules. Since most of the cases that have been proposed so far in tip-enhanced spectroscopy are based on an AFM to control a metallic tip at the close proximity of a sample surface, metallic atoms at the end of the tip would affect molecular vibrations. In the case of SERS experiments in which metallic atoms are attached to molecules, the spectral shift of the vibration mode has been reported for adenine molecules [97–99]. In SERS, adenine is adsorbed in equilibrium onto silver surfaces, whereas in TERS, the silver probe tip with a constant atomic force, which is schematically depicted in Fig. 19.15, presses nanocrystalline adenine molecules. Assuming that the atomic force is applied only to the contraction of the bond between the silver atom of the tip and the adjacent nitrogen of the adenine molecule, we would expect the bond distance to shrink and the vibrational frequencies may then shift. In the experiment, for example, a cantilever with a spring constant of $0.03\,\mathrm{N\,m^{-1}}$ is used, and the silver-coated tip apex diameter of this cantilever is 5–10 nm. The atomic force was kept constant at 0.3 nN by the feedback loop. Deduced from the unit cell parameters of single crystal of 9-methyladenine [100, 101], a couple of adenine molecules exist in a rectangle of approximately 0.77 nm lateral dimension by 0.85-nm wide. Under the assumption that the force is equally applied to all the molecules which are adjacent to the tip apex, the adenine molecules are subjected to a pressure of approximately 1–5 pN per molecule by the silver atoms attached on the surface of the tip. For further understanding of the tip-enhanced Raman-active species of adenine molecules, Watanabe et al. [102] investigated the transition states of possible optimized geometries of Ad-N3 and Ad-N7 isomers (Fig. 19.15, right) by changing the bond distance between the nitrogen of the adenine molecule and the silver atom. The calculated frequency shifts of the Raman bands of both the ring-breathing mode of a whole molecule ($\sim 720\,\mathrm{cm^{-1}}$) and the ring-breathing mode of a diazole ring ($\sim 1360\,\mathrm{cm^{-1}}$) are shown in Fig. 19.16. The calculated potential curves are also plotted as a function of the bond distance in Fig. 19.16. As the metallic tip approaches the surface of the adenine nanocrystal, the tip is subject to van der Waals attractive forces at first and after passing through the equilibrium position (approximately 2.5 Å), the tip receives a repulsive force. In the experiment, the atomic force which is balanced with the repulsive force is set at 1–5 pN per molecule as described before. When the bond distance of the Ag–N linkage is reduced by 10%, a repulsive force of 7 pN per molecule is derived from a harmonic oscillation of the displacement by 0.025 nm and the energy difference is $1.7\,\mathrm{kcal\,mol^{-1}}$ in the case of the Ad-N3. The repulsive force coincides with the atomic force obtained with the system. The ring-breathing mode of Ad-N3 shows a significant shift towards a higher frequency as a function of the contracted bond distance between the silver atom and the nitrogen of adenine. The ring-breathing mode of an adenine molecule is shifted upwards by $5\,\mathrm{cm^{-1}}$

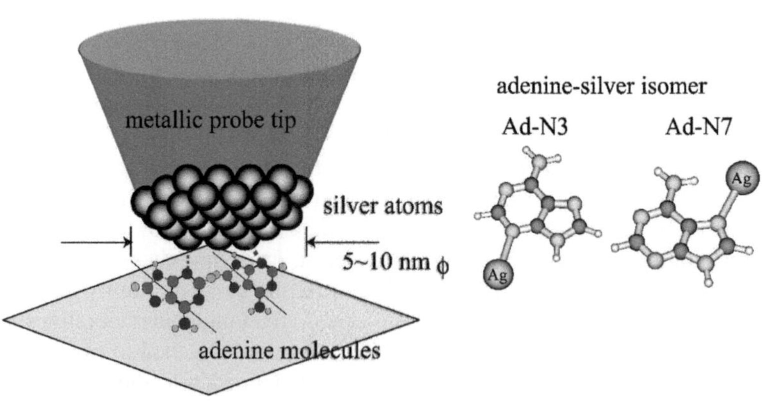

Fig. 19.15. The concept of tip-pressurized effect of a metallic tip, and possible model of adenine-silver isomers

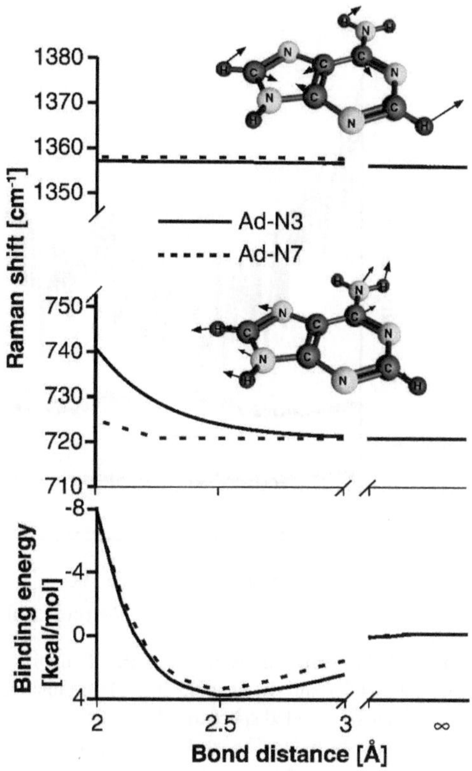

Fig. 19.16. Calculated frequency shifts of the Raman bands of both the ring-breathing mode of a whole molecule (approximately 720 cm^{-1}) and the ring-breathing mode of a diazole ring (approximately 1360 cm^{-1}). Calculated potential curves of isomers are also plotted as a function of the bond distance between the nitrogen atom and the silver atom [102]

when the bond distance is reduced by 10% and by 17 cm^{-1} when the bond distance is reduced by 20%. The frequency shift of the ring-breathing mode (approximately 720 cm^{-1}) agreed well with TERS experiments (Fig. 19.17), while the ring-breathing mode of the diazole ring (approximately 1360 cm^{-1}) was not shifted as predicted by the calculation [102] and observed in the experiments [79, 89]. In addition to the higher-frequency shift, Raman band broadening is also observed in the tip-enhanced Raman spectra (Fig. 19.17). The line broadening as well as the Raman frequency shift have been reported in high-pressure induced Raman spectroscopy [103–105]. Tip-enhanced Raman spectra caused by the pressurized tip could be thought to observe the phenomenon which was similar to thet seen in the high-pressure Raman study. The line broadening occurs not only as a result of surface interaction, but also as a result of pressures caused by the silver tip. However, unlike the isotropic hydraulic pressure studied in high-pressure Raman spectroscopy, the tip pressure is unidirectional, and hence it is expected to change the bond lengths uniaxially, resulting in modifications of molecular vibrations. Verma et al. [106] have recently proved this by observing vibration modes of C_{60} molecules. With a cagelike spherical hollow structure, C_{60} molecules are suitable for studying the effects of uniaxial pressure.

While the quantum chemical vibrational calculation can provide quantitative estimates of frequency shifts, its accuracy is dependent on the physical model. Here, the silver tip is treated as a single atom, leading to some discrepancies between the calculation and the experiment. More practical models are required for getting

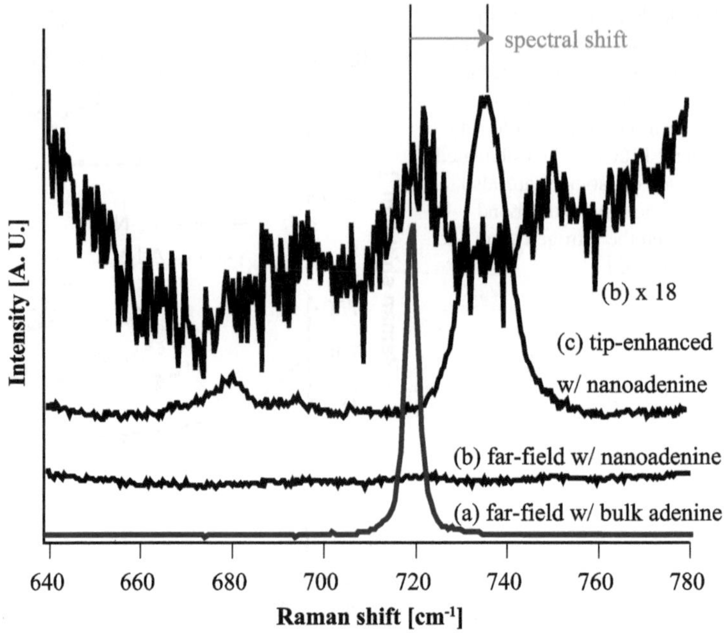

Fig. 19.17. Tip-enhanced Raman spectra of adenine molecules. (**a**) Far-field Raman spectrum of adenine powder and (**b**) far-field and (**c**) tip-enhanced Raman spectra of an adenine nanocrystal. An 18-times expanded spectrum (**b**) is also shown. The ring-breathing mode at 720 cm^{-1} in (**a**) and (**b**) was shifted to 738 cm^{-1} in (**c**) [89]

a precise picture. The phenomenon of shifting tip-enhanced Raman spectra caused by pushing molecules with an AFM is used for a novel spectroscopic instrumentation of molecular analysis and identification at nanoscale. Spatial resolution of the proposed tip-pressurized method should be given by an AFM (not the size of the enhanced electric field but the area directly in contact with molecules), which perturbs individual molecules. Hence, the technique has the possibility to achieve molecular resolution in vibrational spectroscopy, such as of short fragments of DNA sequencing lying flat on a surface.

19.7.2 Nonlinear Effect

In nanoscale signal sensing, the higher the spatial resolution becomes, the weaker the signal is because of the smaller number of molecules observed. In order to circumvent the possible extremely small signal, the combination of nonlinear optical process [107] with the tip-enhanced electric field is quite promising. Figure 19.18 shows the efficiency of the linear and nonlinear optical processes at the enhanced electric field of a metallic tip. The efficiency of first-order optical processes such as Raman spectroscopy exactly reflects the electric field distribution of the tip-enhanced field. On the other hand, nonlinear optical process, second- or third-order optical

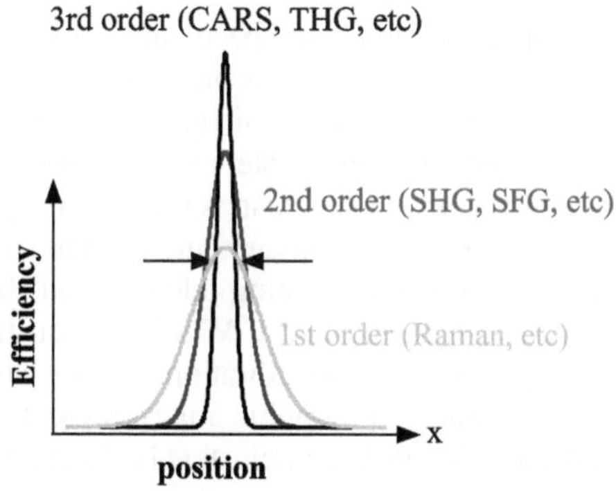

Fig. 19.18. Efficiency of linear and nonlinear optical processes under a tip-enhanced field. The first-order profile exactly reflects the electric intensity profile of the tip-enhanced field. CARS coherent anti-Stokes Raman scattering, THG third-harmonic generation, SHG second-harmonic generation, SFG sum-frequency generation

processes, can be induced selectively at the position of higher electric field intensity. This is based on the same reason as how nonlinear optical microscopy using a conventional objective lens goes beyond the diffraction limit [2, 3]. Only the difference here is the source of the light field, a tip-enhanced electric field in tip-enhanced microscopy and a diffraction-limited focused spot in conventional microscopy. In this section, discussions are focused on a third-order nonlinear optical process with the tip enhancement of a metallic tip, i.e., a technique for vibrational nanoimaging with tip-enhanced coherent anti-Stokes Raman scattering (CARS) [108–110], one of third-order coherent nonlinear Raman scattering [107]. CARS is one of the most powerful four-wave mixing spectroscopic methods, which includes three pump fields with the frequency of two ω_1 and ω_2 ($\omega_1 > \omega_2$), and the resulting anti-Stokes field with frequency $2\omega_1 - \omega_2$. Figure 19.19 describes the energy diagram of the CARS process. When the frequency difference between ω_1 and ω_2 corresponds with a specific vibration level of a target molecule, the molecule is excited to the vibrationally excited state. Accordingly, anti-Stokes Raman scattering ($\omega_1 - \omega_2 + \omega_1 = 2\omega_1 - \omega_2$) from the vibrationally excited state is coherently and resonantly enhanced. This is so-called CARS. Tuning the frequency of ω_2 while fixing the frequency of ω_1 can provide us with the same information of spontaneous Raman spectroscopy. The details and advantages of the CARS process can be found in [107].

The tip-enhanced CARS system is based on the same concept as the tip-enhanced Raman system (Fig. 19.4) using a silicon cantilever tip coated with silver films. Different from the tip-enhanced Raman system, in which we used a continuous-wave laser as an excitation light source, near-IR pulsed lasers, two mode-locked picosecond Ti:sapphire lasers (5 ps, 800 kHz), are used to induce the nonlinear optical process CARS. The ω_1 and ω_2 beams from the two lasers are collinearly

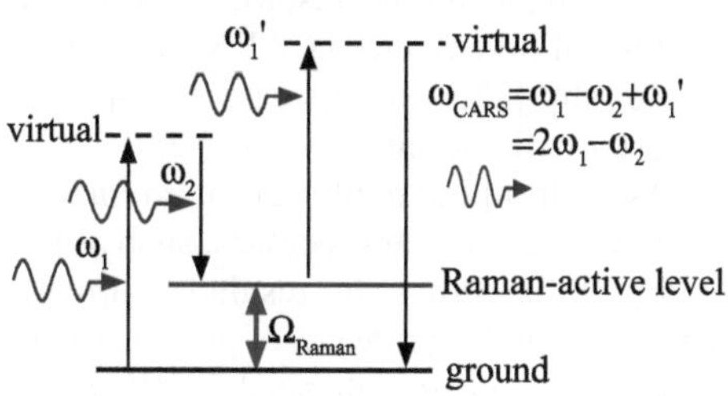

Fig. 19.19. Energy diagram of the CARS process

overlapped in time and space, and are introduced into the microscope and focused by a 1.4 NA objective lens onto the sample surface.

As a demonstration of tip-enhanced CARS, DNA molecules of poly(dA-dT) aggregated into nanoclusters were used for tip-enhanced CARS imaging [110]. The dimension of the nanoclusters is typically approximately 20 nm in height and approximately 100 nm in diameter, and the nanoclusters consists of stranded DNA double-helix structures containing adenine and thymine molecules. Figure 19.20a and b shows the tip-enhanced CARS image at the on-resonant frequency ($1336.9\,\mathrm{cm}^{-1}$) corresponding to the vibrational mode shown in Fig. 19.16 and the simultaneously obtained topographic image, respectively. The ring-breathing mode of adenine molecules in DNA nanoclusters is clearly discernible in the tip-enhanced CARS image, which corresponds well with the topographic image. On the other hand, at the off-resonant frequency ($1278.3\,\mathrm{cm}^{-1}$) in Fig. 19.20c, the tip-enhanced CARS signals have mostly vanished. The residual weak tip-enhanced CARS signals from nanoclusters are due to the contribution of nonresonant CARS [107] of not only adenine molecules but also thymine molecules and DNA backbones in the nanoclusters. These nonresonant contributions are insensitive to the frequency difference of the two lasers $\omega_1 - \omega_2$. Figure 19.20a and c verified that the tip-enhanced CARS signal obtained originated from a vibrationally resonant state, in this case the ring-breathing mode of adenine molecule. Figure 19.20d is the far-field CARS image at the on-resonant frequency without the tip. The line profiles of all the images indicated by arrows in the figure are shown in Fig. 19.20e-1 and e-2. The left axis of Fig. 19.20e-1 is expanded in Fig. 19.20e-2. It is noticed in Fig. 19.20e-1 that the full width at half-maximum (FWHM) of the tip-enhanced CARS signal is smaller than that of the topographic image. This is attributed to the fact that the CARS intensity is proportional to the square of the number of molecules. In Fig. 19.20e-2, the residual background signal of approximately 15 counts/100 ms in the presence of the tip is emitted at $2\omega_1 - \omega_2$ from the silver tip itself [108] owing to the third-order nonlinear susceptibility of silver. This undesirable emission may be encountered in such tip-enhanced spectroscopy combined with nonlinear optical spectroscopy, however, the nonlinear light emission from a metallic tip has the potential to be actively utilized as a nanoscale light source (e.g., second-harmonic generation at a metallic prove tip [111–113]). In the case of tip-enhanced CARS, one can suppress the nonlinear emission from a metallic tip by the means of time-resolved CARS [114]. Owing to the third-order nonlinearity of the CARS process, which dramatically emphasizes the discrimination between with and without the tip, it is impossible to detect any CARS signal from the nanoclusters in the absence of the surface-enhancing tip in Fig. 19.20d. Consequently, there is no background signal from the tip and no CARS signal from the sample, resulting in a zero count at each pixel. Figure 19.21 is another example of tip-enhanced CARS imaging on a DNA double-helix structure grown on a cover slip [115]. In this case, the spatial resolution of approximately 15 nm was achieved with the same system [109].

Aside from the nonlinear vibrational process, this concept is also applicable to any nonlinear optical processes, such as two-photon-excited fluorescence [36, 116]. Figure 19.22 shows the results of tip-enhanced two-photon-excited fluorescence carried out on J-aggregates of pseudo-isocyanine dye in a poly(vinyl sulfonate) film. As seen also in tip-enhanced CARS imaging, the optical image shows a slightly

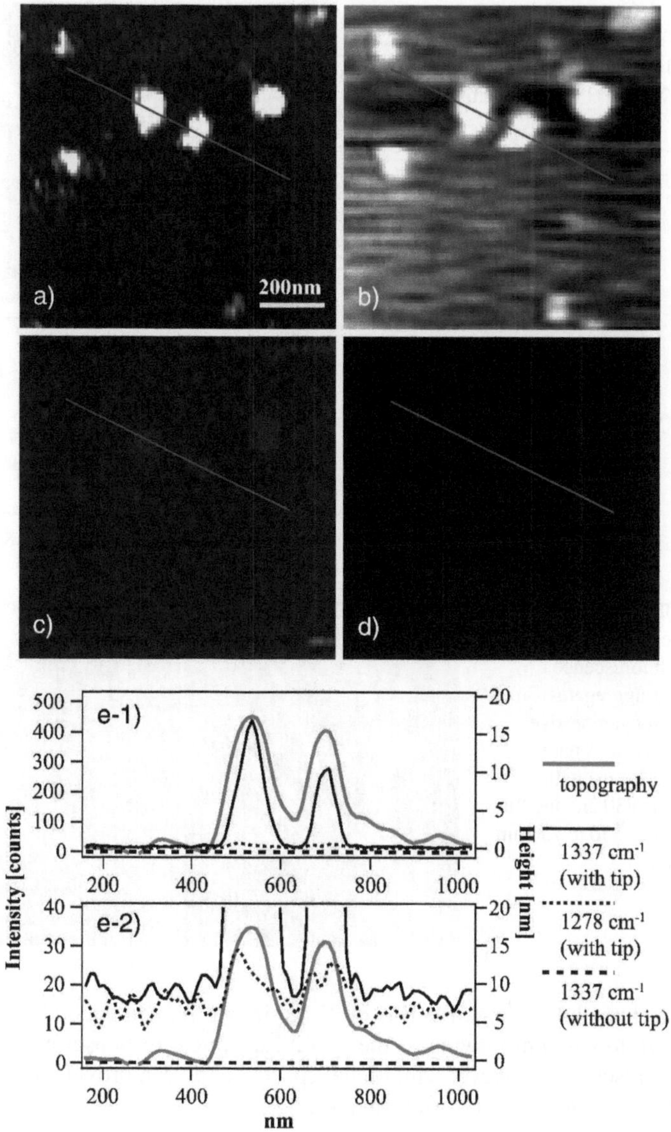

Fig. 19.20. Tip-enhanced CARS images of the DNA nanoclusters. (**a**) Tip-enhanced CARS image at on-resonant frequency 1336.9 cm^{-1}. (**b**) The simultaneously obtained topographic image. (**c**) Tip-enhanced CARS image at the off-resonant frequency 1278.3 cm^{-1}. (**d**) Far-field CARS image of the corresponding area obtained without the silver tip. (**e-1**,**e-2**) Cross-sectional line profiles at the *solid lines* in the images. The scanned area is 1 μm × 1 μm consisting of 100 pixels in a column by 50 pixels in a row. The number of photons counted in 100 ms was recorded for each pixel [110]

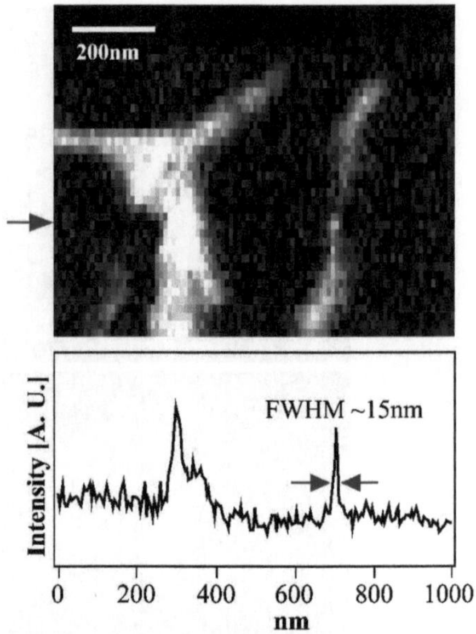

Fig. 19.21. Tip-enhanced CARS image of adenine molecules contained in a DNA double-helix structure and its cross-sectional profile [109]. FWHM full width at half maximum

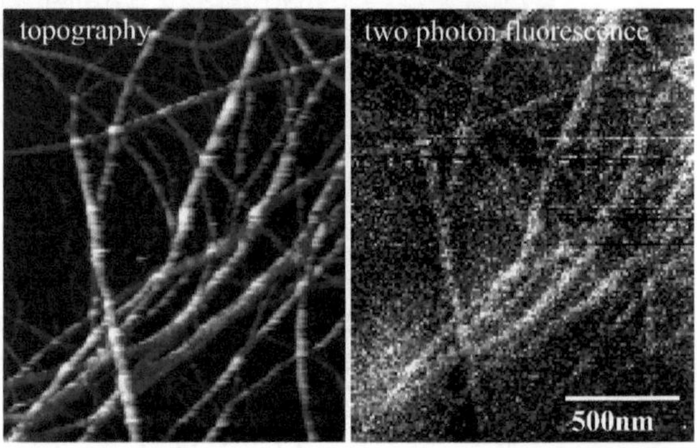

Fig. 19.22. Topography and tip-enhanced two-photon-excitation fluorescence images of J-aggregates of pseudo-isocyanine dye in apoly(vinyl sulfonate) film [116]. The excitation wavelength is 830 nm and fluorescence from 550 to 750 nm was detected

better spatial resolution (e.g., FWHM of 25 nm in two-photon-excited fluorescence vs. 30 nm in topography) owing to the nonlinear optical response of two-photon-excited fluorescence. However, vibrational spectroscopy can provide us with much rich information on molecular conformations.

19.8 Conclusion

Owingto the enhancement virtue of a metallic probe tip as a surface enhancer for the SERS effect, TERS allows us to obtain localized molecular vibrational information without averaging the signal inside the diffraction-limited focused spot, and we have not only high spatial resolution but also high signal sensitivity compared with far-field spectroscopy, which requires many more molecules to compensate for the small Raman scattering cross-section. The combination of the aforementioned techniques, including tip-enhancement, polarization control, tip-pressurized effect, and nonlinear processes, is quite promising for higher spatial resolution and sensitivity. In addition, fabrications of specific tip shapes [117–119] for higher tip enhancement or findings of new near-field effects are also very important factors for future development of tip-enhanced spectroscopy.

References

1. Born M, Wolf E (1999) *Principles of Optics 7th edition*, Cambridge University Press, Cambridge
2. Kawata S, Sun H-B, Tanaka T, Takada K (2001) Nature 412, 697
3. Hell SW, Wichmann J (1994) Opt Lett 19, 780
4. Pohl DW, Denk W, Lanz M (1984) Appl Phys Lett 44, 651
5. Harootunian, Betzig E, Isaacson M, Lewis A (1986) Appl Phys Lett 49, 674
6. Kawata S (ed) (2001) *Near-Field Optics and Surface Plasmon Polaritons*, Springer, Berlin Heidelberg New York
7. Synge EH (1928) Philos Mag 6, 356
8. Abbe E (1873) Archiv Miroskop Anat 9, 413
9. Binnig G, Rohrer H (1982) Helv Phys Acta 55, 726
10. Zenhausern F, O'Boyle MP, Wickramasinghe HK (1994) Appl Phys Lett 65, 1623
11. Inouye Y, Kawata S (1994) Opt Lett 19, 159
12. Sugiura T, Okada T, Inouye Y, Kawata S (1997) Opt Lett 22, 1663
13. Betzig E, Chichester RJ (1993) Science 262, 1422
14. Xie XS, Dunn RC (1994) Science 265, 361
15. Ambrose WP, Goodwin PM, Martin JC, Keller RA (1994) Science 265, 364
16. Tsai DP, Othonos A, Moskovits M, Uttamchandani D (1994) Appl Phys Lett 64, 1768
17. Jahncke CL, Paesler MA, Hallen HD (1995) Appl Phys Lett 67, 2483
18. Zeisel D, Dutoit B, Deckert V, Roth T, Zenobi R (1997) Anal Chem 69, 749
19. Deckert V, Zeisel D, Zenobi R (1998) Anal Chem 70, 2646
20. Nakano T, Kawata S (1993) Optik 94, 159
21. Nakano T, Kawata S (1994) Scanning 16, 368
22. Dragnea B, Preusser J, Schade W, Lecne SR (1999) J Appl Phys 86, 2795
23. Michaels CA, Stranick SJ, Richter LJ, Cavanagh RR (2000) J Appl Phys 88, 4832
24. Masaki T, Goto K, Inouye Y, Kawata S (2004) J Appl Phys 95, 334
25. Masaki T, Inouye Y, Kawata S (2006) Chem Phys Lett 417, 410
26. Fleischmann M, Hendra PJ, Mc Quillan AJ (1974) Chem Phys Lett 26, 163
27. Jeanmaire DI, Van RP Duyne (1977) J Electroanal Chem 84, 1
28. Albrecht MG, Creighton JA (1977) J Am Chem Soc 99, 5215
29. Kellner R, Mizaikoff B, Jakusch M, Wanzenbock HD, Weissenbacher N (1997) Appl Spectrosc 51, 495
30. Osawa M, Yoshii K (1997) Appl Spectrosc 51, 512
31. Hayazawa N, Inouye Y, Kawata S (1999) J Microsc 194, 472
32. Azoulav J, Debarre A, Richard A, Tchenio T (1999) J Microsc 194, 486
33. Hamann HF, Gallagher A, Nesbitt DJ (2000) Appl Phys Lett 76, 1953
34. Lessard TJ, Lessard GA, Quake SR (2000) Appl Phys Lett 76, 378
35. Hamann HF, Kuno M, Gallagher A, Nesbitt DJ (2001) J Chem Phys 114, 8596
36. Sánchez EJ, Novotny L, Xie XS (1999) Phys Rev Lett 82, 4014
37. Knoll B, Keilmann F (1999) Nature 399, 134
38. Wessel J (1985) J Opt Soc Am B 2, 1535
39. Fischer UC, Pohl DW (1989) Phys Rev Lett 62, 458

40. Bachelot R, Gleyzes P, Boccara AC (1995) Opt Lett 20, 1924
41. Martin YC, Hamann HF, Wickramasinghe HK (2001) J Appl Phys 89, 5774
42. Chang RK, Furtak TE (ed) (1981) *Surface Enhanced Raman Scattering*, Plenum, New York
43. Hayazawa N, Inouye Y, Sekkat Z, Kawata S (2000) Opt Commun 183, 333
44. Stöckle RM, Suh YD, Deckert V, Zenobi R (2000) Chem Phys Lett 318,131
45. Hartschuh A, Sánchez EJ, Xie XS, Novotny L (2003) Phys Rev Lett 90, 095503
46. Anderson MS, Pike WT (2002) Rev Sci Instrum 73, 1198
47. Roy D, Leong SH, Welland ME (2005) J Korean Phys Soc 47, S140
48. Sun WX, Shen ZX (2003) Ultramicroscopy 94, 237
49. Mehtani D, Lee N, Hartschuh RD, Kisliuk A, Foster MD, Sokolov AP, Maguire JF (2005) J Raman Spectrosc 36, 1068
50. Saito Y, Motohashi M, Hayazawa N, Iyoki M, Kawata S (2006) Appl Phys Lett 88:143109
51. Inouye Y, Hayazawa N, Hayashi K, Sekkat Z, Kawata S (1999) Proc SPIE Int Soc Opt Eng 3791, 40
52. Okamoto T (2001) *Near-Field Optics and Surface Plasmon Polaritons*, Springer, Berlin Heidelberg New York
53. Stöckle RM, Deckert V, Fokas C, Zenobi R (2000) Appl Spectrosc 54, 1577
54. Saito Y, Hayazawa N, Kataura H, Tsukagoshi K, Inouye Y, Kawata S (2005) Chem Phys Lett 410, 136
55. Anderson MS (2000) Appl Phys Lett 76, 3130
56. Novotny L, Bian RX, Xie XS (1997) Phys Rev Lett 79, 645
57. Saito Y, Wang JJ, Smith DA, Batchelder DN (2002) Langmuir 18, 2959
58. Saito Y, Wang JJ, Batchelder DN, Smith DA (2003) Langmuir 19, 6957
59. Li XL, Xu WQ, Jia HY, Wang X, Zhao B, Li BF, Ozaki Y (2004) Appl Spectrosc 58, 26
60. Ren B, Picardi G, Pettinger B (2004) Rev Sci Instrum 75, 837
61. Wang HJ, Saito Y, Batchelder DN, Kirkham J, Robinson C, Smith DA (2005) Appl Phys Lett 86, 263111
62. Saito Y, Murakami T, Inouye Y, Kawata S (2005) Chem Lett 34, 920
63. Bachelot R, Gleyzes P, Boccara AC (1997) Appl Opt 36, 2160
64. Vasile MJ, Grigg DA, Russel JE (1991) Rev Sci Instrum 62, 2167
65. Freeman RG, Grabar KC, Allison KJ, Bright RM, Davis JA, Guthrie AP, Hommer MB, Jackson MA, Smith PC, Walter DG, Natan MJ (1995) Science 267, 1629
66. Bar G, Rubin S, Cutts RW, Taylor TN, Zawodzinski TA (1996) Langmuir 12, 1172
67. Jia HY, Zeng JB, Song W, An J, Zhao B (2006) Thin Solid Films 496, 81
68. Furukawa H, Kawata S (1998) Opt Commun 148, 221
69. Novotny L, Sánchez EJ, Xie XS (1998) Ultramicroscopy 71, 21
70. Hayazawa N, Inouye Y, Sekkat Z, Kawata S (2002) J Chem Phys 117, 1296
71. Chance RR, Prock A, Silbey R (1978) Adv Chem Phys 37, 1
72. Nie S, Emory SR (1997) Science 275, 1102
73. Hildebrandt P, Stockburger M (1984) J Phys Chem 88, 5935
74. Sunder S, Bernstein HJ (1981) Can J Chem 59, 964
75. Watanabe T, Pettingger B (1982) Chem Phys Lett 89, 501
76. Watanabe H, Hayazawa N, Inouye Y, Kawata S (2005) J Phys Chem B 109, 5012
77. Xu H, Bjerneld EJ, Käll M, Börjesson L (1999) Phys Rev Lett 83, 4357
78. Hayashi S, Konishi T (2005) Jpn J Appl Phys 44, 5313
79. Hayazawa N, Yano T, Watanabe H, Inouye Y, Kawata S (2003) Chem Phys Lett 376, 174
80. Anderson N, Hartschuh A, Cronin S, Novotny L (2005) J Am Chem Soc 127, 2533
81. Yano T, Verma P, Inouye Y, Kawata S (2006) Appl Phys Lett 88, 093125
82. Saito R, Fujita M, Dresselhaus G, Dresselhaus MS (1992) Appl Phys Lett 60, 2204
83. Kataura H et al. (1999) Synth Met 103, 2555

84. Dresselhaus MS, Eklund PC (2000) Adv Phys 49, 705
85. Rao M et al. (1997) Science 275, 187
86. Henrard L, Hernandez E, Bernier P, Rubio A (1999) Phys Rev B 60, R8521
87. Alvarez L, Righi A, Guillard T, Rols S, Anglaret E, Laplaze D, Sauvajol JL (2000) Chem Phys Lett 316, 186
88. Turrel G, Raman J (1984) Spectosc 15, 103
89. Hayazawa N, Saito Y, Kawata S (2004) Appl Phys Lett 85, 6239
90. Grüneis A, Saito R, Jiang J, Samsonidze GG, Pimenta MA, Jorio A, Filho AGS, Dresselhaus G, Dresselhaus MS (2004) Chem Phys Lett 387, 301
91. Moon P, Spencer DE (1961) *Field Theory for Engineers*, Van Nostrand Reinhold, New York
92. Tsang JC, Mooney PM, Dacol F, Chu JO (1994) J Appl Phys 75, 8098
93. Atkinson A, Jain SC (1999) J Raman Spectrosc 30, 885
94. J. McCarthy, Bhattacharya S, Perova TS, Moore RA, Gamble H, Armstrong BM (2004) Scanning 26, 235
95. Goodman GG, Pajcini V, Smith SP, Merrill PB (2005) Mater Sci Semicond Process 8, 225
96. Hayazawa N, Motohashi M, Saito Y, Kawata S (2005) Appl Phys Lett 86, 263115
97. Otto C, van den Tweel TJJ, de Mul FFM, Greve J (1986) J Raman Spectrosc 17, 289
98. Giese B, Mc DJNaughton (2002) J Phys Chem B 106, 101
99. Kneipp K, Kneipp H, Kartha VB, Manoharan R, Deinum G, Itzkan I, Dasari RR, Feld MS (1998) Phys Rev E 57, 6281
100. Hoogsten K (1959) Acta Crystallogr 12, 822
101. Stewart RF, Jensen LH (1964) J Chem Phys 40, 2071
102. Watanabe H, Ishida Y, Hayazawa N, Inouye Y, Kawata S (2004) Phys Rev B 69, 155418
103. Ferraro JR (1984) *Vibrational Spectroscopy at High External Pressures*, Academic, New York
104. Crowell RA, Chronister EL (1992) Chem Phys Lett 195, 602
105. Hambir SA, Franken J, Hare DE, Chronister EL, Baer BJ, Dlott DD (1997) J Appl Phys 81, 2157
106. Verma P, Yamada K, Watanabe H, Inouye Y, Kawata S (2006) Phys Rev B 73, 045416
107. Shen YR (1984) *The Principles of Nonlinear Optics*, Wiley, New York
108. Hayazawa N, Ichimura T, Hashimoto M, Inouye Y, Kawata S (2004) J Appl Phys 95, 2676
109. Ichimura T, Hayazawa N, Hashimoto M, Inouye Y, Kawata S (2004) Phys Rev Lett 92, 220801
110. Ichimura T, Hayazawa N, Hashimoto M, Inouye Y, Kawata S (2004) Appl Phys Lett 84, 1768
111. Zayats V, Sandoghdar V (2000) Opt Commun 178, 245
112. Takahashi S, Zayats AV (2002) Appl Phys Lett 80, 3479
113. Bouhelier M, Beversluis A, Hartschuh A, Novotny L (2003) Phys Rev Lett 90, 013903
114. Volkmer, Book LD, Xie XS (2002) Appl Phys Lett 80, 1505
115. Tanaka S, Cai LT, Tabata H, Kawai T (2001) Jpn J Appl Phys 40, L407
116. Hartschuh, Beversluis MR, Bouhelier A, Novotny L (2004) Philos Trans R Soc Lond Ser A, 362, 807
117. Krug T II, Sánchez EJ, Xie XS (2002) J Chem Phys 116, 10895
118. E. J. Sánchez, Krug JT II, Xie XS (2002) Rev Sci Instum 73, 3901
119. Farahani N, Pohl DW, Eisler H-J, Hecht B (2005) Phys Rev Lett 017402

20 Investigating Individual Carbon Nanotube/Polymer Interfaces with Scanning Probe Microscopy

Asa H. Barber · H. Daniel Wagner · Sidney R. Cohen

Abbreviations

SPM	scanning probe microscope
SEM	scanning electron microscope
TEM	transmission electron microscope
SWCNT	single-walled carbon nanotube
MWCNT	multiwalled carbon nanotube
E_{comp}	Young's modulus of a composite
E_f	Young's modulus of a fiber
E_m	Young's modulus of a matrix
G_m	shear modulus of a matrix
V_f	volume fraction of fibers in a composite
V_m	volume fraction of matrix in a composite
η	viscosity of a solvent
ν^*	viscosity of a colloid suspension
c	particulate concentration in a solvent
ℓ	fiber length
ℓ_c	critical fiber length
L_{emb}	length of nanotube embedded within a polymer
r	fiber radius
A_f	cross-sectional area of a fiber
R	distance from a fiber at which stress falls to an equilibrium value
σ_f	stress in a fiber
τ_i	interfacial shear stress
ε_m	strain in a matrix
x	position along the principle fiber axis
β	shear lag parameter
τ	liquid contact angle with a solid substrate
γ_{lv}	surface tension at the liquid–vapor boundary
γ_{sv}	surface tension at the solid–vapor (typically air) boundary
γ_{sl}	surface tension at the solid–liquid boundary
r_d	droplet radius
T	line tension of a droplet at the solid–liquid–vapor triple junction
F	force
ρ	perimeter (radial) of a fiber
d	nanotube diameter

kPa	kiloPascal
MPa	megaPascal
GPa	gigaPascal
TPa	teraPascal

20.1 Mechanical Properties of Carbon-Nanotube Composites

The outstanding mechanical properties of carbon nanotubes have promoted considerable research into the development of carbon nanotube–polymer composites. These composites are predicted to have strength to failure and stiffness far in excess of conventional fiber-reinforced polymer composites, given adequate nanotube dispersion, surface quality, and orientation. Fundamental to the reinforcement of polymers with nanotubes is the role of the interface. As the interface in these nanocomposites is extremely small, numerous challenges exist in understanding interfacial phenomena at this boundary. This chapter reviews current understanding in carbon nanotube-interfacial behavior and how experimental techniques using scanning probe microscopy (SPM) have been fundamental in obtaining quantitative information on these interfaces. The implications from experimental methods on mechanisms of adhesion and stress transfer from the matrix to the nanotube reinforcement during composite loading are considered.

20.1.1 Introduction

Technologically, polymers filled with a reinforcing phase remain the most used commercial composite material and thereby remain the focus of this text. First, the properties of the carbon nanotubes are briefly described, followed by a summary of the mechanical properties of nanocomposites. This should serve as an introduction for the main emphasis: that of the role of the interface between carbon nanotubes and polymer matrices.

20.1.2 Mechanical Properties of Carbon Nanotubes

The motivation for nanocomposite research began with the discovery of synthetically produced carbon nanotubes. The first production process was developed for multiwalled carbon nanotubes (MWCNT) in 1991 by Iijima [1], followed soon after by the synthesis of single-walled tubes (SWCNTs) [2, 3]. While typical fibers used in conventional composites such as graphite, glass and Kevlar have diameters on the order of 10 μm, MWCNTs are considerably smaller at around 20 nm and SWCNTs possess diameters typically less than 2 nm. This size effect is interesting from mechanical considerations as fibers have been known to increase in strength as the fiber length, or diameter, gets smaller until the material can be considered as a defect free unit such as a single crystal. It is because of their size and almost perfect structure that carbon nanotubes are ideal candidates for high-strength materials. Furthermore, the similarity of carbon nanotubes to graphite sheets indicates that other mechanical properties such as the Young's modulus may be high.

Due to the considerable challenges that exist in measuring the mechanical properties of such small materials individually, the first studies of carbon nanotubes were theoretical. Tersoff [4], recognizing that carbon nanotubes shared similar properties to inplane graphite, calculated the energy required to deform a nanotube using elasticity theory. Additional simulation work on bending of nanotubes was related to high-resolution microscopy observations of bent nanotubes [5]. As experimental techniques became more sophisticated it was possible to bend nanotubes using SPM. Wong et al. [6] pinned a single MWCNT at one end and bent the opposite free length of the nanotube with an SPM tip whereas Falvo et al. [7] showed successive bending of a single MWCNT on a surface, again with an SPM tip. All of these studies highlighted how durable nanotubes could be, with flexibility and elastic behavior reported at high bending angles equivalent to local strains of up to 16%.

The stiffness of individual nanotubes has been extensively studied using molecular dynamics simulations. Yakobson et al. [8] predicted a Young's modulus of 5.5 TPa and strain to failure of around 40% for a single wall of a carbon nanotube [9]. A far more conservative estimate [10], based on the simple assumption that the inplane stiffness of the nanotube wall is the same as graphite, gave a Young's modulus value of 0.8 TPa. Lu [11] calculated elastic properties and strengths of a variety of different nanotubes, both MWCNTs and SWCNTs, using an empirical force-constant model. This model revealed a Young's modulus of around 1 TPa for MWCNTs. Interestingly, Lu also found that the elastic modulus of a MWCNT should be independent of the number and spacing of the walls in the tube, so that SWCNTs will exhibit behavior similar to the outer wall of a MWCNT. Treacy and coworkers first calculated the Young's modulus of both MWCNTs [12] and SWCNTs [13] from the mean-square vibration amplitudes directly visualized using transmission electron microscopy (TEM). This yielded modulus values above 1 TPa. Bending of single MWCNTs with an SPM tip also gave similar calculated modulus values [6]. Further experimental methods have deflected SWCNT bundles suspended over porous alumina membranes in an SPM to measure the elastic modulus of the SWCNT bundle [14]. This modulus was found to decrease as the diameter of the bundle increased, indicating that significant intertube sliding occurs during deformation of increasingly larger SWCNT bundle diameters.

While the measurement of stiffness of carbon nanotubes has been the focus of numerous theoretical and experimental studies, the tensile strength properties of carbon nanotubes have been more difficult to quantify due to limits in predicting nonelastic behavior or experimentally straining a nanotube to failure. A powerful technique for measuring the tensile strength, as well as Young's modulus, of a carbon nanotube was first carried out by the group of Ruoff [15] for individual MWCNTs followed soon after in that year by SWCNT ropes [16] as shown in Fig. 20.1. The experiment requires an individual nanotube to be held between two silicon SPM tips and pulled apart while measuring the applied tensile force due to bending of the SPM cantilever. Cantilever bending is viewed directly in an electron microscope. This experimental setup is analogous to the micromechanical tests typically carried out to examine materials in tension. While the spread in data was quite large, the experiments found maximum strength and modulus values of up to 63 GPa and 950 GPa, respectively,

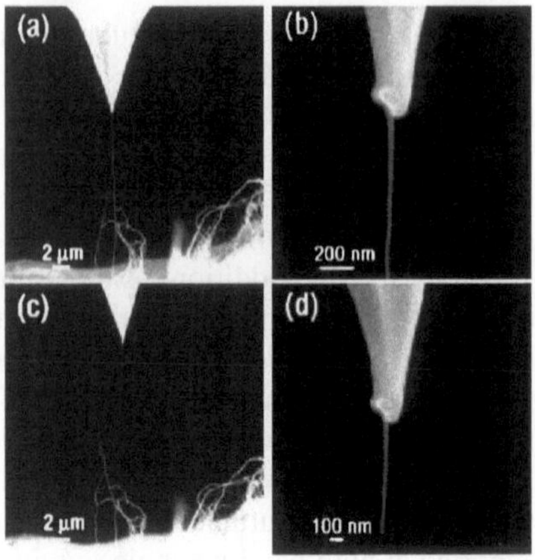

Fig. 20.1. Direct testing of SWCNT bundles can be performed by nanomanipulation techniques within the chamber of an SEM. Reprinted figure with permission from [16]. Copyright by the American Physical Society (2000)

for MWCNTs. Evaluation of the mechanical properties of SWCNTs proved to be more difficult due to nanotube bundling but maximum strength and modulus values of 52 GPa and 1330 GPa were recorded. Finally, additional tensile tests on individual MWCNTs using a similar approach to the group of Ruoff have demonstrated tensile strengths of up to 150 GPa [17] as well as highlighting the effect of defects on the distribution of the carbon nanotube tensile strength [18, 19].

20.1.3 Carbon-Nanotube Composites

From the above studies it has been established that carbon nanotubes are extremely flexible with a Young's modulus and strength of the order of a TPa and tens of GPa, respectively. Comparisons can be made between the mechanical properties of carbon nanotubes and conventional engineering fibers, as shown in Table 20.1. Clearly the strength and stiffness of carbon nanotubes are considerably higher than other fibers typically used to reinforce polymers. A simple rule of mixtures approach can be used to emphasis the potential of nanotubes reinforcing polymers. Thus, the Young's modulus (E_{comp}) of a composite is given by:

$$E_{\mathrm{comp}} = E_{\mathrm{f}} V_{\mathrm{f}} + E_{\mathrm{m}} V_{\mathrm{m}} \,, \tag{20.1}$$

where E_{f} and E_{m} are the Young's moduli of (fibrous) carbon nanotubes and the polymer matrix, and V_{f} and V_{m} are the volume fractions of carbon nanotubes and polymer, respectively, in the composite. Considering most polymer matrices have a Young's modulus on the order of 5 GPa or lower [21] and a reasonable estimate for the carbon nanotube Young's modulus (from Table 20.1) is ~ 1 TPa, the overall modulus of the composite can be above 100 GPa at relatively small (10% by volume) carbon nanotube content in the composite. Therefore, carbon nanotubes would appear to be the "ultimate" fiber for reinforcing polymer composites and increasing the composite's strength and stiffness.

Early efforts to produce carbon nanotubes–polymer composites were initially disappointing. First attempts to produce MWCNT–epoxy composites [22] showed

Table 20.1. Mechanical properties of engineering fibers and nanotubes

Fiber	Strength (GPa)	Young's modulus (GPa)	Reference
MWCNT	–	1300	[6]
SWCNT/MWCNT	–	800	[9]
MWCNT	–	∼1000	[10]
SWCNT/MWCNT	–	1000	[11]
SWCNT	–	1250	[13]
SWCNT	–	∼1300	[14]
MWCNT	18–68	270–950	[15]
SWCNT	13–52	320–1470	[16]
MWCNT	150	–	[17]
Kevlar	2.3	143	[20]
Carbon (PAN)	2.2–2.7	250–390	[21]
E-glass	1.4–3.5	76	[21]

that, while the compression modulus of the composite did show a reasonable increase with 5 wt % of nanotubes, the composite tensile Young's modulus showed very little increase. The authors attributed this small nanotube reinforcing effect in tension to a poor nanotube distribution within the epoxy matrix as shown in Fig. 20.2. Similar composites using SWCNTs incorporated within an epoxy matrix were also tested [23], with the deformation of the nanotubes both in tension and compression shown to be very small using Raman spectroscopy. Better nanotube reinforcement of polymers in tension was achieved when MWCNTs were mixed into the hardener of the epoxy and then the complete epoxy mix injected into sample moulds [24], with 400% increases in the composite Young's modulus (as compared to the pure epoxy) obtained with 4 wt % additions of MWCNTs. Thin MWCNT-epoxy films fabricated by first sonicating and dispersing the nanotubes in a solvent before addition to an epoxy have also shown mechanical property improvements due to good distribution of the carbon nanotubes within the polymer [25, 26]. Progress in processing conditions have been achieved by producing fibers of isotropic petroleum pitch [27], polystyrene [28], polypropylene [29] and polymethyl-methacrylate [30] incorporating carbon nanotubes. The benefit of producing these fibers, usually by extruding the polymer containing the carbon nanotubes, is to cause good dispersion and alignment of the nanotubes in the polymer fiber composite. This dispersion is particularly good as shear flow in the molten polymer prior to fiber processing causes nanotubes to separate. Shear mixing of carbon nanotubes with polypropylene, polystyrene and acrylonitrile-butadiene-styrene (ABS) has also been shown to be effective in producing composite films [31], with a resultant effective reinforcement of the polymer by the nanotubes. Despite the initially poor results from solvent preparation, refined techniques using nanotubes dispersed within preprepared polymer solutions have produced effectively reinforced polystyrene-co-butyl-acrylate [32], polystyrene [33] and polyvinyl-acetate [34] carbon-nanotube composites as well as allowing the polymer solution to intercalate within SWCNT sheets [35].

In conclusion, the excellent mechanical properties of carbon nanotubes make them suitable as reinforcing fibers in polymer composite materials. While this section has highlighted how effective carbon nanotube polymer composites can be

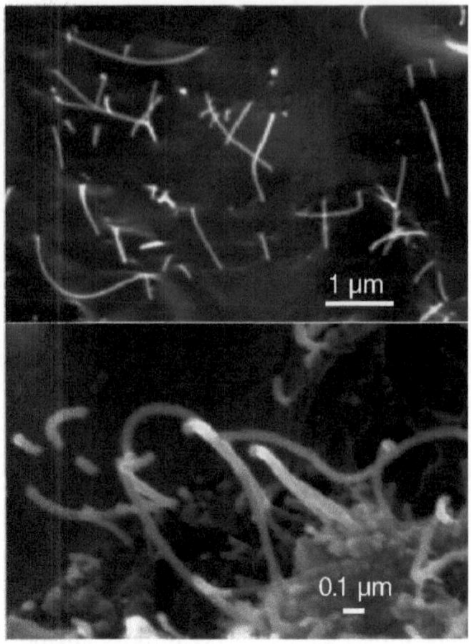

Fig. 20.2. SEM micrograph of a MWCNT-polymer composite showing nanotubes dispersed in the matrix. Note the potential agglomeration of nanotubes in the *right side* of the lower micrograph. Reused with permission from [22]. Copyright 1998, American Institute of Physics

fabricated, the role of the interface in these composites will be expanded upon in the next few sections. It is also important to note that the scarcity of information regarding the nanotube/polymer interface in bulk composite experiments indicates that the nanotube experiments of Sect. 20.2 are fundamental in understanding the overall mechanical behavior of carbon-nanotube-reinforced polymer composites.

20.2 Interfacial Adhesion Testing

20.2.1 Historical Background

The concept of one phase modifying or reinforcing a second phase has been the subject of numerous studies. Indeed the process of composite reinforcement was, in principle, first initiated by the efforts of Einstein [36]. This work considered how particles in a solvent would modify the overall viscosity of a colloidal suspension. Relative to the solvent itself, the addition of particles to the solvent would increase the energy required to perturb the system. Thus, this increase in energy corresponds to an increase in the viscosity of the colloid solution. An equation was derived for this phenomenon relating the viscosity of the colloid suspension, η^* with volume concentration c, to the viscosity of the solvent, η, by:

$$\eta^* = \eta(1 + 2.5c) \ . \qquad (20.2)$$

Equation (20.2) accurately describes both mono- and polydisperse suspension of particulates in solvent despite its simplicity. There are also strong similarities with the rule-of-mixtures equation, essentially linking the overall property of a system, in this case the viscosity of a colloidal suspension, to the property of its constituents and a concentration of the modifying phase. Subsequent work of both a theoretical [37] and experimental nature [38] took the viscosity relationship of Einstein and applied it to a two-phase system of a particulate within a second rubbery solid, as opposed to a solvent. The external force perturbing the system was therefore causing a solid deformation instead of setting up a viscous flow, thus enabling the properties of the composite to be evaluated in terms of stiffness or Young's modulus. While even the basic rule-of-mixtures can sometimes be used to predict composite properties sufficiently, the interface between the two or more phases in a composite is yet to be fully explored. Proper treatment of these systems must acknowledge the fact that properties of composite materials are more complex than the sum of those of the two mixed materials.

20.2.2 Shear-Lag Theory

A mathematical approach for evaluating the mechanical properties of fibrous composites was developed in 1952 by Cox [39] and added to by Kelly [40] to produce the classical equations known as shear-lag theory. One of the breakthroughs in this approach was to accurately model the tensile stress profile σ_f along a single fiber within a loaded composite as well as that of the interfacial shear stress τ_i at a fiber–matrix boundary when the fiber is aligned parallel to the load. These equations are shown below and relate the mechanical properties of the constituents; Young's modulus of the fiber E_f, Young's modulus of the matrix material, E_m, shear modulus of the matrix G_m, and the strain in the matrix ε_m, to geometric parameters such as the length of the fiber, ℓ, and the radius of the fiber, r, thus:

$$\sigma_f = E_f \varepsilon_m \left\{ 1 - \frac{\cosh \beta(\ell/2 - x)}{\cosh \beta(\ell/2)} \right\} , \tag{20.3}$$

$$\tau_i = E_f \varepsilon_m \left\{ \frac{G_m}{2E_f \ln(R/r)} \right\}^{1/2} \frac{\sinh \beta(\ell/2 - x)}{\cosh \beta(\ell/2)} . \tag{20.4}$$

Therefore, any position, x, along the principle axis of the fiber can be used to calculate the tensile stress in the fiber or interfacial shear stress. The constant β in the above equations is a stress-transfer constant equal to:

$$\beta = \left\{ \frac{2G_m}{E_f A_f \ln(R/r)} \right\}^{1/2} , \tag{20.5}$$

where A_f is the cross-sectional area of the fiber and G_m is the shear modulus of the polymer at the interface. While shear-lag theory can be used to accurately predict the stress conditions in the fiber and at the fiber/matrix interface, fitting is required based on the ratio R/r. The constant R is somewhat abstract; it derives from relations of the properties of a bulk composite properties, i.e. a system with many fibers, from which the mechanical environment of a single-fiber system is subsequently calculated. Specifically, R is the interfiber distance originally used in the derivation of Cox. However, for single-fiber systems, this description is inadequate and is more accurately expressed as the distance from the fiber beyond which the matrix feels no contribution from its presence [41]. For good fiber–matrix adhesion, the stress transfer from the matrix to the fiber will be efficient and the R value will be large whereas poor interfacial adhesion will result in small R values. These shear-lag equations therefore identify a quantity, β, that is dependent on the quality of the interface and extends the mechanical behavior of composites to include the role of the interface. The assumptions of shear-lag theory have been proven experimentally using Raman spectroscopy for composites that behave elastically [42].

20.2.3 Kelly–Tyson Approach

It should be stressed that purely elastic behavior is used to predict the mechanical properties of a composite with shear-lag theory. Kelly and Tyson made the second critical study into the role of the interface in composite materials in 1965 [43]. Composites of molybdenum or tungsten wires reinforcing a copper matrix were fabricated and tested in tension to failure. The fracture cross sections of the composite were then examined so that the number of fibers that had broken at the fracture surface or pulled out from this surface could be counted. These observations showed how the ratio of broken to pulled-out fibers could change with the length of the reinforcing wire. Importantly, the work also revealed how shear stress at the fiber–matrix reinforcement was limited by the shear yield stress (or failure stress in shear) of the matrix or the interface. To test this theory, further experimental work was performed using a single fiber of molybdenum partially embedded within a matrix of copper. Pulling of this fiber at a large embedded length caused the fiber to fracture, whereas smaller fiber-embedded lengths resulted in the fiber pulling out, thus recreating the fiber fracture and pull-out observed in the bulk composite tensile testing. The single fiber pull-out test was preferred over bulk testing due to the direct information obtained on the fiber/matrix interfacial properties. The calculated interfacial failure stress during pull-out was found to decrease as the embedded length decreased. An empirical relationship was then derived from single fiber pull-out to relate the critical fiber length, ℓ_c, at which there is a transition from fiber pull-out to failure, to the interfacial shear strength by:

$$\ell_c = \frac{\sigma_f^r}{\tau_i}\,, \tag{20.6}$$

where σ_f is the tensile strength of the fiber. Equation (20.6) can be used to evaluate the quality of the interface as weak interfaces will give relatively large critical fiber lengths, whereas strong interfacial shear strength between the fiber and matrix will result in a smaller critical fiber length.

20.2.4 Single-Fiber Tests

The theories developed by Cox as well as Kelly and Tyson have been crucial in a large area of work dedicated to the study of interfacial properties within composite materials, specifically by single-fiber testing. These tests have been important in evaluating fundamental changes at the interfacial region, without considering other parameters such as fiber orientation that can complicate the interpretation of interfacial changes in bulk composites. The single-fiber pull-out test has been particularly important in evaluating hundreds of different modifications applied to the surface of the reinforcing fiber, mostly for the benefit of improving the transfer of stress from the matrix to the fiber under an applied external load through chemical bonding at the interface. While stress-based analysis is usually employed to evaluate interfacial strength, work of fracture methods have also been devised [44, 45].

Two additional methods have also been developed to ascertain the interfacial shear strength between a fiber and matrix; the single-fiber fragmentation test [46] and the microdebond test [47]. The single-fiber fragmentation test is perhaps the simplest for sample preparation as the whole fiber length is embedded within a matrix. This technique relies on loading the single-fiber composite in tension until the fiber fractures. This will continue until all of the fiber fragments are no longer at a length sufficient to fracture. A statistical relationship for the average fragment length [48] can then be used to calculate a critical fiber length, thus giving an interfacial shear strength value from (20.6). Unfortunately, the technique is less useful for composites where the matrix has a relatively low strain to failure as the matrix may fail before the fiber fragments.

While fragmentation testing has been one of the principle tests for the evaluation of single engineering fiber–polymer composite interfaces, recent work has observed the fragmentation of individual carbon nanotubes in a loaded polymer composite using a transmission electron microscope [49–51]. These studies provided some of the first experimental evidence in understanding the mechanical properties at the nanoscale, using a modified Kelly–Tyson approach [50] to estimate an interfacial shear strength as high as 500 MPa under certain conditions. However, a Kelly–Tyson approach to evaluating interfacial shear strength through (20.6) requires knowledge of the strength of the reinforcing fiber that is being broken. The real strength of a fibrous material is often not a single value and will increase or decrease as the length of the fiber decreases or increases, respectively. As the strength of carbon nanotubes has already been shown to have a considerably large statistical spread [18, 19], the direct proportionality between the interfacial shear strength τ_i and variable strength in (20.6) leads to a wide variation in estimated τ_i.

The microdebond, or microdroplet test [47] is similar to the pull-out test and consists of a droplet of polymer being placed on a fiber. After curing/solidification, the fiber is pulled though a sheet with a hole slightly larger than the fiber diameter. The fiber will slide through the hole unimpeded until the droplet contacts the edge of the hole. Further pulling increases the shear stress at the interface between the fiber and matrix until, at a critical stress, the interface fails.

20.3 Single Nanotube Experiments

20.3.1 Rationale and Motivation

The power and directness of evaluating the interface via single-fiber testing has been so successful that the study of fiber–matrix adhesion from single-fiber testing has been prevalent for many tens of years. This principle is so persuasive that testing of carbon-nanotube/polymer interfacial strength should be best examined using single nanotube tests, for which scanning probe microscopy is gaining acceptance as an ideal tool.

A number of benefits are to be had from examining both failure and wetting on single nanotubes rather than on a macroscopic sample. The first and foremost is that since no method has been devised yet to synthesize nanotubes of uniform size, the measurements made at a macroscopic level will include a statistical range of sizes, so that any dimensional effect (e.g., strength with diameter) could not be directly probed. Furthermore, any dynamic effects related, for instance, to friction of a nanotube being pulled from a polymer matrix or viscous flow as it is removed from a liquid, would be lost in an experiment where one measurement is made on a large number of "events". Finally, unique behavior that may occur around a single tube such as unsheathing of a multiwalled nanotube, rupture of a surrounding polymer matrix, or the influence of specific defects on wetting can only be examined when the events are monitored one at a time.

With the advent of recent advances in manipulation and measurement at nanometric scales, mechanical measurements on single nanoparticles (including carbon nanotubes) have become possible, as described in Sect. 20.1. The field of composites presents unique technical challenges, especially in the understanding of events occurring at the boundary between carbon nanotubes and their surrounding environment, i.e. a polymer matrix. While Sect. 20.1.3 examines the properties of carbon-nanotube–polymer composites, the experimental methods discussed here for direct evaluation of individual carbon nanotube interfacial properties are particularly important at bridging the knowledge gap between the nanoscale and the highly developed macroscopic studies in Sect. 20.2.

Interfacial strength testing of nanotube composites by SPM can be separated into two categories: tests involving ex situ formation of the composite and those involving in situ formation. The latter refers specifically to an in situ preparation of a composite unit of a single carbon nanotube embedded in a polymer matrix in the SPM, whereby one end of the nanotube is attached to the SPM tip. This enables both performing the mechanical testing and imaging the result of the detachment using one instrument in serial fashion, with no need for transfer between instruments. In ex situ testing, the composite is formed according to a standard procedure and then transferred to an SPM setup only for the mechanical strength testing. Imaging for evaluation purposes can therefore be carried out using alternative techniques to SPM. However, the transfer of the sample from SPM to this alternative imaging method requires the development of sample preparation/handling as well as a registry method so that a specific location can be found.

20.3.2
Drag-out Testing (Ex Situ Technique)

The drag-out interfacial adhesion method tests samples that originate as a bulk mixture of carbon nanotubes within an epoxy, subsequently microtomed for electron microscopy observation. As described in [52], macroscopic samples of carbon nanotubes in an epoxy resin are prepared so as to include voids in the sample interior. Statistically, some of these voids are bridged by the carbon nanotube intercalants. Microtoming the epoxy composites to thin slices of approximately 70 nm yields a surface that contains holes, some of which are bridged by nanotubes. These thin slices can be readily imaged by transmission electron microscopy (TEM) so that a suitable area containing a bridging nanotube can be located. The thin composite slice is then transferred to a SPM, where the sample is imaged to relocate the area of interest. The tip of the SPM then engages the free length of the bridging carbon nanotube and applies a lateral force perpendicular to the principle nanotube axis, while recording the force that develops as a result of the resistance encountered. To compute the interfacial strength, the embedded area must also be known. This is not readily evaluated from the SPM image because the embedded regions are subsurface and not generally observed. Furthermore, the image quality on these thin slices near holes is not suitable to accurately determine important parameters like nanotube diameter. For this reason, TEM micrographs are used to provide the relevant image information. Figure 20.3, which compares SPM and TEM images of the same region, illustrates this point. Since the two different microscopies must be applied to the same sample, precise registration of a specific area (typically with dimensions on the order of 100–200 nm) is performed in order to allow accurate transfer between the instruments. This is done by microtoming the composite slice directly onto a labeled TEM grid that has features and dimensions that overlap between the lower magnifications of the two high-resolution microscopes, and an optical microscope integrated with the SPM. The sample must be mounted so as to allow transfer from one microscope to the other without use of undue force (as may occur while securing it with an adhesive), while still affixing it strongly enough to allow noise-free imaging in the SPM. These two requirements were satisfied by

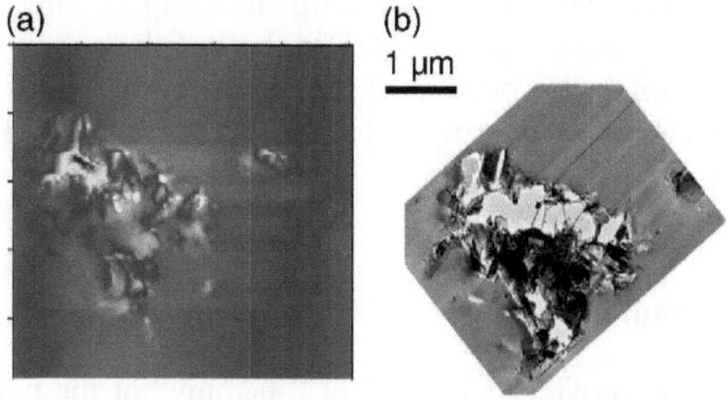

Fig. 20.3. Comparisons of images of a sample as used for drag-out experiments in (**a**) SPM and (**b**) TEM. Scale is same in the two parts. Note that although the corresponding images show features that clearly represent the same region, features are much sharper in the TEM, in addition to the exposure of subsurface features

Fig. 20.4. TEM micrographs before (*left*) and after (*right*) pull-out experiment with SPM tip. In (**a**), two MWCNTs (one short, one long) are initially bridging the gap. After pull-out, the short one has hopped to a position, to the upper right of the hole, whereas the longer one has not only broken, but the inner parts have become unsheathed from the outer shell, which can be seen at the bottom of the picture (*arrow*) angled to the left. Additionally, plastic damage induced on the polymer is observed (*arrow* 2). In (**b**), four MWCNTs that were originally bridging the gap have been broken, and one long one has been partially withdrawn, leaving a large curvature. Here also, plastic damage has occurred (*arrow, upper left*). In (**c**), the pull-out is "clean" in that no damage to the polymer matrix is seen. The pull-out has partially removed the MWCNT from the polymer, leaving a void at the bottom (*arrow*), and completely detached it from its upper part

placing the grid on a polyimide surface, polymer side down, so that the electrostatic force between polymer and polyimide secures the grid and sample for the SPM measurement, yet can be easily slid off for transfer to the TEM.

The beauty of performing single-nanotube measurements is shown in Fig. 20.4. Each drag-out is unique, and the different events such as unsheathing of the inner tubes (which requires much smaller forces as shown in elegant TEM experiments [53]), rupture of the surrounding polymer, clean extraction of a nanotube segment, or breakage of the nanotube are directly observed. The lateral force recorded corresponds to the nanotube detachment and hence is directly correlated with a specific mechanical process. In order for this to be used in the computation of interfacial strength, we must then be assured that all of the applied force is directed into rupture of the interfacial bond. Some events may occur during the drag-out that are not associated with the debonding process. In these cases the reported interfacial strength represents a maximum value describing the situation where the work exerted goes both into rupturing the intimate nanotube–polymer bond, and into other events within the surrounding polymer. A further question that arises in evaluating these experiments is how to treat the fact that the vector describing the force applied by the cantilever torsion is not directed along the tube axis, but nearly perpendicular to it. Simple geometrical considerations yield the resolution of this force along the nanotube axis. However, any slack in the free nanotube could lead to a bent structure that would result in a change in the resultant forces. Comparison of the measured tip deflection with the actual sample motion allows exact determination of the extent of bending of the NT (relative travel between tip and sample). Thus, it was seen that for all data included in the set, at least 90% of the sample motion was transformed directly into tip motion so that bending of the nanotube was small. Similar drag-out experiments conducted at the microscopic scale with polymer fibers also exhibit this phenomenon [54].

20.3.3 Pull-out Testing (In Situ)

The ex situ technique provides an excellent "anatomy" of the pull-out due to the recording of high-resolution TEM images before and after the mechanical test, as well as direct measuring of the nanotube detachment forces by SPM. Nonetheless, the experiments are technically tedious largely due to the delicate nature of the samples and the necessity to transfer them and reregister position in different instruments.

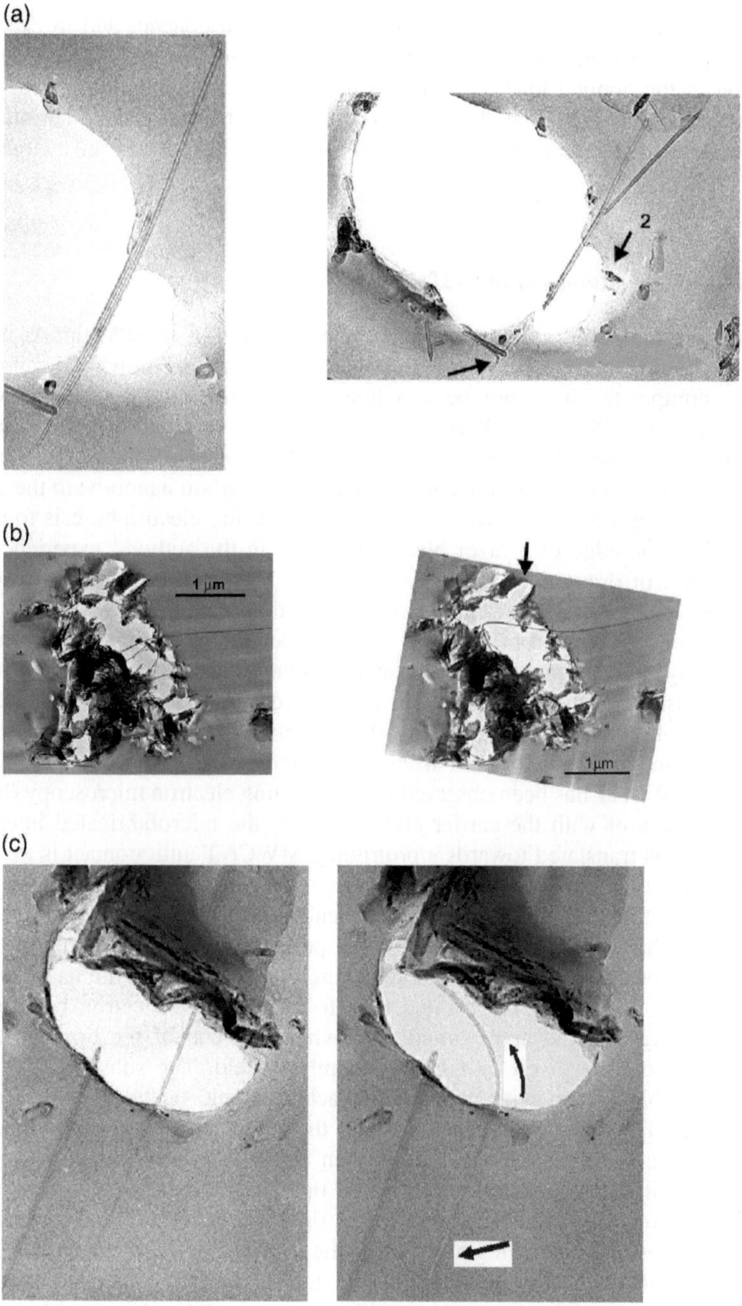
(a)
2
(b)
1 μm
1μm
(c)

In this respect, the pull-out experiments are simpler in concept and in execution since they involve only a single instrument and a more robust sample [55]. The direction of the applied load along the principle axis of the nanotube also allows for more direct interpretation of the forces acting at the nanotube/polymer interface. Another outcome from this technique is that better statistics result from a larger data set.

20.3.3.1 Carbon-Nanotube Attachment to an SPM Tip

The first step in the pull-out procedure is the preparation of an SPM tip. As the experiment should directly measure the forces acting on a simple individual nanotube–polymer composite, the nanotube was first attached to the end of the SPM tip. In this case, the SPM tip itself is the nanoscopic equivalent to the grip of a tensile testing machine often used to conduct microscopic pull-out as described in Sect. 20.2. The procedure for attaching an individual carbon nanotube to the end of an SPM tip is based on a previous method [56] of using electrophoresis to attract nanotubes to the edge of a razor blade. However, in the authors' experience, the simple action of dragging a razor blade through MWCNT powder is also effective for attaining a substrate sample where individual nanotubes, as well as some bundles, are protruding from the edge of the blade. This technique can be further simplified by taking a metallic block with carbon adhesive tape attached to one side and brushing this side through nanotube powder. Large carbon nanotube bundles will be observed on the tape using electron microscopy but some individually resolvable nanotubes sticking out from these bundles can also be used. Once an individual MWCNT has been observed using scanning electron microscopy (SEM), a manipulator arm with the carrier chip including the microfabricated integrated cantilever/tip is translated towards a protruding MWCNT until contact is made, as shown in Fig. 20.5.

Often, the translation and contact between the SPM tip and MWCNT can be difficult to achieve due to the lack of depth perception from SEM imaging. The best method for overcoming this and achieving contact between the nanotube and SPM tip is to first center the tip and then focus the SEM electron beam on its apex. This gives a good approximation of where the end of the tip is in the z-plane, with the error given by the SEM depth of field. The substrate, i.e. razor blade, metal block, etc. with nanotubes attached should be fixed to the sample stage of the SEM within the chamber, with the stage set to a low position. This sample stage containing the substrate is then translated using the SEM x-y manipulators until it is underneath the in-focus tip. The sample stage is then moved up in the z-direction towards the end of the tip, with small x-y movements used to stop the substrate contacting any part of the carrier chip. The nanotubes on the edge of the substrate will slowly come into focus, indicating that the nanotubes are close to the tip. The substrate should be moved continually up towards the tip until contact is achieved, usually observed by an individual MWCNT snapping into contact with the SPM tip due to van der Waals forces. Generally, some thermal noise and vibrations cause both the cantilever/tip and the nanotube to oscillate during these experiments. The damping of these oscillations, especially of

Fig. 20.5. Manipulation of MWCNTs can be viewed inside the chamber of the SEM. For nanotube attachment to a Si tip, the substrate with nanotubes protruding from the edge is moved into contact with the Si tip (*top*). At this stage gluing and alignment are performed. The substrate is then separated from the tip (*middle*), with the resultant nanotube–Si tip (*bottom*) usable for pull-out experiments

the nanotube, can be useful for determining when this contact is achieved, especially as the exact position of the tip relative to the nanotube is not precisely known.

Small adjustments of either the SPM tip or the substrate can be used to control and manipulate the orientation of the nanotube so that it protrudes from the tip at an angle nearly normal to the cantilever long axis. Once correctly aligned, the nanotube can be fixed to the SPM tip by concentrating the electron beam at the carbon nanotube–tip junction to deposit amorphous carbon by electron-beam-induced deposition. When using an FEI XL-30 ESEM this is typically performed by increasing the accelerating electron beam voltage from the usual imaging voltage of 10 kV to 25–30 kV while increasing the spot size by 30% from a regular imaging value of 3 to 4. The tube–tip contact is then imaged under high magnifications of 500 k to 1000 k that decomposes local hydrocarbon contamination under the intense electron beam, with the contamination then forming a strong glue of amorphous carbon at the nanotube–tip interface. The mechanical properties of this glue have been evaluated recently, showing it to possess outstanding clamping capability [57]. Indeed, SEM micrographs obtained after the pull-out revealed that failure never occurred at this contact.

20.3.3.2
Preparing and Performing Individual Nanotube Pull-Out

The single nanotube composite is prepared by dipping the carbon nanotube into a liquid polymer that must then undergo a phase change to solidify around the nanotube. The approach of the nanotube to contact with, and later separate from, the polymer can be easily controlled by the SPM system itself. The phase change must be induced so that the polymer solidification can occur with the SPM maintaining the position of the carbon nanotube. Various means to cast and cure a polymer around the nanotube inside the SPM were attempted in our lab. Two possible approaches were considered using a UV-curable polymer, which is exposed to UV light after the nanotube has been submersed in the film, and heating a thermoplastic above its softening point, followed by rapid cooling after immersion of the nanotube. In either case, mechanical testing is performed by retracting the solid polymer sample from the nanotube-SPM tip while recording the cantilever deflection. This cantilever deflection will increase until, at a maximum force required to fail the interface, debonding occurs and the carbon nanotube slides out of the polymer. Subsequent imaging of the polymer can then be performed with the nanotube-SPM tip to reveal the presence of a cavity in the polymer surface corresponding to the embedded position of the nanotube, as highlighted in Fig. 20.6 for a thermoplastic matrix.

Problems are encountered when using both UV-curable and thermoplastic polymers. UV curing requires directing sufficient UV power to the liquid polymer region directly below the SPM tip. This is most conveniently done by an open SPM system which allows optical access to the tip region, for instance from underneath. Unfortunately, UV lamps can generate considerable heat that needs to be avoided so as not to induce significant drift that may pull the nanotube away from the desired position. In the second preparation method, the required heating of a polymer to give a liquid phase, followed by cooling after carbon-nanotube contact also causes potential drift between the SPM tip and sample surface. This is due to the mechanical loop of the SPM setup, as well as changes in the volume of the polymer sample.

Practically, the drift can be split into two components: lateral (x-y) and vertical (z). The latter can be controlled within the scan range of the z-piezo element through feedback control – that is, the deflection (force) initially applied between tip and surface can be maintained throughout the cooling cycle by extension or retraction of the piezo tube in order to keep this value constant. The lateral drift is more problematic; to monitor and control this motion, position-sensitive detectors would need to record the sideways motion of the sample surface relative to the cantilever base, which could then be corrected by a separate feedback system. An alternative approach taken by us is to note the extent of lateral drift by imaging the surface repetitively at incremental temperatures within the range studied. Since this drift was small, on the order of 10% of the carbon nanotube free length, it was neglected, and the vertical feedback alone was applied to control and monitor the force.

It is clear that these experiments require precise control of the tip position in all three directions. An SPM closed-loop control system, where the piezo elements of the SPM are monitored and corrected to reduce unwanted movement from piezo hysteresis and creep, enhances the accuracy of the manipulations. In the absence of closed-loop control, these artifacts can be minimized by proper care. For instance, piezo creep decays logarithmically with time, so that in general if one zooms or pans to a new region, repeat scanning over several minutes of the same area while adjusting for offset allows the creep to settle to a negligible level.

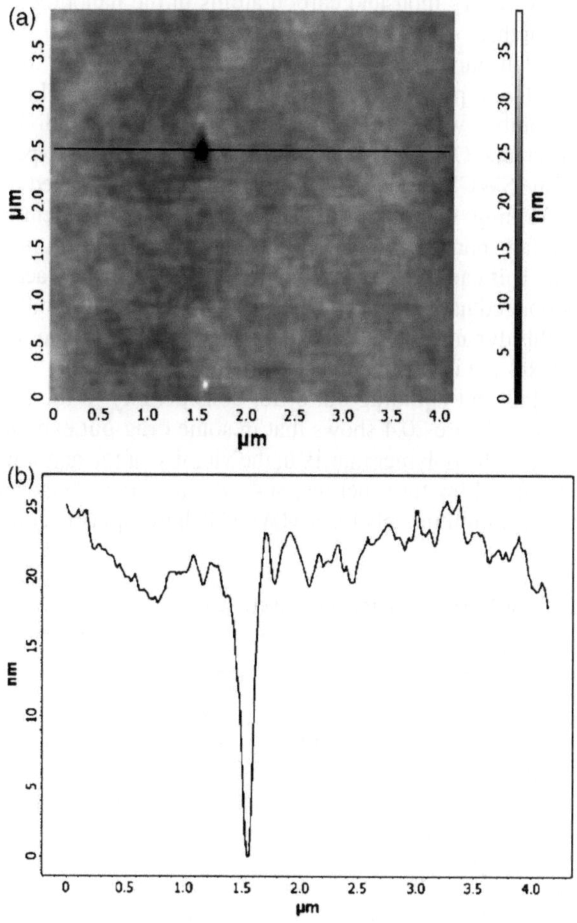

Fig. 20.6. SPM topography scan (**a**) of a nanotube pull-out hole and the corresponding cross-sectional scan along the indicated line (**b**)

20.3.4 Comparison of In Situ and Ex Situ Experiments

Table 20.2 shows the results of measured and calculated values resulting from the drag-out experiments. The maximum force observed in the SPM is divided by the interfacial area measured by TEM to give the interfacial strength. Work done is obtained from integrating the force vs. distance curves. Several conclusions can be drawn from these results. First, the scatter in the results is quite large, a consequence of the statistical nature of the events. Despite this, the interfacial strengths are significantly higher than those measured for carbon fibers under comparable conditions. Although theoretical considerations are presented below in Sect. 20.3.2, it is of interest to note that simple van der Waals forces arising only from the total interaction area would lead to strengths on the order of 5 MPa. The values measured indicate either a unique conformation that enhances the interaction, or stronger chemical bonding. Presuming the force required to break a chemical bond is 5 nN, a "back of the envelope" calculation shows that bonding arising from a defect at only one of every thousand carbon atoms in the nanotube can account for the observed strengths.

Pull-out experiments show a similar trend to the drag-out results. Although a different polymer, a thermoplastic, was used, the interfacial shear strengths using nanotubes were about 5 times greater than for individual carbon fibers in similar polymers. Only MWCNTs were suitable for these experiments. The small diameters of the SWCNTs render them too flexible to be pushed into the polymer melt without deforming. One interesting feature of the experimental plots is that the force does not fall sharply from its peak as the debonding occurs, but rather exhibits a gradual tail. This can be assigned to a small friction process accompanying the pulling of the debonded nanotube through the polymer. The drag-out experiments are not sensitive to this dynamic effect since the relaxation of the nanotube at the instant of debonding relaxes the intimate contact with the probing tip.

Pull-out and drag-out differ considerably in the total extent of sample area affected. Figure 20.4 shows that in some drag-out experiments, noticeable damage is done to the polymer matrix in the vicinity of the nanotube. For pull-out experiments, interfacial fracture energies of 4–70 $J\,m^{-2}$ have been calculated using energy-based conservation models for a MWCNT-thermoplastic composite [58]. This range con-

Table 20.2. Data from drag-out experiments [52]

Specimen #	1	2	3	4	5	6
Diameter (nm)	8.2	11.0	24.0	13.4	13.4	24
Embedded length (nm)	484	256	2570	379	708	1870
Interfacial area ($\times 10^{-14}\,m^2$)	1.01	0.88	19.4	1.60	2.99	14.07
Debond force (μN)	3.8 ± 0.5	2.8 ± 0.6	6.8 ± 1.7	0.6 ± 0.04	2.3 ± 0.6	12.8 ± 2.1
Work of debond ($\times 10^{-13}$ J)	2.9	3.3	16	1.3	1.6	7.8
Pullout energy ($J\,m^{-2}$)	26.4	36.9	8.2	0.9	5.35	5.54
Interfacial strength (MPa)	376 ± 40	318 ± 16	35 ± 9	38 ± 2	77 ± 20	91 ± 15

forms well to the range of fracture energies measured in the drag-out experiments as listed in Table 20.2. Similar values can be calculated from the pull-out traces as shown in Fig. 20.7. While we would expect interfacial fracture energies to vary for the different polymers, the consistency of the range of values between two independent experiments and modeling calculation highlights the quantitative nature of these individual nanotube–polymer interfacial tests. The direct nature of the nanotube pull-out test, i.e. load applied directly to the nanotube, makes them more reliable than the drag-out test, for which the SPM tip is coupled with the nanotube bridging a hole in the nanotube. Figure 20.8 also shows how a large number of data points can be attained using the pull-out test. The delicate nature of the thin polymer slice for the drag-out testing does lead to some rupture of adjacent areas of the polymer. However, TEM micrographs revealed that these cracks were generally small relative to the debonded area, indicating that the integral of the force–distance plot used to calculate the work done during the drag-out corresponds predominantly to fracture of the carbon-nanotube/polymer interface.

A final, but central point is the effect of carbon-nanotube diameter. Whereas the nanotubes have been shown to have significantly higher interfacial strengths than carbon fibers, the difference in diameters between these two structures is orders of magnitude. The single nanotube testing experiments show conclusively that the trend of increased strength with smaller size continues down to the smallest nan-

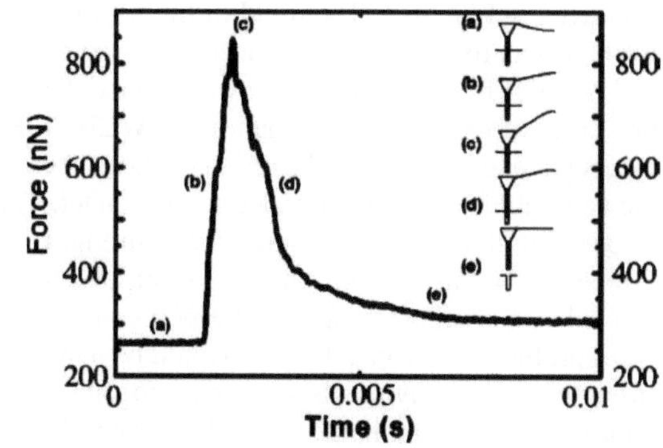

Fig. 20.7. Plot of measured force with progression of experiment (time) for pull-out of a MWCNT from a thermoplastic polymer matrix. Reused with permission from [52]. Copyright 2003, American Institute of Physics

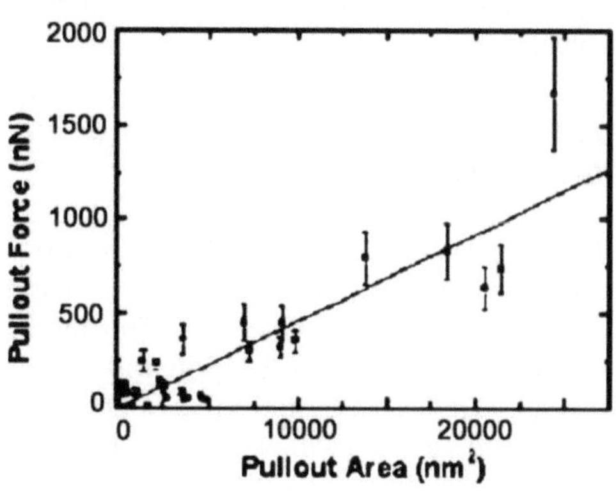

Fig. 20.8. Plot of measured pull-out force vs. interfacial area determined from nanotube diameter and embedded length. The slope of the fitted line is 47 MPa, which represents the interfacial shear strength. Reused with permission from [52]. Copyright 2003, American Institute of Physics

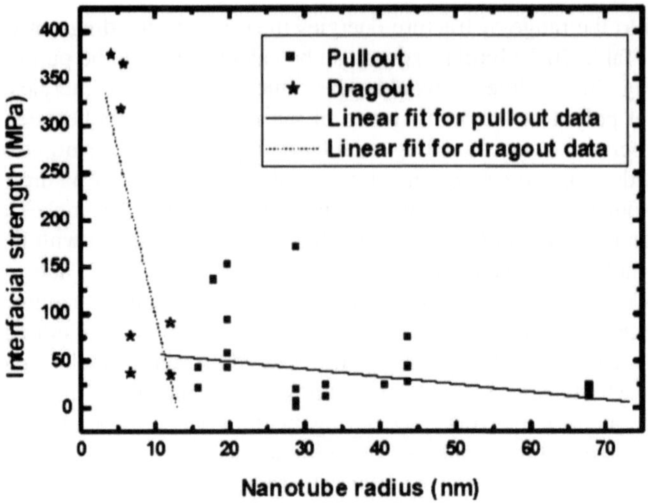

Fig. 20.9. Dependence of interfacial strength on the MWCNT radius, measured from both individual nanotube pull-out and drag-out experiments. Reused with permission [96]. Copyright 2003, American Institute of Physics

otube diameters tested. A compilation of both drag-out and pull-out experiments is displayed in Fig. 20.9. The trend of higher strength with smaller radius is indicated for both sets of experiments. Further support comes from comparison of SWCNTs and MWCNTs in the drag-out experiments. The SWCNTs have diameters at least one order of magnitude less than the MWCNTs. Here, only one of the four experiments yielded a pull-out. The rest (not shown here, see [54]) resulted in breakage of the nanotube, leaving the bonded region intact. The breakage values thus represent a lower limit for the interfacial shear strength. These values range between 120 to over 700 GPa, by taking the relevant area as the outer diameter of the bundle formed in the braided rope. If the polymer interacts only with the tightly wound outer profile of the rope, the area would be a minimum, giving maximum computed strength. If the polymer is presumed to interact with each individual tubule, the total interaction area is the individual nanotube cross-sectional area multiplied by their total number. Due to the larger area in this case, a lower strength would be calculated.

20.3.5 Wetting Experiments

The first, and indeed essential, step in formation of the carbon-nanotube/polymer interface is wetting of the nanotube by the polymer, a phenomenon controlled by the state of the nanotube surface. This initial wetting in polymer systems is driven by the van der Waals forces, primarily London dispersion forces [59,60]. Whereas Fig. 20.10 indicates good wetting of the NT by the polymer, this qualitative observation remains as such without a means for quantitative evaluation of the wetting angle. van der Waals forces have historically been evaluated through wetting properties.

Fig. 20.10. SEM micrograph of an individual nanotube, protruding from a bundle of MWCNTs, being spontaneously pulled into a liquid PEG surface. Reprinted figure with permission from [69]. Copyright (2005) by the American Physical Society

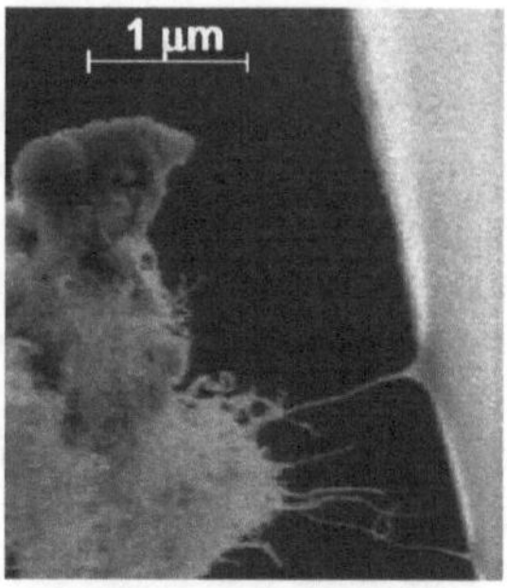

Wetting measurements, furthermore, provide a quantitative evaluation of the surface properties, notably the surface free energy as described by Young's equation:

$$\gamma_{\mathrm{LV}} = \cos\theta\,(\gamma_{\mathrm{SV}} - \gamma_{\mathrm{SL}}) \;, \tag{20.7}$$

where the γ's represent liquid–vapor (LV), solid–vapor (SV), and solid–liquid (SL) surface tensions, respectively, and θ is the contact angle of the droplet. Under proper conditions, the liquid–vapor surface tension can be directly determined by wetting measurements [61]; however the properties of the solid surface are less accessible. Fundamental work by Fox and Zisman [62], made a direct relation between the critical surface tension corresponding to the surface tension of a liquid that wets a particular surface, and the surface tension of that surface. Fowkes [63] suggested that the surface tension is mathematically equivalent to the geometric mean of those van der Waals forces arising purely from dispersion, and those arising purely from polar (e.g., hydrogen bonding) interactions. Building on these concepts, Owens and Wendt [64] developed a method to directly deduce the polar and dispersive components of surface tension for a solid by measurement of the contact angles for a homologous series of liquids, namely

$$\frac{\gamma_{\mathrm{LV}}\,(1+\cos\theta)}{2\sqrt{\gamma_{\mathrm{LV}}^{d}}} = \sqrt{\gamma_{\mathrm{SV}}^{p}}\left(\frac{\sqrt{\gamma_{\mathrm{LV}}^{p}}}{\sqrt{\gamma_{\mathrm{LV}}^{d}}}\right) + \sqrt{\gamma_{\mathrm{SV}}^{d}} \;, \tag{20.8}$$

where the superscripts d and p refer to dispersive, and polar components, respectively. The attractiveness of this expression in this linear form as displayed is as follows: Since the polar and dispersive components of the surface tension are known for many liquids, knowledge of the contact angle then allows determination of the polar and dispersive components of the surface directly from the slope and intercept respectively of the line indicated by (20.8).

A complication in this conventional analysis is the influence of line tension. Since the influence of line tension on the contact angle is predicted to vary inversely with droplet radius [65] the very small meniscus curvatures that would form at the nanotube–liquid interface could well be governed by such effects. A recent SPM work [66] suggests that most deviations from Young's equation are, in fact due to surface heterogeneities, rather than line tension. This work deduced the value for the line tension T derived for van der Waals liquids to be approx. -2×10^{-12} N.

Since deviations from Young's equations due to this term are given by $T/\gamma_{LV}.r_d$, using typical vdW liquid surface tensions of $\gamma_{LV} = 40\,\mathrm{m\,J\,m^{-2}}$ shows that this term is negligible for radii greater than a few nm. We therefore can conclude that any size-related effects observed in wetting of nanotubes of larger radii are due to changes at the highly curved solid nanotube surface rather than the liquid curvature.

There are many experimental considerations in choosing the method for contact-angle measurement. First, it must be recognized that all of the relevant equations derive from thermodynamic considerations, so that liquid–vapor equilibrium must be attained during the course of the experiment. Secondly, nanotube wetting occurs on a size scale that is not accessible to optical measurements. Although these droplets can, and have been imaged in the electron microscope, not only do the vacuum conditions hinder achieving equilibrium, but the electron beam itself can interact strongly both with the liquid and the nanotube [67]. For this reason, the Wilhelmy plate balance technique for contact-angle measurement, which is based on a force, rather than angular measurement, is an attractive approach [68,69]. The experimental setup is shown in Fig. 20.11. The tube (or plate) is dipped into a liquid, and the developing meniscus results in a downward force, which is monitored, and used to compute the contact angle. This method was traditionally used for determining the surface tension of liquids, although more recently it was applied to wetting measurements on microscopic carbon fibers. The compact relationship

$$F = \gamma_{LV}\rho\cos\theta \tag{20.9}$$

reveals that a measurement of wetting force gives directly the contact angle, provided that the nanotube perimeter ρ and liquid surface tension are known. Thus, the height of the meniscus, which is not measured in this experiment, is not required.

As discussed previously, lateral bending of a nanotube could occur at much lower forces than its buckling stiffness would indicate. A nanotube that bends so that the long axis lies nearly parallel to the liquid surface would exert a very different wetting force than the configuration indicated by Fig. 20.11. This complication was directly tested by observing the nanotube dipping in an environmental scanning electron microscopy (SEM) using a testing liquid (PEG) that is compatible with this environment. As seen in Fig. 20.10, no bending occurs under these conditions, even for the nanotube entering the liquid at an angle of about 20° from the normal. In fact, it appears that the wetting force helps to align the nanotube with respect to the liquid surface.

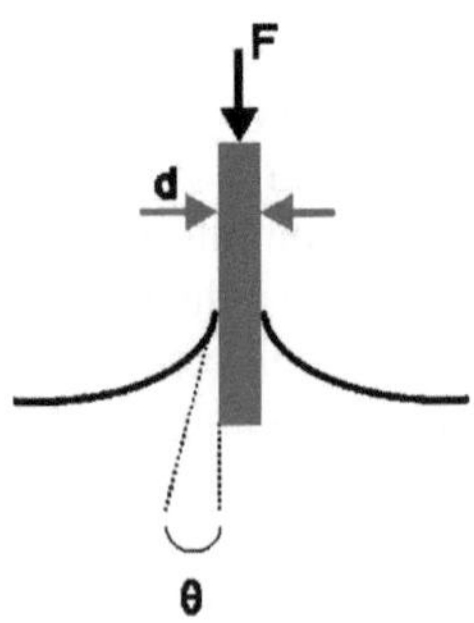

Fig. 20.11. Schematic of the force-based measurement of contact angle, based on the Wilhelmy balance technique. The downward wetting force leads to a deflection of the cantilever holding the NT and is directly detected and converted to force. If d is the diameter of the NT, then $F = \gamma_{LV}\pi d\cos\theta$

In the single-nanotube wetting experiments performed in our laboratory, MWCNTs attached to the SPM tip were dipped into droplets of various liquids, and the equilibrium wetting force measured from the deflection of the cantilever upon wetting. In order to obtain results on "virgin" nanotubes that had not been previously exposed to liquid, the cantilever was modulated at its resonance frequency, and the drop in amplitude was monitored as an indication of approach to the liquid surface. By successively controlling the approach motor and ramp on the z-control piezo element, it was possible to record the initial wetting of the nanotube. No significant differences were observed between initial and subsequent immersions, so that it is sufficient to consider the easier experiment of change in force upon pulling the nanotube out of the liquid. Results are shown in Fig. 20.12 and Table 20.3 for unmodified arc-discharge and chemical vapor deposition grown carbon nanotubes. According to (20.9), the wetting force is independent of immersion depth. Indeed, pushing the nanotube to larger depths into the liquid had no effect on the equilibrium pull-out force.

The data can further be analyzed using (20.8) to determine the relative influence of polar and dispersive components on the surface free energy. Dispersive and polar components of the surface free energy for the liquids used are tabulated in Table 20.4. Additionally, the slope and intersect of the resulting linear plot generated using (20.8) give values of 10.2 mJ m^{-2} and 17.6 mJ m^{-2} for the polar and dispersive components of the carbon nanotube as indicated in the table. Complementary values for planar graphite are listed for comparison. These results show that whereas the overall surface free energy of the MWCNT is similar to that of planar graphite, the

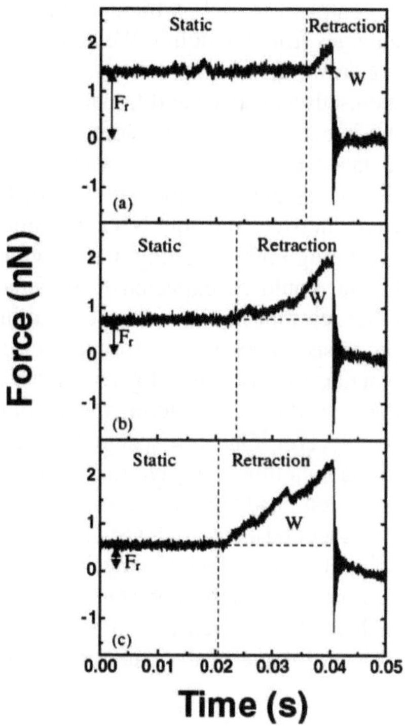

Fig. 20.12. MWCNTs were partially immersed in three liquids (**a**) PEG, (**b**) glycerol, (**c**) water. The forces acting on the nanotubes (y-axis) are shown againts time of the experiment (x-axis). Initially the nanotube was static in the liquid (*left side* of plot), followed by retraction from the liquid (*right*) causing an increase of force acting on the nanotubes. Seperation of the nanotubes from liquid caused a rapid force drop to zero. Reprinted figure with permission from [68]. Copyright (2004) by the American Physical Society

Table 20.3. Initial wetting forces for arc-discharge and CVD grown MWCNTs partially immersed in various organic liquids, normalized for a nanotube diameter of 20 n

Liquid	Initial wetting force (nN) Arc-discharge grown MWCNTs	CVD grown MWCNT
Polyethyleneglycol (PEG)	1.64	0.86
Glycerol	1.10	1.44
Water	0.78	4.23

Table 20.4. Polar and dispersive components of organic liquids used for wetting experiments

Probe liquid	γ_{LV} (mJ m^{-2})	γ_{LV}^{d} (mJ m^{-2})	γ_{LV}^{p} (mJ m^{-2})
Polydimethylsiloxane (PDMS)	25.1	22.7	2.4
Polyethyleneglycol (PEG)	48.3	29.3	19.0
Glycerol	64.0	34.0	30.0
Water	72.8	21.8	51.0

polar component of the curved structure is twice that of the flat one. This property will strongly control the interaction of nanotubes with a surrounding liquid or other (e.g., polymer) matrix material.

Figure 20.12 exhibits a dynamic feature that cannot be explained by the equilibrium treatment presented above: From the inception of the pull-out until the nanotube disengages from the liquid, a steady rise in the attractive force is observed. This is in contrast to (20.9), which predicts a flat force trace until the nanotube completely emerges from the liquid. We therefore must conclude that the liquid meniscus is being drawn up with the tube, so that the force observed is that required to stretch the meniscus and create additional liquid surface area. The resultant work W, taken from the integral of the area under the retraction in Fig. 20.12 (after converting time to distance on the abscissa), is a strong function of liquid surface free energy, indicating that the shape and ultimate stretched configuration of the meniscus varies significantly between the liquids. On the other hand, the work done actually rises for liquids exhibiting higher wetting angles with the nanotube, for which the initial meniscus height would be expected to be smaller. Hence, the effect likely correlates with other relevant liquid properties that influence energy dissipation, such as frictional processes. Invoking macroscopic viscosity values cannot explain the observations: whereas the viscosity of glycerol is 3 orders of magnitude higher than that of water, the work done in extracting the nanotube from the water is several times higher than for glycerol. One can speculate as to a unique ordering of water molecules close to the nanotube that give rise to the effect. Unique organization of water has been predicted inside of nanotubes, however, this is a confinement effect [70].

A second anomaly is seen in the calculated contact angles for water with the CVD nanotubes [69]. The nonphysical value obtained indicates that the model presented does not accurately represent the wetting process. If, however, (20.9) is modified to include internal wetting of the nanotube, with perimeter taken from the known ratio of inner to outer wall diameter, this contradiction can be resolved. Thus,

$$F = \gamma_{LV} \left(p_{out} \cos\theta_{out} + p_{in} \cos\theta_{in} \right) , \qquad (20.10)$$

where subscripts out and in refer to inner and outer. In order to satisfy the data, the inner contact angle must approach zero, which has been qualitatively observed in MWCNTs by electron microscopy [71,72]. Theoretical work suggests that imbibing of liquids into the nanotube should be a very rapid and facile process [73]. Interestingly, the liquid rise into the nanotube interior is only implicated in our experiments for the case of water, which suggests the possibility of the unique orientation of water molecules in the confined space may lead to such behavior. Such structuring, as noted above, has been predicted theoretically [70].

Finally, the complete force balance must include the pressure that would develop in the nanotube interior as the liquid rises, compressing the air trapped inside. The meniscus height can be initially estimated from the SEM micrographs of PEG wetting on nanotube exteriors as achieving a maximum of 200 nm. For a typical nanotube length of 800 nm, and 8 nm inner diameter, this corresponds to a pressure increase of 30 kPa, which exerts a force of 2 pN. This is insignificant compared with the wetting forces of several nN. Even if the inner meniscus were to fill most of the inner tube, leading to pressures ten times this estimate, the force exerted would amount to only 10% of the measured wetting force. The effect of such a force would be to increase the buoyancy of the nanotube, thus decreasing the apparent wetting angle.

Little experimental quantitative work exists that considers the wetting of nanotubes with liquids. A previous study [72] has shown how various metals, when liquefied by heating, would appear to wet carbon nanotube powders. Subsequent cooling and imaging of these samples using TEM was used to evaluate contact angles between the probe and nanotube exterior. As expected, higher surface tension metals gave larger contact angles at the nanotube surface. Although there are various errors associated with observation and measurement of these contact angles from visual inspections, contact angle data was used to extrapolate a critical liquid surface tension of 100–200 $\mathrm{mJ\,m^{-2}}$. This values suggests that liquids with a higher surface tension will not wet carbon nanotubes, although liquids below these values may not fully wet a carbon nanotube surface as suggested from the organic liquid wetting of individual carbon nanotubes described in this section.

20.4 Implication of Results and Comparison with Theory

20.4.1 Interfaces in Engineering Composites

The interfacial strengths measured for nanotubes dragged and pulled out of polymers are of the order of tens to hundreds of MPa. These results should not be considered in isolation, but instead questions such as "how do these values compare to other interfacial strength values?" and "how does the bonding between the nanotube and polymer relate to these interfacial strength values?" should be raised and, ideally, answered.

A comparison of carbon-nanotube/polymer interfacial strength values with other nanofibrous materials is difficult to make due to the lack of data. However, the most relevant comparison to make is with other fibrous carbon materials, of which typical engineering fibers of graphite have been most widely studied.

Testing of single carbon fiber–epoxy interfaces has produced a large and varied body of data. Interfacial shear strength determination using a pull-out configuration gives values ranging from 26–100 MPa [74–76]; similar values result from microdebond testing [76,77]. This range is large and reflects the parameters that can affect adhesion at the interface. In particular, promotion of chemical bonding between the fiber and polymer can cause increases in the measured interfacial shear strength. The wide range of observed values is indicative of the complexity of the pull-out mechanism, a factor that in principle could be controlled better for single-nanotube experiments, as the nanotube is a more well-defined physicochemical species. The mechanical properties of the fiber and matrix constituents also play a prominent role at the interface and it is the strength of the polymer matrix around the interface, usually in shear, that is often critical in controlling the interfacial strength. Thus, similar pull-out and microdebond experiments performed using graphite fibers but with thermoplastic polymers of a relatively low shear yield strength [77, 78] give lower interfacial shear strengths due to the matrix failing preferentially, or in addition to the interface.

Carbon nanotube pull-out and drag-out span a range of values as with their carbon-fiber counterparts. The stiffer epoxy matrix used in the drag-out experiments give maximum values of many hundreds of MPa, and indicate that the interfacial strength for nanotubes is higher than microscopic values measured for single-carbon-fiber composites. An interfacial strength of around 50 MPa measured between carbon nanotubes and a thermoplastic is comparable to the highest strength interfaces recorded for carbon-fiber–epoxies and again indicate the resilience of interfaces at the nanoscale. Epoxies used in previous studies [74–77] are considerably stiffer than the thermoplastic used in the nanotube pull-out [78,79], which suggests that unique interfacial-bonding mechanisms may occur between nanotubes and polymers.

20.4.2 Simulation of Carbon-Nanotube/Polymer Interfacial Adhesion Mechanisms

One of the earliest and most striking proposals for polymer adhesion to nanotubes was initiated by Lordi and Yao [80]. Molecular simulations were used to calculate binding energies of various polymers to the surface of a carbon nanotube. These binding energies play a minor role in nanotube–polymer adhesion, as do frictional forces. While hydrogen bonding and van der Waals interactions were more significant bonding mechanisms, substantial increases in the adhesion between polymers and nanotubes were obtained if the polymer chain could wrap around the carbon nanotube surface. The binding energies from this wrapping process gave an order of magnitude increase when compared to the other binding mechanisms, with maximum adhesion values equivalent to interfacial strengths of the order of 130 MPa. However, these simulations used very specific helical polymer conformations required to wrap around the nanotube surface, indicating that polymer wrapping may not always occur. Similar strong noncovalent bonding by wrapping was also simulated for a polymer chain, although the polymer conformation at the nanotube surface was not helical [81,82] as shown in Fig. 20.13.

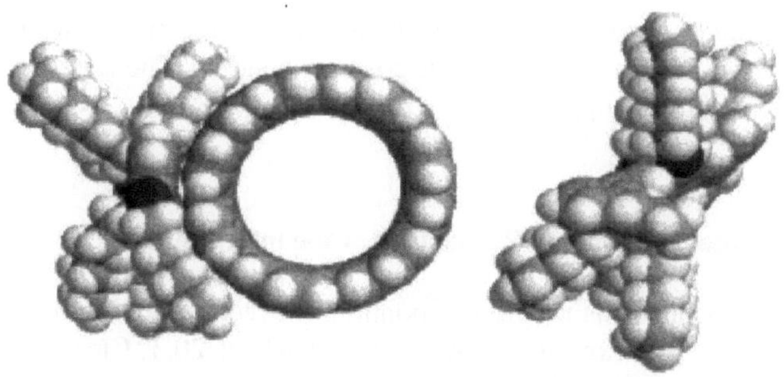

Fig. 20.13. Molecular simulation of a polymer (*right*) that changes its configuration upon wrapping at a carbon nanotube surface. Reprinted with permission from [81] (Fig. 20.8). Copyright 2002 American Chemical Society

Morphology of the polymer has been invoked to explain augmented carbon-nanotube-composite strength when crystalline polymers are used as a matrix in preference to amorphous polymers [83]. Indeed, direct evidence has been produced to show that distinctive crystalline polymer morphology can exist around a carbon-nanotube surface [84]. The formation of crystalline layers at the interphase region around the nanotube surface could be partially responsible for the improvement of mechanical properties at the interface relative to those of the bulk. Higher failure stresses at the interface during composite loading would therefore result. However, the magnitude of any potential increase in the mechanical properties in the interphase region is still unknown and no direct evidence of unique polymer morphology around a nanotube, which differs from the bulk, has been observed despite research showing carbon nanotubes disrupting the overall crystal morphology of semicrystalline polymers [85]. More significantly, the direct experiments used to measure interfacial strength in the previous section do not use crystalline polymers and would indicate that special crystalline polymer morphologies are not responsible for these high interfacial strengths.

Liao and coworkers [86, 87] have suggested additional mechanisms of adhesion between carbon nanotubes. Molecular simulations indicate that high interfacial strengths of 138 MPa and 160–186 MPa can be obtained in carbon nanotube/epoxy and carbon-nanotube/polystyrene composites. Radial compression of the polymer matrix around the nanotube and the subsequent deformation of the nanotube surface were highlighted as being responsible for this improved adhesion. This scenario shares similarities with increased interfacial adhesion observed due to thermal residual stresses in microcomposite samples [88, 89]. Finally, molecular dynamics were employed to elucidate the different interfacial shear strength values between an ideal ethylene polymer nonbonded and covalently bonded to the surface of a carbon nanotube using a pull-out configuration [90]. A nonbonded interface gave a low interfacial strength of about 3 MPa, which is expected when considering van der Waals interactions only and not other mechanisms such as polymer wrapping around the nanotube. Covalent crosslinking from the polymer to the nanotube resulted in significant increases in the interfacial shear strength. Values of 30 MPa and 110 MPa were simulated for disordered (amorphous) and ordered (crystalline) polymers despite barely 1 in 100 carbon atoms from the nanotube being covalently bonded to the polymer matrix. For smaller-diameter nanotubes, the strain and hence defect density increases, which would enhance covalent bonding. The experimental results of Fig. 20.9 are thus in support of these models.

20.4.3 Discussion of Potential Bonding Mechanisms at the Interface

So what can we deduce on the nature of bonding between carbon nanotubes and polymers based on the SPM experiments summarized in Sect. 20.3? Clearly, weak van der Waals forces do not control the interfacial bonding in a nanotube–polymer composite as evidenced by the relatively high nanotube-separation forces calculated in the pull-out and drag-out experiments. As formation of the nanotube–polymer composites for individual nanotube experiments involve thermal processing, potential radial stresses acting on the nanotube due to shrinkage of the polymer around the nanotube will certainly contribute to some degree towards the experimentally measured high interfacial strengths. Chemical bonding may also play some role in such recorded interfacial strength values, with the simulation work previously reported [90] suggesting that if 1% of the carbon atoms bond with the thermoplastic matrix used in the pull-out experiment, then τ values would approach 100 MPa. The possibility of polymer chains wrapping around the nanotube is useful in interpretation of individual nanotube pull-out and drag-out data, as detailed in Fig. 20.9, which shows that carbon-nanotube/polymer interfacial strength varies with nanotube diameter. The trend of higher interfacial strength at the smallest nanotube diameters implies that potential wrapping mechanisms could be easier as the nanotube diameter gets smaller.

As discussed previously, the polymers used in the experiments of Sect. 20.3 are exclusively amorphous and would not be expected to form failure-resistant crystalline interphase layers around the nanotube. However, while the interface at the nanotube/polymer boundary could be very strong, it is still unclear as to why the adjacent polymer, which is not intimately bound to the nanotube, and not the interface itself, is so resistant to stress relative to the bulk polymer properties, which is the limiting factor in conventional composite materials.

20.5 Complementary Techniques

20.5.1 Raman Spectroscopy

Raman spectroscopy is a nondestructive optical technique that can provide unique information about the vibrational and electronic properties of carbon nanotubes. The Raman spectrum of carbon nanotubes is unique and has been previously used as an analytical tool to provide evidence of carbon nanotubes presence in a carbonized product [91]. The strong peaks in the SWCNT spectrum have been particularly valuable in understanding the local environment around the carbon nanotube itself. For example, relative to the SWCNT Raman spectra in air, carbon nanotubes within matrices of polystyrene (PS) and styrene-isoprene (PI) have a peak related to their radial breathing mode shifted to lower frequencies [92]. This peak shift is attributed to the radial deformation applied to the nanotube by the surrounding polymer matrix. Compressive hydrostatic pressures applied to SWCNTs in solution can also be resolved using Raman spectroscopy, with increased peak shifts being related to the cohesive energy density of the liquid [93, 94].

As with previous research using Raman spectroscopy to record the stresses in Kevlar fibers processed within polymer matrices [95], the stress-sensing capability of carbon nanotubes using Raman spectroscopy has been extensively utilized within polymer-composite research. The D^* peak present in the Raman spectrum of MWCNTs within an epoxy has been shown to shift when a 1% compressive load is applied to the composite [96]. This shift was attributed to stress transfer from the loaded polymer matrix to the outer wall of the MWCNT. Tensile and compressive loads applied to SWCNT–polymer composites initially showed little, if any, shift in the Raman peaks due to bundling and poor dispersion of SWCNTs within the matrix [22]. Better carbon nanotube Raman peak shifts were recorded in a polymer composite when tested under differing temperatures, showing how thermal shrinkage of a polymer around carbon nanotubes could alter the Raman spectrum [97]. Improvements in sample preparation and the application of strain in the elastic regime of a SWCNT composite highlighted how a significant shift in the D^* Raman peak occurred due to an externally applied load [98]. This was formalized by Wood et al. [99] to indicate how the change in the D^* peak position was linearly dependent on the applied strain when the polymer behaved elastically. The magnitude of this dependence of the D^* peak shift with applied strain varies with the type of polymer used, indicating that different levels of adhesion will be present between a carbon nanotube and polymer. Thus, improving the adhesion between a carbon nanotube and polymer will result in a larger D^* peak shift per unit of applied strain.

The work of Cronin et al. [100], although not directly related to carbon-nanotube/polymer interfacial adhesion, is interesting in the context of the Raman work presented above. In this work an individual SWCNT was directly strained using an SPM tip and the Raman spectrum taken. The results showed that the D^* peak shifted by about $40\,\mathrm{cm}^{-1}$ when a tensile strain of 1.65% was applied. As the Raman work described above can currently only be used to rank adhesion between different polymers and nanotubes [101], this result is important for the precise calculation of the nanotube strain with applied strain. For example, [102] gives a D^* peak shift of $9\,\mathrm{cm}^{-1}$ per 1% of strain when applying an external load to a SWCNT-polyurethane acrylate (PUA) composite. It should be noted that the nanotubes in this sample were not particularly well aligned, suggesting that tensile loading could place some misaligned nanotubes in compression and give an underestimate for the D^* peak shift [103]. As perfect interfacial adhesion between carbon nanotubes and the polymer would lead to the nanotube being strained identically to the matrix (or applied) strain, which is not the case for PUA, a D^* peak shift of $9\,\mathrm{cm}^{-1}$ per 1% strain corresponds to 0.225% strain in the SWCNT relative to a 1% applied strain. A stiffer epoxy matrix reinforced with SWCNTs was also examined using Raman spectroscopy [104] and gave a D^* peak shift of around $20\,\mathrm{cm}^{-1}$ per 1% applied strain, indicating that the adhesion at the interface is better than that in SWCNT-PUA composites and furthermore highlighting how significant straining of nanotubes (~50%) can occur due to very good adhesion between the nanotube and epoxy matrix.

20.5.2 Scanning Electron Microscopy

The technique of electron microscopy is appealing for the study of most materials due to the rapid and precise acquisition of visual information on a sample. Scanning electron microscopy (SEM) is the most flexible of all electron microscopy techniques requiring little or no special preparation of the sample to be studied. On first examination, performing carbon-nanotube/polymer interfacial adhesion studies using SEM appears to show no additional benefit compared with the individual carbon nanotube pull-out from Sect. 20.3. This is because individual carbon nanotube pull-out with SPM gives all of the parameters required for adhesion calculations; namely the nanotube embedded length (or contact area) and the force required to pull the nanotube out of the polymer matrix. Critically, the carbon-nanotube probe used for imaging in SPM is also the sample to be tested and can therefore gives no information on the specific condition of the probe after a pull-out experiment. This is of fundamental importance if the classic experiments of Kelly and Tyson for pull-out of microscopic fibers can be reproduced at the nanoscale for carbon nanotubes. In particular, the critical embedded length of the carbon nanotube within a polymer matrix at which fracture occurs in preference to pull-out cannot be found using an SPM system alone. This was the motivation that led to the development of an individual carbon nanotube pull-out experiment observed within the chamber of an SEM [105].

One carrier chip with appropriate integrated cantilever/tip is used with an individual MWCNT attached to the end of the Si tip using the attachment technique discussed in Sect. 20.3.3.1. For these experiments, a long carbon nanotube is much easier to use. The second probing technique used in the experiment was the electron beam of the SEM that imaged the cantilever bending associated with forces acting on the nanotube tip within the SEM, thus eliminating the need for the optical detection of cantilever bending used in SPM. A liquid epoxy polymer was then introduced onto a solid substrate fixed to the sample holder of the SEM. This epoxy is a two-part resin and crosslinker which spontaneously solidifies upon mixing, although initial progression is slow, thus allowing time for the nanotube to contact the compliant (liquid) epoxy surface.

Positioning and contact of the nanotube with the liquid epoxy uses exactly the same procedure as the attachment of a nanotube to a Si tip; except that here the carbon nanotube powder on the edge of a substrate is replaced with the liquid epoxy. Contact is considerably easier to achieve since the relatively long nanotube has significantly more flexibility to snap into contact with the liquid than a shorter nanotube. Wetting between the epoxy and nanotube develops (Fig. 20.14), which results in an increase in the embedded length. The initial wetting is further modified by the uncontrolled drift of the nanotube into the epoxy or gentle manipulation of the nanotube to push it into the liquid, although in both cases some correction is required (by manipulation and movement of the liquid droplet using the sample stage positioning) to maintain a nanotube direction perpendicular to the epoxy surface. While the epoxy resin fully crosslinks to produce a solid, drift naturally occurs between the carbon nanotube and epoxy polymer. Again, the long length of the carbon nanotube ensures that some "slack" in the system is present and nullifies any problems with drift that can occur

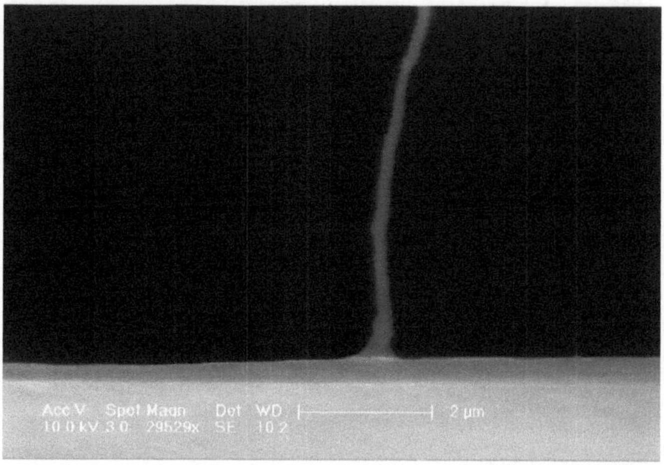

Fig. 20.14. SEM micrograph of an individual MWCNT manipulated to show contact with a polymer. *Scale bar* is 2 µm. Note the presence of a wetting meniscus at the nanotube–polymer contact

for the SPM pull-out experiments where the nanotube may prematurely drift away and detach from the polymer matrix prior to the pull-out.

Pull-out of the nanotube from the solid polymer is achieved simply by translating the SEM sample stage away from the carrier chip along the principle axis of the carbon nanotube. This results in progressive bending of the cantilever (Fig. 20.15) until the stresses developed at the nanotube/polymer interface are sufficient to cause nanotube detachment. Image analysis of the cantilever bending on recorded SEM pictures during and post pull-out can then be used to calculate the forces. Carbon nanotube embedded length can also be calculate from these SEM pictures taken from before nanotube contact with the polymer as well as just before the pull-out procedure, with the difference in the nanotube free length giving the nanotube embedded length in the polymer. The interfacial adhesion is simply the recorded maximum pull-out force divided by the calculated contact area between nanotube and polymer.

Not surprisingly, the total force required for nanotube pull-out from the matrix increases as the embedded length (or contact area) increases although the pull-out force will not simply increase with embedded length ad infinitum. Instead, the force applied to the system may eventually cause a build up of stress in the nanotube that is sufficient to fracture the nanotube (Fig. 20.16) if the embedded length is sufficient to allow this stress build up. The transition from nanotube pull-out to failure is not well defined due to the statistical nature of the nanotube strength but can still be estimated by observing the fracture of the reinforcement and recording the nanotube embedded length. Plots of nanotube pull-out to failure [105] have been used to calculate a critical fiber length, ℓ_c as defined in (20.6) using unmodified (pristine) carbon nanotubes as well as nanotubes chemically modified to induce strong bonding with the epoxy matrix (Fig. 20.17). ℓ_c values of 1400 nm and 400 nm for pristine and modified carbon nanotubes therefore correspond to interfacial shear strengths of 75 MPa

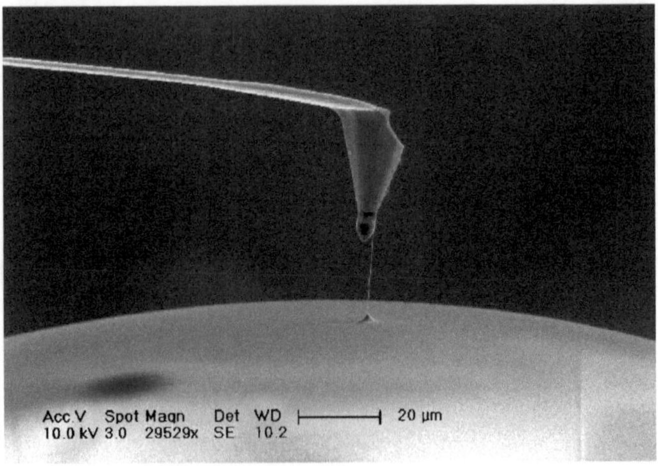

Fig. 20.15. SEM micrograph of a partially embedded nanotube being pulled from a solid epoxy polymer. *Scale bar* is 20 µm. The bending of the cantilever, clearly shown in the figure, can be used to calculate the forces acting on the nanotube

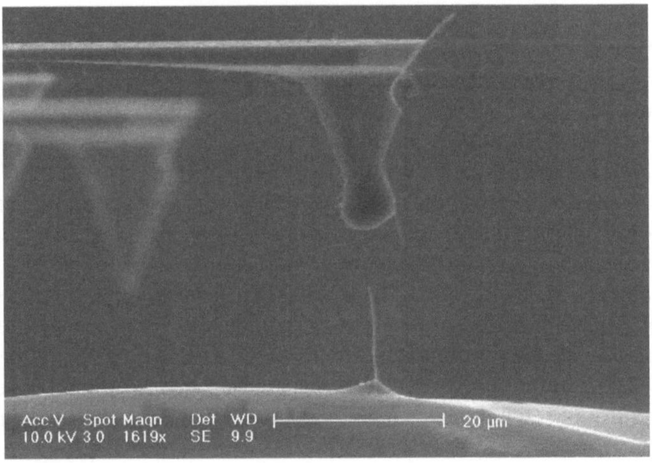

Fig. 20.16. SEM of micrograph of a MWCNT that has been embedded within a polymer matrix at sufficient length to cause nanotube fracture, in preference to nanotube pull-out. Scale bar is 20 µm

and 500 MPa, respectively, using a Kelly–Tyson approximation. These values are certainly very large and corroborate the previous findings of the potentially strong interfaces that can form at carbon-nanotube/polymer interfaces, especially when inducing strong chemically covalent bonds at the interface. However, the Kelly–Tyson approach has been questioned when applied at the nanoscale due to the approximation of the nanotube as a filled cylinder [106]. An alternative view of the stresses generated at the interface results from applying the model of Cox, as defined in (20.3) and (20.4), which avoids use of a critical fiber length. Equation (20.4) can be modified and simplified to relate the average interfacial shear strength, $\tau_{average}$ recorded from experiments to the maximum shear strength occurring at the interface, τ_{max} by:

$$\tau_{average} = \tau_{max} \frac{\tanh(\beta L_{emb})}{\beta L_{emb}} . \tag{20.11}$$

A plot of (20.11) using the experimental data from Fig. 20.17 is shown in Fig. 20.18. Thus, the average interfacial shear strength shown in Fig. 20.18 falls from a maximum value at very small embedded lengths to a more constant value at larger L_{emb}.

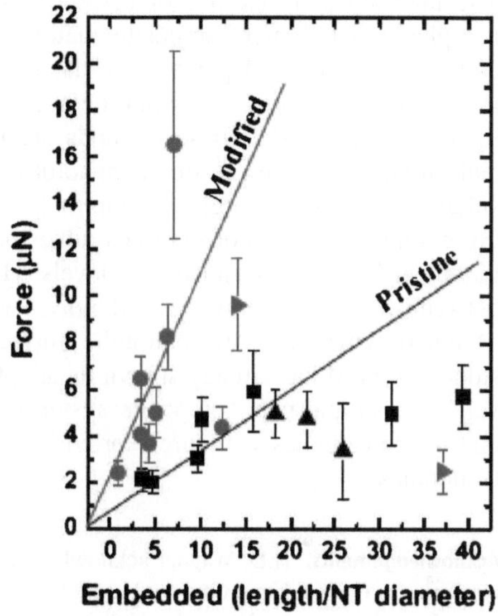

Fig. 20.17. Plot of the pull-out force applied to nanotube against embedded length normalized against the nanotube diameter for the chemically modified carbon nanotubes (*light symbols*) compared to pullout from pristine carbon nanotubes (*black*). Nanotube pull-out is shown by *squares* (for pristine) and *circles* (for modified). Nanotube fracture occurs at larger embedded lengths, indicated by *triangle* data points. Reprinted with permission from [105], Wiley-VCH Verlag (2005)

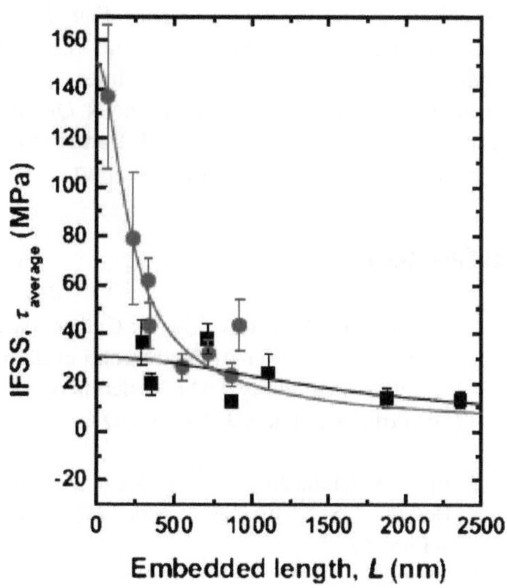

Fig. 20.18. Plot of the average interfacial shear strength, τ_i, against embedded nanotube length for the pull-out of pristine carbon nanotubes (*black*) or chemically modified carbon nanotubes (*gray*) from an epoxy matrix. The *lines* indicate a shear-lag fit, of the form indicated by (20.11), to the experimental data. Reprinted with permission from [105], Wiley-VCH Verlag (2005)

The τ_{max} values extrapolated from the fit in Fig. 20.18 also reflect the quality of the adhesion at pristine (30 MPa) and modified (150 MPa) carbon nanotube/epoxy interfaces. It is significant to note that the τ_{max} values from (20.11) are less than the average $\underset{\cdot}{\tau}$ values obtained from (20.6). This leads to the conclusion that the large variability in the strength of carbon nanotubes [18, 19] makes defining critical length values far too difficult for the calculation of a single interfacial strength number.

20.5.3 Overall Conclusions

The mechanical properties of carbon nanotubes make them ideal as a reinforcing phase in polymer-composite materials, with the overall performance of the composite critically dependent on the interfacial properties. Tests to evaluate the mechanical properties at a nanotube/polymer interface using SPM have been highlighted in this work. These methods are powerful in directly testing the interfacial strength between a single nanotube and the surrounding polymer. The resultant interfacial strength is extremely high relative to other typical engineering composites, and corroborates adhesion measured using complimentary techniques at both nano- and macro-levels. However, the reason for strong adhesion between a carbon nanotube and polymer, as well as for the general resilience of the polymer near the nanotube surface to high stresses is not fully understood. As SPM has already shown its adaptability in measuring interfacial strength values in simple nanocomposite systems, future research into composite interfacial nanomechanics will surely provide fruitful research opportunities using SPM techniques.

Acknowledgements. H.D. Wagner acknowledges the support of the (CNT) Thematic European network on "Carbon Nanotubes for Future Industrial Composites" (EU), the NOESIS European project on "Aerospace Nanotube Hybrid Composite Structures with Sensing and Actuating Capabilities," and the Israel Science Foundation (Grant # 290/02). H.D. Wagner and S.R. Cohen acknowledge the support of the G.M.J. Schmidt Minerva Centre of Supramolecular Architectures. H.D. Wagner is the recipient of the Livio Norzi Professorial Chair. A.H. Barber wishes to thank M. Whitby, T. Peijs and N. Quirke for their various comments and proof reading in helping produce this chapter. A.H. Barber thanks the Department of Materials, Queen Mary, University of London for logistical support.

References

1. Iijima S (1991) Nature 354:56; Carbon-based nanotubes were independently discovered several years earlier by Nesterenko et al. Their work remained largely unknown to the scientific community. [AM Nesterenko, NF Kolesnik, YS Akhmatov. VI Sukhomlin, and OV Prilutski, Metals 3 (1982), UDK 869.173.23, News of the Academy of Science USSR, 12–16.]
2. Iijima S, Ichihashi T (1993) Nature 363:603
3. Bethune DS, Clang CH, de Vries MS, Gorman G, Savoy R, Vazquez J, Breyers R (1993) Nature 363:605
4. Tersoff J (1992) Phys Rev B 46:15546

5. Iijima S, Brabec C, Maiti A, Bernholc JJ (1996) J Chem Phys 104:2089
6. Wong EW, Sheehan PE, Lieber CM (1997) Science 277:1971
7. Falvo MR, Clary GJ, Taylor RM II, Chi V, Brooks FP Jr, Washburn S, Superfine R (1997) Nature 389:582
8. Yakobson BI, Brabec CJ, Bernholc J (1996) Phys Rev Lett 76:2511
9. Yakobson BI, Campbell MP, Brabec CJ, Bernholc J (1997) Comp Mater Sci 8:341
10. Ruoff RS, Lorents DC (1995) Carbon 33:925
11. Lu JP (1997) Phys Rev Lett 79:1297
12. Treacy MMJ, Ebbesen TW, Gibson JM (1996) Nature 381:678
13. Krishnan A, Dujardin E, Ebbesen TW, Yianilos PN, Treacy MMJ (1998) Phys Rev B 58:14013
14. Salvetat J-P, Briggs GAD, Bonard J-M, Bacsa RR, Kulik AJ, Stöckli T, Burnham NA, Forró L (1999) Phys Rev Lett 82:944
15. Yu MF, Lourie O, Dyer MJ, Moloni K, Kelly TF, Ruoff RS (2000) Science 287:637
16. Yu MF, Files BS, Arepalli S, Ruoff RS (2000) Phys Rev Lett 84:5552
17. Demczyk BG, Wang YM, Cumings J, Hetman M, Han W, Zettl A, Ritchie RO (2002) Mater Sci Eng A 334:173
18. Barber AH, Andrews R, Schadler LS, Wagner HD (2005) Appl Phys Lett 87:203106
19. Barber AH, Kaplan-Ashiri I, Cohen SR, Tenne R, Wagner HD (2005) Compos Sci Tech 65:2380
20. Yang HH (1993) Kevlar Aramid Fiber, John Wiley & Sons Ltd., NY, p 26
21. Hull D (1981) An introduction to composite materials, 2nd edn Cambridge University Press UK
22. Schadler LS, Giannaris GC, Ajayan PM (1998) Appl Phys Lett 73:3842
23. Ajayan PM, Schadler LS, Giannaris GC, Rubio A (2000) Adv Mater 12:750
24. Allaoui A, Bai S, Cheng HM, Bai JB (2002) Compos Sci Tech 62:1993
25. Xu X, Thwe MM, Shearwood C, Liao K (2002) Appl Phys Lett 81:2833
26. Wong M, Paramsothy M, Xu XJ, Ren Y, Li S, Liao K (2003) Polymer 44:7757
27. Andrews R, Jacques D, Rao AM, Rantell T, Derbyshire F, Chen Y, Chen J, Haddon RC (1999) Appl Phys Lett 75:1329
28. Thostenson ET, Chou T-W (2002) J Phys D 35:L77
29. Kearns JC, Shambaugh RL (2002) J Appl Polym Sci 86:2079
30. Haggenmueller R, Gommans HH, Rinzler AG, Fischer JE, Winey KI (2002) Chem Phys Lett 330:219
31. Andrews R, Jacques D, Minot M, Rantell T (2002) Macromol Mater Eng 287:395
32. Dufresne A, Paillet M, Putaux JL, Canet R, Carnoa F, Delhaes P, Cui S (2002) J Mater Sci 37:3915
33. Safadi B, Andrews R, Grulke EA (2002) J Appl Polym Sci 84:2660
34. Liu L, Barber AH, Nuriel S, Wagner HD (2005) Adv Funct Mater 15:975
35. Coleman JN, Blau WJ, Dalton AB, Muñoz E, Collins S, Kim BG, Razal J, Selvidge M, Viero G, Baughman RH (2003) Appl Phys Lett 82:1682
36. Einstein A (1911) Ann Phys 34:591
37. Rehner J (1943) J Appl Phys 14:638
38. Guth E (1945) J Appl Phys 16:20
39. Cox HL (1952) Brit J Appl Phys 3:72
40. Kelly A (1973) Strong Solids, Oxford University Press, Oxford, 5, 121
41. Nayfeh AH (1977) Fiber Sci Technol 10:195
42. Galiotis C (1991) Compos Sci Tech 42:125
43. Kelly A, Tyson WR (1965) J Mech Phys Solids 13:329
44. Chua PS, Piggott MR (1985) Compos Sci Technol 22:107
45. Chua PS, Piggott MR (1985) Compos Sci Technol 22:33

46. Fraser WA, Ancker FH, DiBenedetto AT (1975) Proc 13th Anniversary Technical and Management Conf 22-A1
47. Agarwal BD, Broutman LJ (1980) Analysis and performance of fiber composites, Wiley (New York) p 72–78
48. Ohsawa T, Nakyama A, Miwa M, Hasegawa A (1978) J Appl Polym Sci 22:3203
49. Lourie O, Wagner HD (1999) Compos Sci Technol 59:975
50. Wagner HD, Lourie O, Feldman Y, Tenne R (1998) Appl Phys Lett 72:188
51. Lourie O, Cox DM, Wagner HD (1998) Phys Rev Lett 81:1683
52. Cooper CA, Cohen SR, Barber AH, Wagner HD (2002) Appl Phys Lett 81:3873
53. Cumings J, Zettl A (2000) Science 289:602
54. Nuriel S, Katz A, Wagner HD (2005) Compos A 36:33
55. Barber AH, Cohen SR, Wagner HD (2003) Appl Phys Lett 82:4140
56. Nishijima H, Kamo S, Akita S, Nakayamaa Y, Hohmura KI, Yoshimura SH, Takeyasu K (1999) Appl Phys Lett 74:4061
57. Ding W, Dikin DA, Chen X, Piner RD, Ruoff RS, Zussman E, Wang X, Li X (2005) Appl Phys Lett 98:014905
58. Barber AH, Cohen SR, Kenig S, Wagner HD (2004) Compos Sci Tech 64:1915
59. London F (1937) Trans Faraday Soc 33:8
60. Israelachvili JN (1985) Intermolecular and Surface Forces, Academic Press, London
61. Adamson AW (1990) Physical Chemistry of Surfaces, 5th edn, John Wiley & Sons, New York
62. Fox HW, Zisman WA (1952) J Coll Sci 7:109
63. Fowkes FM (1964) Ind Eng Chem 1956:40
64. Owens DK, Wendt RC (1969) J Appl Polym Sci 13:1741
65. Gaydos J, Nuemann AW (1987) J Colloid Interf Sci 120:76
66. Checco A, Guenoun P, Daillant J (2003) Phys Rev Lett 91:186101
67. Reimer L (1985) Scanning electron microscopy: physics of image formation and analysis, Springer-Verlag, Berlin
68. Barber AH, Cohen SR, Wagner HD (2004) Phys Rev Lett 92:186103
69. Barber AH, Cohen SR, Wagner HD (2005) Phys Rev B 71:115443
70. Noon WH, Ausman KD, Smalley RE, Ma J (2002) Chem Phys Lett 355:445
71. Gogotsi Y, Libera JA, Yoshimura M (2000) J Mater Res 15:259167
72. Dujardin E, Ebbesen TW, Krishnan A, Treacy MJ (1998) Adv Mater 10:1472
73. Supple S, Quirke N (2003) Phys Rev Lett 90:214501
74. Ramanathan T, Bismarck A, Schulz E, Subramanian K (2001) Compos Sci Tech 61:599
75. Marieta C, Schulz E, Mondragon I (2002) Compos Sci Tech 62:299
76. Mak M, Lowe A, Jar B, Stachurski A, Fitzgerald J (1998) J Mater Sci Lett 17:645
77. Miller B, Muri P, Rebenfeld L (1987) Compos Sci Tech 28:17
78. Ramanathan T, Bismarck A, Schulz E, Subramanaian K (2001) Compos Sci Tech 61:1703
79. Bismarck A, Schulz E (2003) J Mater Sci 38:4965
80. Lordi V, Yao N (2000) J Mater Res 15:2770
81. in het Panhuis M, Maiti A, Dalton AB, van den Noort A, Coleman JN, McCarthy B, Blau WJ (2003) J Phys Chem B 107:478
82. McCarthy B, Coleman JN, Czerw R, Dalton AB, in het Panhuis M, Maiti A, Drury A, Bernier P, Nagy JB, Lahr B, Byrne HJ, Carroll DL, Blau WJ (2002) J Phys Chem B 106:2210
83. Cadek M, Coleman JN, Barron V, Hedicke K, Blau WJ (2002) Appl Phys Lett 81:5123
84. Barber AH, Cohen SR, Wagner HD (2004) Nano Lett 4:1439
85. Assouline E, Lustiger A, Barber AH, Cooper CA, Wachtel E, Klein E, Wagner HD (2003) J Polym Sci B 41, 520
86. Liao K, Li S (2001) Appl Phys Lett 79:4225

87. Wong M, Paramsothy M, Xu XJ, Ren Y, Li S, Liao K (2003) Polymer 44:7757
88. Wagner HD (1994) Compos Interfaces 2:321
89. Wagner HD (1996) Phys Rev B 53:5055
90. Frankland SJV, Caglar A, Brenner DW, Griebel M (2002) J Phys Chem B 106:3046
91. Kim K, Jin L (2001) Nano Lett 1:631
92. Barraza HJ, Pompeo F, O'Rear EA, Resasco DE (2002) Nano Lett 2:797
93. Wood JR, Frogley MD, Meurs ER, Prins AD, Peijs T, Dunstan DJ, Wagner HD (1999) J Phys Chem B 103:10388
94. Wood JR, Zhao Q, Frogley MD, Meurs ER, Prins AD, Peijs T, Dunstan DJ, Wagner HD (2000) Phys Rev B 62:7571
95. Galiotis C, Young RJ, Yeung PHJ, Batchelder DN (1984) J Mater Sci 19:3640
96. Barber AH, Cooper CA, Cohen SR, Wagner HD (2003) In: Koenrad PK, Kemerink M (eds) AIP Conf Proc 696:580
97. Lourie O, Wagner HD (1998) J Mater Res 13:2418
98. Cooper CA, Young RJ, Halsall M (2001) Compos A 32:401
99. Wood JR, Zhao Q, Wagner HD (2001) Compos A 32:391
100. Cronin SB, Swan AK, Ünlü MS, Goldberg BB, Dresselhaus MS, Tinkham M (2005) Phys Rev Lett 93:167401
101. Barber AH, Zhao Q, Wagner HD, Baillie CA (2004) Compos Sci Tech 64:1915
102. Zhao Q, Wagner HD (2003) Compos A 34:1219
103. Frogley MD, Zhao Q, Wagner HD (2002) Phys Rev B 65:113413
104. Zhao Q, Frogley MD, Wagner HD (2002) Compos Sci Technol 62:147
105. Barber AH, Cohen SR, Eitan A, Schadler LS, Wagner HD (2006) Adv Mater 18:83
106. Wagner HD (2002) Chem Phys Lett 361:57

Subject Index

MIX
Papier aus verantwortungsvollen Quellen
Paper from responsible sources
FSC® C105338

If you have any concerns about our products,
you can contact us on
ProductSafety@springernature.com

In case Publisher is established outside the EU,
the EU authorized representative is:
Springer Nature Customer Service Center GmbH
Europaplatz 3, 69115 Heidelberg, Germany

Printed by Libri Plureos GmbH
in Hamburg, Germany